Alice Heath

KEY MATHS

Edexcel

Summary and Practice

HIGHER

- Paul Hogan
- Barbara Job
- Diane Morley

Published in 2002 by:
Nelson Thornes Ltd
Delta Place
27 Bath Road
CHELTENHAM
GL53 7TH
United Kingdom

02 03 04 05 06 / 10 9 8 7 6 5 4 3 2 1

A catalogue record for this book is available from the British Library.

ISBN 0 7487 6772 X

Illustrations by Oxford Designers & Illustrators
Page make-up by Tech Set Ltd

Printed and bound in China by Midas Printing International Ltd.

Acknowledgements

The publishers wish to thank the following for permission to reproduce copyright material: Corel (NT): 68, 78, 155, 178, (top); Digital Vision: 59, 178 (bottom); Photodisc: 107; D. Phillips/Science Library: 122 (left); Eye of Science/Science Photo Library: 122 (right); Air Traffic Control Tower, Denver, CO, USA: 110. All other photographs Nelson Thornes Archive.

The publishers have made every effort to contact copyright holders but apologise if any have been overlooked.

Contents

Introduction

Key Maths GCSE Edexcel Summary and Practice Higher is designed to support you as you work through the Higher Edexcel specification GCSE course. It has been designed to be used throughout your course or for practice and revision in your final preparation for the examinations.

It contains the following features to support you in your work:

- Summaries in every chapter to focus you on the essential concepts and skills that you need to cover
- A range of questions to provide comprehensive practice of all these core areas
- Two pages per chapter of worked examination questions with hints and tips, for comprehensive examination preparation
- Further examination questions to give you practice at answering full examination standard questions.

All the highly popular features of Key Maths are included throughout the book to assist with your learning and understanding.

Also available from Nelson Thornes:
Key Maths GCSE Edexcel Summary and Practice Foundation 0 7487 6770 3
Key Maths GCSE Edexcel Summary and Practice Intermediate 0 7487 6771 1

1 Transformations

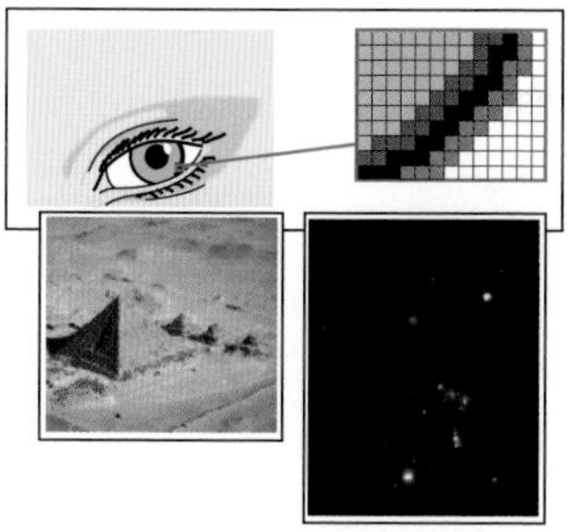

You need to know about:

- describing a translation
- describing a reflection
- describing a rotation
- describing an enlargement
- the inverse of a transformation
- combined transformations

Translation

A **translation** is a movement in a straight line.
The best way to describe a translation is to use a column vector.

This is a translation of $\begin{pmatrix} 3 \\ 6 \end{pmatrix}$.

This means a translation of 3 units to the right and 6 units up.
You must measure a translation using corresponding points on the object and on the image.
This translation is shown using the bottom right vertex of the triangle.

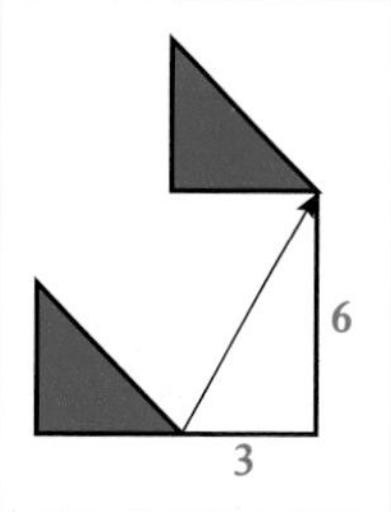

The top number tells you the horizontal movement (+ = right and − = left).
The bottom number tells you the vertical movement (+ = up and − = down).

Reflection

A **reflection** flips a shape over a straight line called the mirror line. To describe a reflection you need to describe the mirror line or give its equation.

This is a reflection in the line $y = x$.
Notice that the image of point A is labelled A′, the image of point B is labelled B′ and so on.

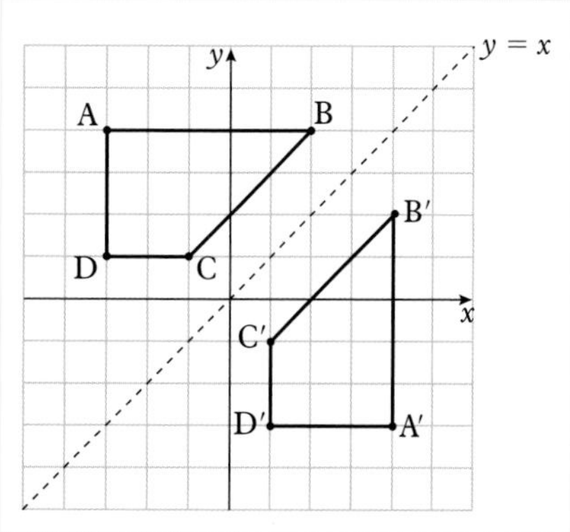

Rotation

A **rotation** turns a shape through a given angle around a fixed point called the centre of rotation. To describe a rotation you need to give: the angle that the shape has turned through, the direction of turn (clockwise or anticlockwise) and the centre of rotation.

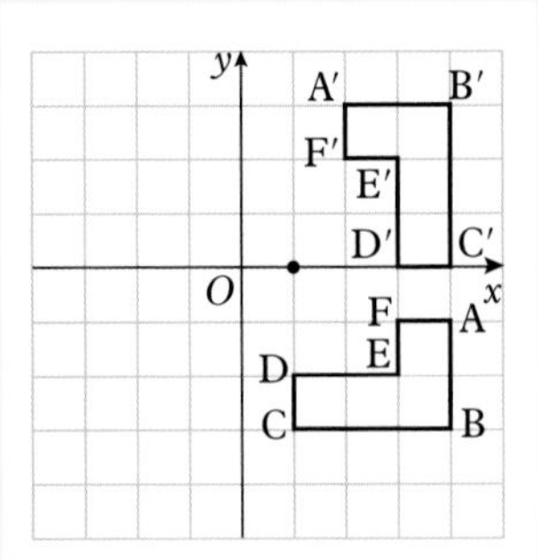

This is a rotation of 90° anticlockwise about (1, 0).

Each of the transformations so far does not change the shape that it transforms. The object and the image are congruent for a translation, a reflection and a rotation.

Enlargement

An **enlargement** changes the size of an object.
To describe an enlargement you need to give the size of the enlargement and the centre of enlargement.

This is an enlargement with scale factor $\frac{1}{2}$, centre (0, 1).

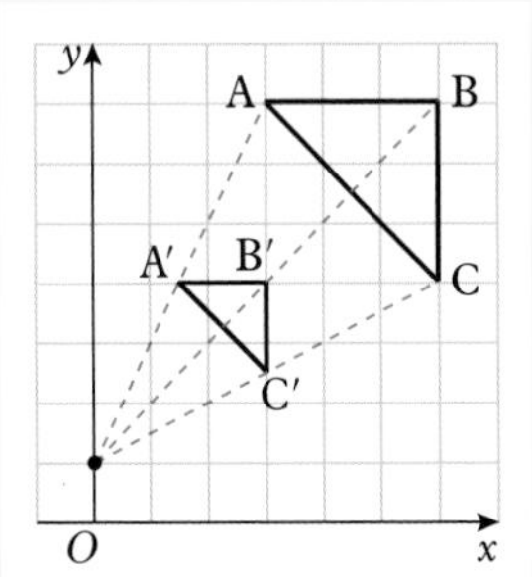

This is an enlargement with scale factor -2, centre O.

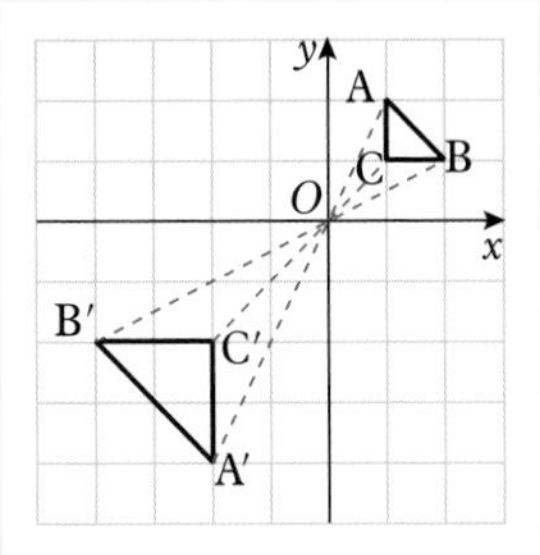

Inverse

The **inverse** of a transformation gets you back from the image to the shape you started with as the object.
The inverse of each transformation is the same type of transformation.

The inverse of a translation of $\begin{pmatrix} -2 \\ 4 \end{pmatrix}$ is a translation of $\begin{pmatrix} 2 \\ -4 \end{pmatrix}$.

The inverse of a reflection is the same reflection.
The inverse of a rotation is the same rotation but in the opposite direction.
The inverse of an enlargement with scale factor k is an enlargement with the same centre with scale factor $\frac{1}{k}$.

You can do more than one transformation on an object. Sometimes the result of doing 2 transformations is the same as doing a single transformation. For example, reflection in the x axis followed by rotation of 180° about the origin is the same as reflection in the y axis.

1 Describe each of these translations using a column vector.

a

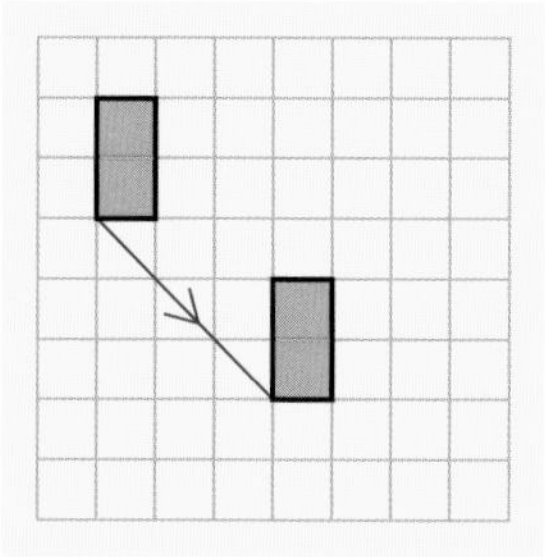

c

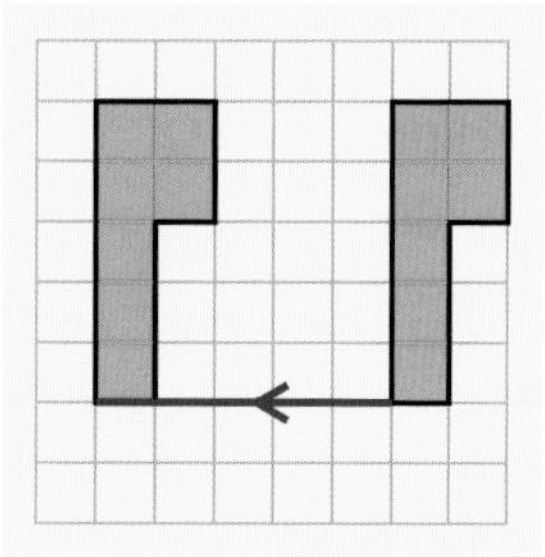

c

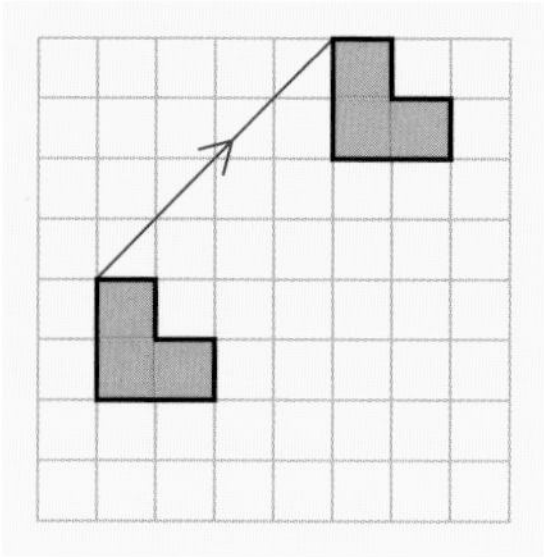

d

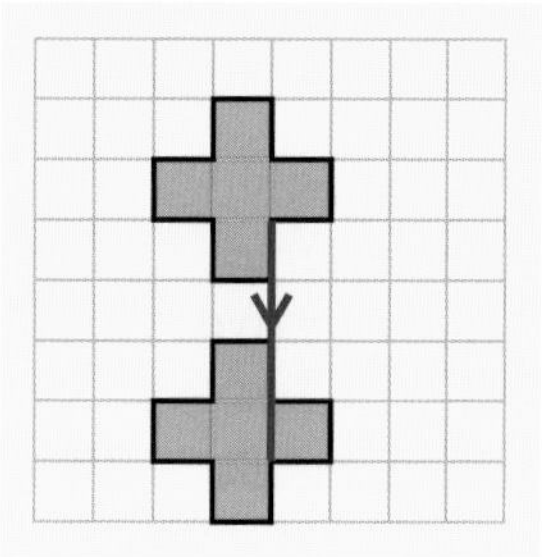

2 Copy this diagram.

a Draw the reflection of **P** in the line $x = 2$.
Label the image **Q**.

b Draw the reflection of **P** in the line $y = 4$.
Label the image **R**.

c Draw the reflection of **P** in the line $y = x$.
Label the image **S**.

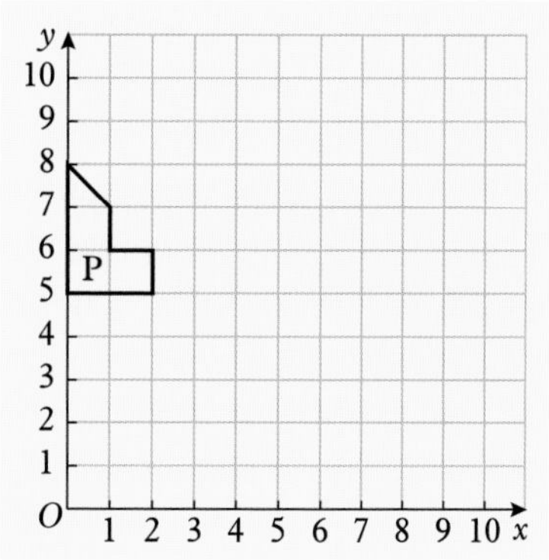

3 Copy this diagram.

a Draw the image of **P** after a rotation through 90° clockwise about O. Label the image **Q**.

b Draw the image of **P** after a rotation through 90° anticlockwise about O. Label the image **R**.

c Draw the image of **P** after a rotation through 180° about O.
Label the image S.

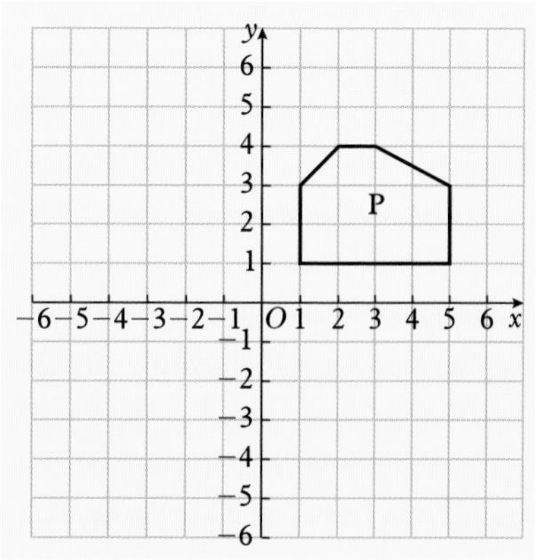

4 Copy this diagram.

a Draw the enlargement of shape **P** with scale factor 2, centre O. Label the image **Q**.

b Draw the enlargement of shape **P** with scale factor $-\frac{1}{2}$, centre O. Label the image **R**.

c Draw the enlargement of shape **P** with scale factor -2, centre O. Label the image **S**.

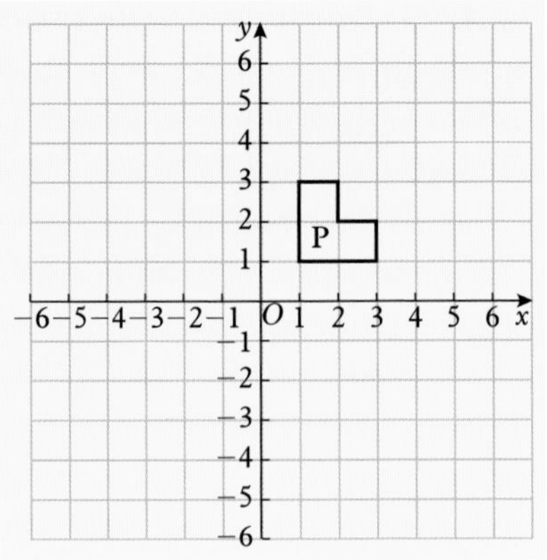

5 Copy the diagram.

a Reflect rectangle ABCD in the line $y = x$ to give rectangle A′B′C′D′.

b Rotate rectangle A′B′C′D′ 90° clockwise about O to get rectangle A″B″C″D″.

c Reflect rectangle A″B″C″D″ in the line $y = -2$ to get rectangle A‴B‴C‴D‴.

d Write down the transformation needed to map rectangle A‴B‴C‴D‴ onto rectangle ABCD.

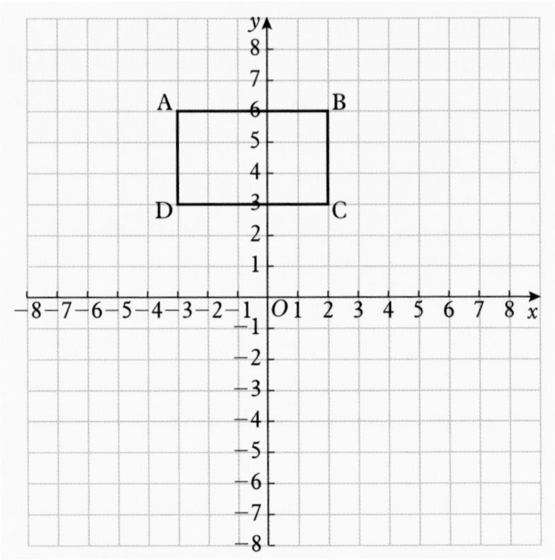

6 Copy the diagram.

a Enlarge shape **A** with scale factor -2 and centre O. Label the image **B**.

b Rotate shape **B** through 180° about O. Label the image **C**.

c Write down the transformation needed to map shape **C** onto shape **A**.

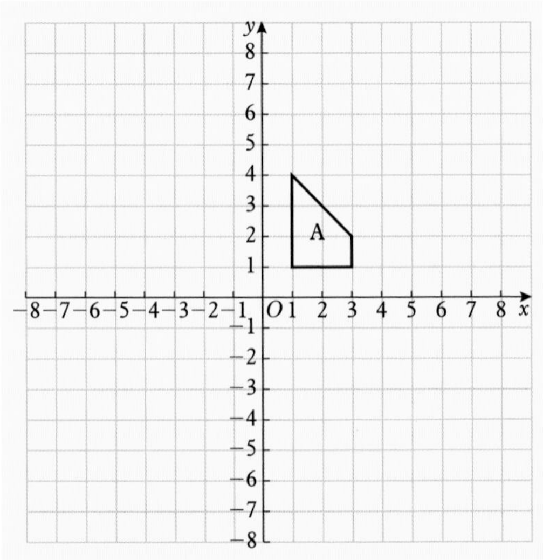

1 Describe the transformation that maps shape **A** onto shape **B**. **(2 marks)**

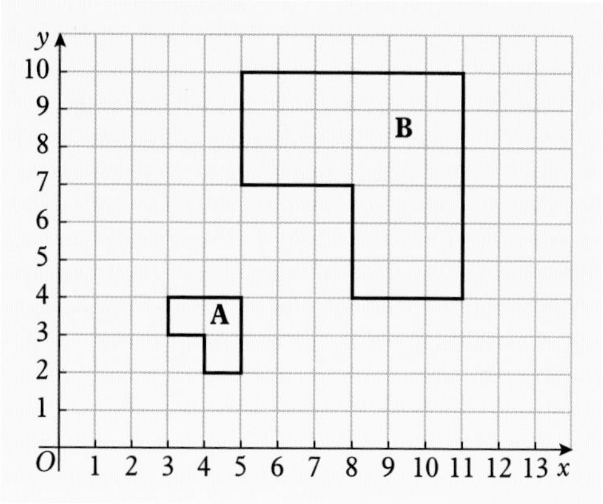

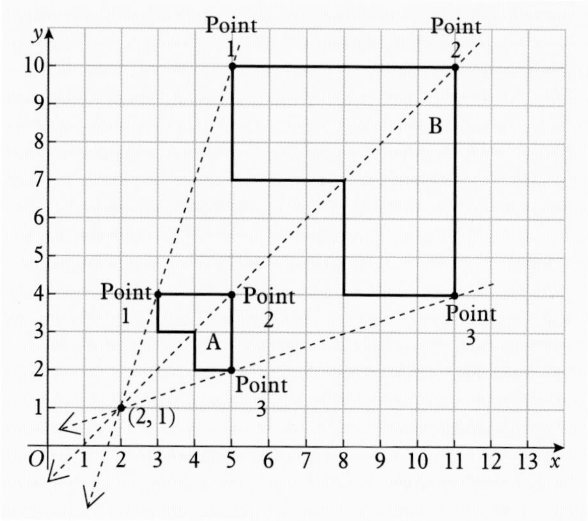

You must first decide what type of transformation you have been asked about. This is an enlargement. The object and image are not the same size. Then you need to give the details about the transformation. For an enlargement, you need the centre of enlargement and the scale factor.

Pick out three points on one shape, and pick out the three corresponding points on the other shape.

Join corresponding points with a line. Use a pencil, as it is easy to correct any mistakes.

Extend the lines until they meet at a single point.

Centre of enlargement = (2, 1). **1 mark**

The height of the large shape (the image) is 6 squares, the height of the small shape (the object) is 2 squares, the scale factor is $6 \div 2 = 3$. **1 mark**

So, the transformation is an enlargement with centre (2, 1) and scale factor 3.

2

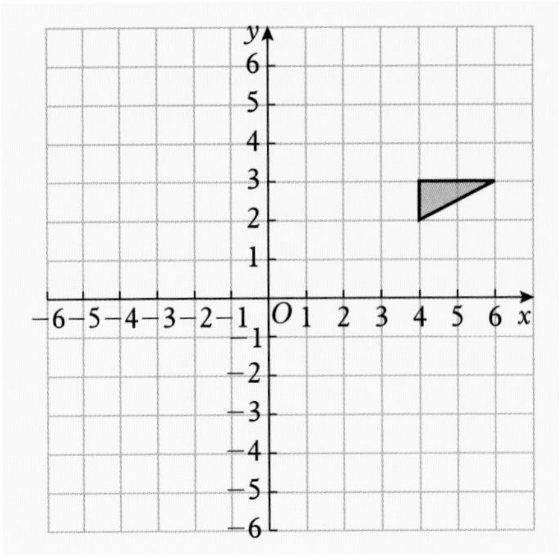

a Reflect the triangle in the line $y = x$. **(1 mark)**

b Rotate the image 90° anticlockwise about the origin. **(1 mark)**

c Describe the single transformation that replaces **a** and **b**. **(2 marks)**

Start by drawing the line $y = x$.

a *Use tracing paper to reflect the triangle across the line.*

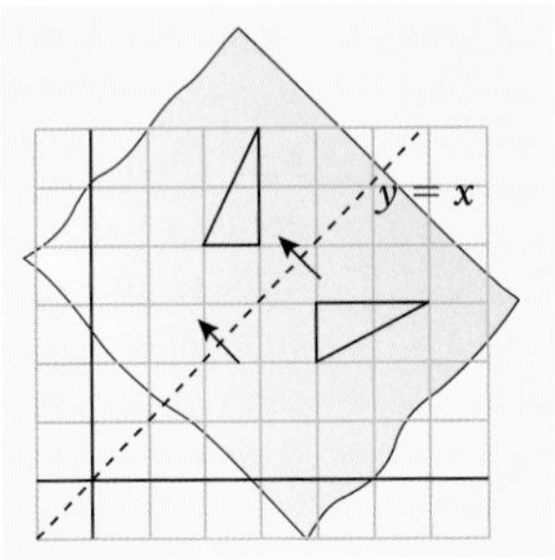

1 mark

b *Use tracing paper to rotate this shape 90°.*

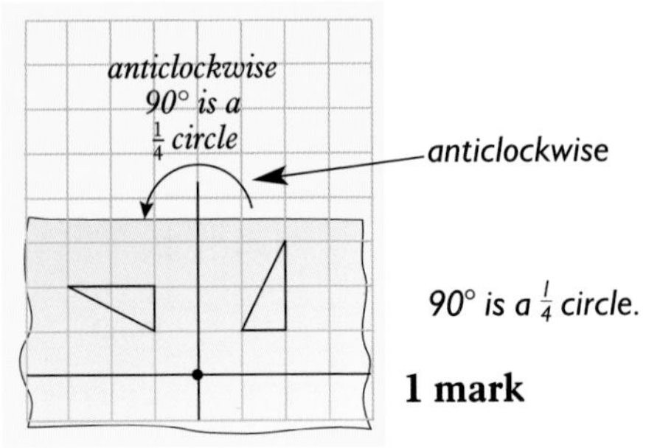

1 mark

c *Now look at the shape you started with and the answer to* **b**. *To get straight to the answer needs a reflection in the y axis.*

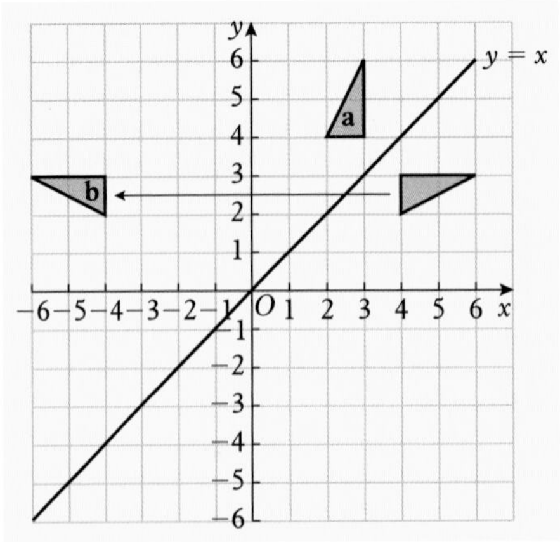

1 mark: reflection
1 mark: y axis

1

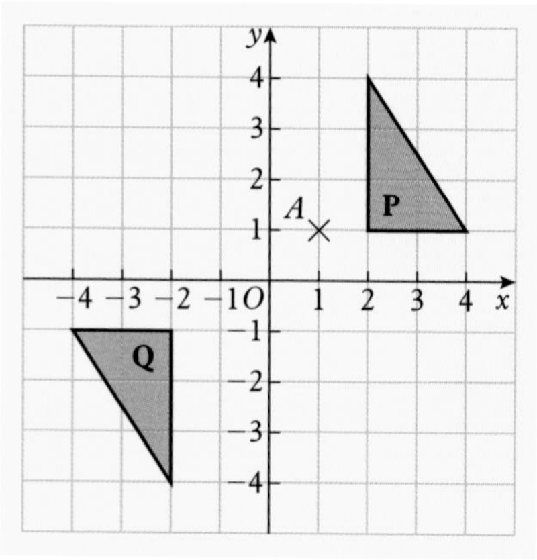

a Describe fully the single transformation that maps shape **P** onto shape **Q**. **(2 marks)**

b Rotate shape **P** 90° anticlockwise about the point $A(1, 1)$. **(2 marks)**

[S2000 P5 Q6]

2

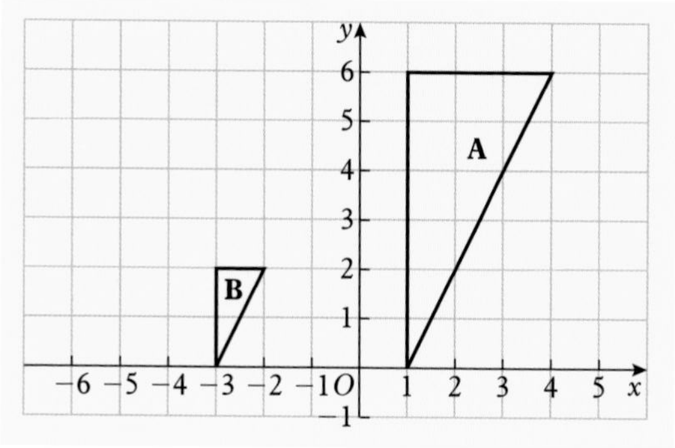

Describe fully the single transformation which maps triangle **A** to triangle **B**. **(3 marks)**

[N1999 P5 Q6]

3

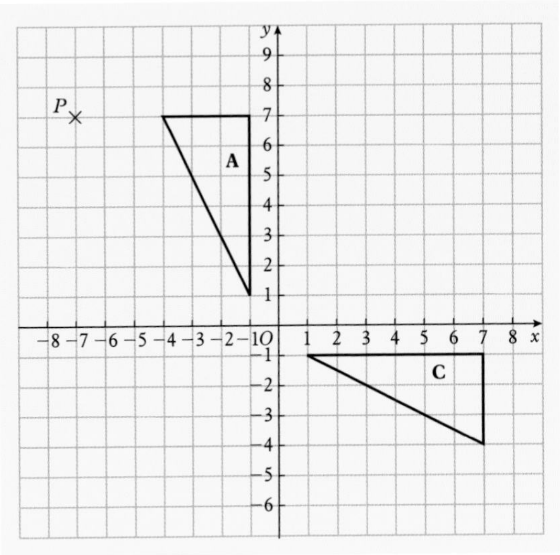

a Enlarge triangle **A** by the scale factor $\frac{1}{3}$ with centre the point $P(-7, 7)$. **(2 marks)**

b Describe fully the single transformation which maps triangle **A** onto triangle **C**. **(2 marks)**

[N2000 P5 Q3]

4 The shape **P** has been drawn on the grid.

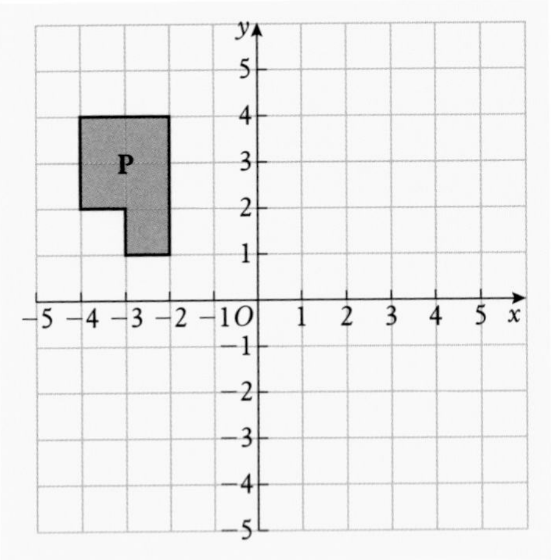

a Reflect the shape **P** in the y axis.
Label the image **Q**. **(1 mark)**

b Rotate the shape **Q** through 180° about (0, 0).
Label this image **R**. **(1 mark)**

c Describe fully the single transformation which maps shape **P** to shape **R**. **(2 marks)**

[N1998 P5 Q4]

5 Shape **A** is shown on the grid.
Shape **A** is enlarged, centre (0, 0), to obtain shape **B**.
One side of shape **B** has been drawn for you.

a On the grid, complete shape **B**. **(2 marks)**

b Enlarge shape **A** by scale factor $\frac{1}{2}$, centre (5, 16).
Label your enlargement **C**. **(2 marks)**

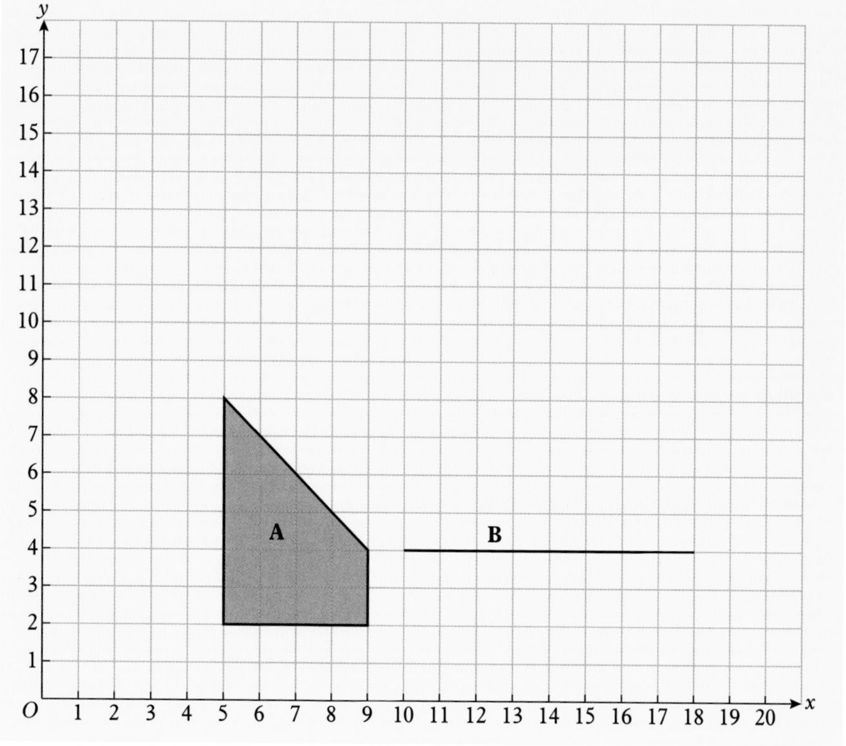

[N2001 P5 Q5]

6 Shape **A** is shown in the diagram.
Shape **A** is enlarged to obtain the shape **B**.
One side of shape **B** has been drawn.

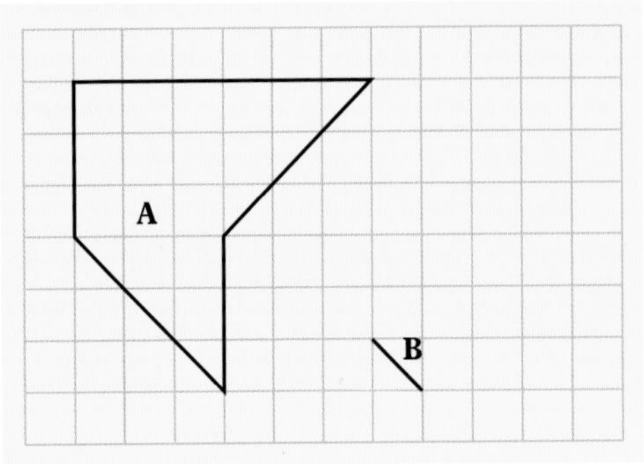

a Write down the scale factor of the enlargement. **(1 mark)**

b Complete the drawing of shape **B** on the diagram. **(2 marks)**

[S1997 P5 Q8]

2 Rational and irrational numbers

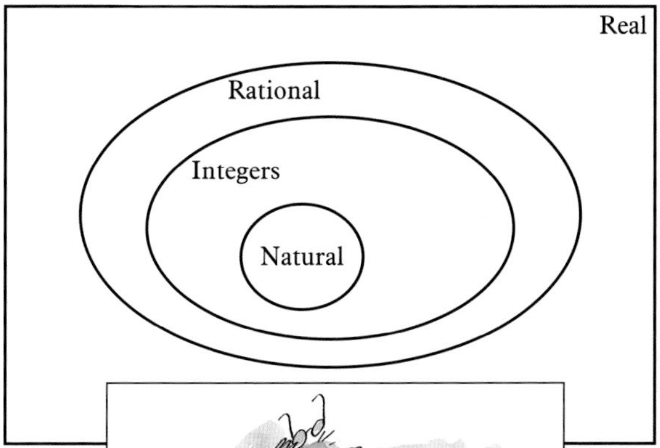

You need to know about:

- types of numbers
- rational and irrational numbers
- factors
- prime factors
- HCFs
- multiples
- common multiples
- LCMs
- recurring decimals
- surds
- rationalising the denominator
- simplifying surds

Rational number

A **rational number** is a number that can be written as a fraction $\frac{a}{b}$, where a and b are integers and $b \neq 0$.

Factor

If a and b are natural numbers, a is a factor of b if $b \div a$ is also a natural number.
For example, 2 is a factor of 10 because $10 \div 2 = 5$.

Prime factor

A factor of a number that is also a prime number is called a **prime factor**. You can write a number as the product of prime factors in only one way.
$12 = 2 \times 2 \times 3 = 2^2 \times 3$
There is no other combination of prime factors of 12.

Highest common factor HCF

The **highest common factor** of two or more numbers is the largest number that is a factor of those numbers.
For example, 8 is the highest common factor of 24 and 32.
$24 = 2 \times 2 \times 2 \times 3$
$32 = 2 \times 2 \times 2 \times 2 \times 2$
Write out all of the common factors
$\text{HCF} = 2 \times 2 \times 2$
$\text{HCF} = 8$

Multiple

A **multiple** of a is any number that has a as a factor. The multiples of 5 are 5, 10, 15, ...

Lowest common multiple LCM

The **lowest common multiple** of two or more numbers is the smallest number that is a multiple of the given numbers.
The LCM of 12 and 15 is 60.
Write out the prime factors of each number
$12 = 2 \times 2 \times 3$
$15 = 3 \times 5$
Write out all of the first list and add any numbers from the second list that aren't already included
$\text{LCM} = 2 \times 2 \times 3 \times 5 = 60$

Writing decimals as fractions

Every terminating decimal can be written as a fraction.
$0.2 = \frac{2}{10}$
$0.31 = \frac{31}{100}$
$0.487 = \frac{487}{1000}$
Every recurring decimal can also be written as a fraction. To do this you need to look at the recurring part and use a denominator of 9, 99, ...
$0.\dot{2} = \frac{2}{9}$
$0.\dot{3}\dot{1} = \frac{31}{99}$
$0.\dot{4}8\dot{7} = \frac{487}{999}$

Adding and subtracting fractions

Write all the fractions with a common denominator.
$\frac{1}{3} + \frac{3}{7} = \frac{7}{21} + \frac{9}{21} = \frac{16}{21}$
$\frac{3}{4} - \frac{2}{5} = \frac{15}{20} - \frac{8}{20} = \frac{7}{20}$

Multiplying fractions

Multiply the numerators and the denominators.
$\frac{2}{3} \times \frac{6}{7} = \frac{12}{21} = \frac{4}{7}$
Change all mixed numbers into improper fractions.
$2\frac{3}{4} \times \frac{2}{5} = \frac{11}{4} \times \frac{2}{5} = \frac{22}{20} = 1\frac{2}{20} = 1\frac{1}{10}$

Dividing fractions

Turn the second fraction over and multiply.
$\frac{1}{8} \div \frac{5}{16} = \frac{1}{8} \times \frac{16}{5} = \frac{16}{40} = \frac{2}{5}$
$2\frac{1}{4} \div 1\frac{1}{2} = \frac{9}{4} \div \frac{3}{2} = \frac{9}{4} \times \frac{2}{3} = \frac{18}{12} = \frac{3}{2} = 1\frac{1}{2}$

Surd

Any square root is irrational unless it is the square root of a square number.
So $\sqrt{2}$ is irrational but $\sqrt{9}$ is rational as $\sqrt{9} = 3$.
An irrational square root is called a **surd**.

Rules for simplifying surds

For any integers m and n:

$$\sqrt{m} \times \sqrt{n} = \sqrt{mn},$$

$$\sqrt{mn} = \sqrt{m} \times \sqrt{n},$$

$$\frac{\sqrt{m}}{\sqrt{n}} = \sqrt{\frac{m}{n}}$$

$$\sqrt{\frac{m}{n}} = \frac{\sqrt{m}}{\sqrt{n}}.$$

Rationalising the denominator

You can simplify a fraction that has a surd in the denominator. You do this by multiplying the top and bottom by the same expression that makes the denominator rational.

To rationalise the denominator of $\frac{2}{\sqrt{3}}$ multiply top and bottom by $\sqrt{3}$.

$$\frac{2}{\sqrt{3}} = \frac{2}{\sqrt{3}} \times \frac{\sqrt{3}}{\sqrt{3}} = \frac{2\sqrt{3}}{3}$$

To rationalise the denominator of $\frac{2}{2+\sqrt{3}}$ multiply top and bottom by $2 - \sqrt{3}$.

$$\begin{aligned}\frac{2}{2+\sqrt{3}} &= \frac{2}{2+\sqrt{3}} \times \frac{2-\sqrt{3}}{2-\sqrt{3}} \\ &= \frac{2(2-\sqrt{3})}{(2+\sqrt{3})(2-\sqrt{3})} \\ &= \frac{4-2\sqrt{3}}{4-3} \\ &= 4-2\sqrt{3}\end{aligned}$$

Further simplification with surds

When you add or subtract surds, it may be possible to collect like terms.

$$\begin{aligned}\sqrt{50} + 2\sqrt{72} + 3\sqrt{98} &= \sqrt{25 \times 2} + 2 \times \sqrt{36 \times 2} + 3 \times \sqrt{49 \times 2} \\ &= 5\sqrt{2} + 2 \times 6 \times \sqrt{2} + 3 \times 7 \times \sqrt{2} \\ &= 5\sqrt{2} + 12\sqrt{2} + 21\sqrt{2} \\ &= 38\sqrt{2}\end{aligned}$$

1 Look at this list of numbers.

3 -7 2.1 $\frac{1}{3}$ 0 16 $\frac{3}{7}$ 5 $0.\dot{7}$ 25 15 $\sqrt{5}$

Write down the numbers in this list that are

a integers

b natural numbers

c rational numbers

d prime numbers

e square numbers

f triangle numbers.

2 Write down all the factors of 30.

3 **a** Write 36 as a product of primes.
b Write 42 as a product of primes.
c Using your answers to **a** and **b**, work out the HCF of 36 and 42.

4 Write down the first 5 multiples of 7.

5 **a** Write 24 as a product of primes.
b Write 14 as a product of primes.
c Using your answers to **a** and **b**, work out the LCM of 24 and 14.

6 Write these decimals as fractions.

a 0.3

b 0.184

c $0.\dot{7}$

d $0.\dot{3}\dot{6}$

7 Is $\sqrt{\frac{25}{4}}$ rational or irrational? Explain your answer.

8 Find:
a a rational number between $\sqrt{8}$ and $\sqrt{10}$
b an irrational number between $\sqrt{8}$ and $\sqrt{10}$.

9 Are these numbers rational or irrational?
Explain your answers.

a $13 + \sqrt{2}$

b $\sqrt{5} \times \sqrt{6}$

c 3.12

d $\pi - 1$

e $\sqrt{12} \times \sqrt{3}$

f $\sqrt{34} \div \sqrt{2}$

g 2π

h $1.2 + \sqrt{3}$

10 **a** Write $\frac{3}{7}$ as a decimal.
b Write $\frac{3}{70}$ as a decimal.
Use your answer to **a** to help you.
c Write 0.2 as a fraction.
d Work out a fraction that is equivalent to $0.2\dot{4}2857\dot{1}$.

11 Simplify these expressions.

a $\sqrt{2} \times \sqrt{10}$

b $\sqrt{32}$

c $\sqrt{3} \times 2\sqrt{5}$

d $\sqrt{2\frac{1}{4}}$

e $\dfrac{3}{\sqrt{5}}$

f $\sqrt{24} \div \sqrt{2}$

g $\dfrac{7}{\sqrt{13}}$

h $\dfrac{2 + \sqrt{2}}{\sqrt{2}}$

12 Expand these brackets.

a $(6 + \sqrt{5})(3 - \sqrt{5})$

b $(3 + \sqrt{2})(3 + \sqrt{2})$

c $(3 + \sqrt{2})(3 - \sqrt{2})$

d $(2 - \sqrt{3})(2 + \sqrt{3})$

13 Simplify these expressions as far as possible.

a $\sqrt{8} + \sqrt{2}$

b $\sqrt{3} + \sqrt{27}$

c $\sqrt{54} - \sqrt{6}$

d $\sqrt{180} - \sqrt{80}$

14 Write down whether each of these numbers can be written as a recurring decimal.

a $\sqrt{2}$

b $\sqrt{\dfrac{4}{81}}$

c $2 + \pi$

d $\sqrt[3]{20}$

e $\dfrac{3}{4 + \sqrt{12}}$

f $\sqrt{5\frac{4}{9}}$

15 **a** Write down the nth even number.
b Write down the mth odd number.
c Add your answers to **a** and **b** together to show that the sum of any even number and any odd number is odd.
d Multiply your answers to **a** and **b** to show that the product of any even number and any odd number is even.

16 **a** Write down the nth even number.
b Write down the $(n + 1)$th even number.
c Write down the $(n + 2)$th even number.
d Use your answers to **a** to **c** to show that the sum of three consecutive even numbers is a multiple of 6.

1 When 56 is written as the product of its prime factors, in index form, we obtain $56 = 2^3 \times 7$.

a Write 126 as the product of its prime factors in index form. **(1 mark)**

b Write down 126×56 as a product of prime factors in index form. **(1 mark)**

c Write down the square root of your answer to **b** as a product of prime factors. **(1 mark)**

It is likely that this type of question will appear on a non-calculator paper.

a *Find the prime factors of 126:*

$$\begin{array}{r|l} 2 & 126 \\ 3 & 63 \\ 3 & 21 \\ 7 & 7 \\ & 1 \end{array}$$

Don't forget to write down the answer:

$126 = 2 \times 3 \times 3 \times 7$

Use powers of the prime numbers as necessary

$= 2 \times 3^2 \times 7$ **1 mark**

b $56 = 2^3 \times 7$

$126 = 2 \times 3^2 \times 7$

So $56 \times 126 = 2^3 \times 7 \times 2 \times 3^2 \times 7$

$= 2^3 \times 2 \times 3^2 \times 7 \times 7$

$= 2^4 \times 3^2 \times 7^2$

Show what you are doing with the numbers and simplify the powers.

1 mark

c $\sqrt{2^4 \times 3^2 \times 7^2} = \sqrt{2^4} \times \sqrt{3^2} \times \sqrt{7^2}$

$= 2^2 \times 3 \times 7$

When you square root you halve the power.

The answer is already in prime factor form.

1 mark

You could work out 56×126 and then find the prime factors but this would be more difficult.

2 Given that $(\sqrt{5} - \sqrt{2})^2 = p - q\sqrt{10}$, find the values of p and q. **(3 marks)**

First work out the brackets.

A common error is to write $(\sqrt{5} - \sqrt{2})^2 = (\sqrt{5})^2 - (\sqrt{2})^2$.

$(\sqrt{5} - \sqrt{2})^2 = (\sqrt{5} - \sqrt{2})(\sqrt{5} - \sqrt{2})$

$= (\sqrt{5})^2 - \sqrt{5}\sqrt{2} - \sqrt{5}\sqrt{2} + (\sqrt{2})^2$ **1 mark**

$= 5 - \sqrt{10} - \sqrt{10} + 2$

$= 7 - 2\sqrt{10}$ **1 mark**

Having worked this out, remember to write down the answers to the question.

$7 - 2\sqrt{10} = p - q\sqrt{10}$

So $p = 7$, $q = 2$. **1 mark**

3 Write $\frac{\sqrt{3}}{2 + 3\sqrt{3}}$ in the form $a + b\sqrt{c}$. **(3 marks)**

You first need to get rid of the common denominator as written with surds.

This is two separate terms and has no common denominator.

The "difference of two squares" can be used to rationalise the denominator.

$$\frac{\sqrt{3}}{2 + 3\sqrt{3}} \times \frac{(2 - 3\sqrt{3})}{(2 - 3\sqrt{3})} = \frac{\sqrt{3}(2 - 3\sqrt{3})}{(2 + 3\sqrt{3})(2 - 3\sqrt{3})}$$

Multiply numerator and denominator by the same expression. Use the denominator with the opposite sign in the middle.

$$= \frac{\sqrt{3}(2 - 3\sqrt{3})}{4 - 6\sqrt{3} + 6\sqrt{3} - (9 \times 3)}$$

1 mark

$$= \frac{2\sqrt{3} - (3 \times 3)}{4 - 27}$$

1 mark

$$= \frac{-9 + 2\sqrt{3}}{-23}$$

$$= \frac{9}{23} - \frac{2}{23}\sqrt{3}$$

1 mark

Always make sure you write the answer in exactly the form asked for in the question.

So $a = \frac{9}{23}$, $b = -\frac{2}{23}$ and $c = 3$.

1

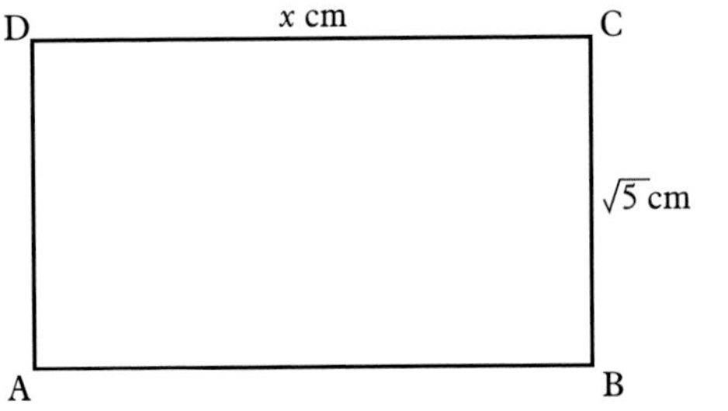

Diagram **NOT** accurately drawn

The diagram shows a rectangle ABCD with width $\sqrt{5}$ and length x cm.
The length and the width of the rectangle are not equal.

a Find a value for x for which the area of the rectangle is a rational number of square centimetres. **(1 mark)**

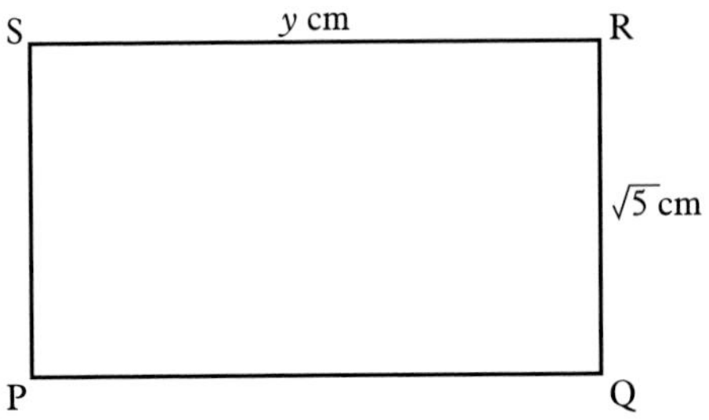

Diagram **NOT** accurately drawn

The diagram shows a rectangle PQRS with width $\sqrt{5}$ and length y cm.

b Find a value of y for which the length of the rectangle is a rational number of square centimetres. **(2 marks)**

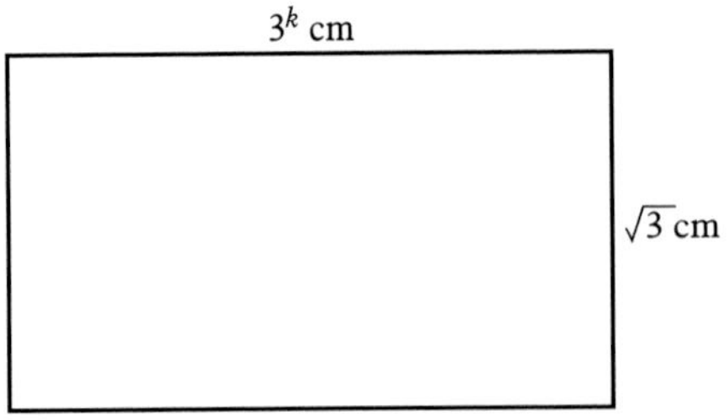

The area of the rectangle is 81 cm^2.

c Work out the value of k. **(3 marks)**

[N2000 P5 Q15]

2 **a** Find the value of

(1) m when $\sqrt{128} = 2^m$ **(2 marks)**

(2) n when $(\sqrt{8} - \sqrt{2})^2 = 2^n$ **(2 marks)**

A rectangle has a length of 2^t cm and a width of $(\sqrt{8} - \sqrt{2})$ cm. The area of the rectangle is $\sqrt{128}$ cm^2.

b Find t. **(2 marks)**

[S2001 P5 Q19]

3 n is an integer greater than 1.

$\sqrt{n}$ is a rational number.

a What type of number is n? **(1 mark)**

b Simplify $(\sqrt{n+4} + \sqrt{n})(\sqrt{n+4} - \sqrt{n})$. **(2 marks)**

c Hence, or otherwise, find a number k, greater than $\frac{1}{4}$, such that the product

$$k(\sqrt{20} + 4)$$

is **rational**. **(1 mark)**

[N1999 P6 Q16]

4 **a** Find the prime factors of 4891 by writing as $70^2 - 3^2$. **(2 marks)**

b By writing the nth term of the sequence 1, 3, 5, 7, . . . as $(2n - 1)$, or otherwise, show that the difference between the squares of any two consecutive odd numbers is a multiple of 8. **(3 marks)**

[S1999 P6 Q16]

5 **a** Find the value of n in the equation

$$2^n = \sqrt{8}$$

(2 marks)

Triangle ABC has an area of 32 cm^2.

b Calculate the value of k.

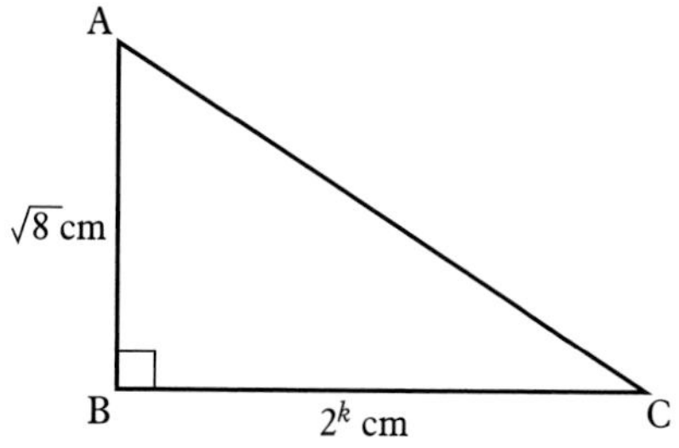

(3 marks)

[S1998 P5 Q9]

6 n is a **rational** number.

$k \times \sqrt{8} = n$

a Write down a non-zero value of k and a non-zero value of n which satisfy this equation. **(2 marks)**

b Write $\sqrt{8}$ in the form 2^c, where c is a rational number. **(1 mark)**

c $4^{d+1} = \sqrt{8}$

Find the value of d. **(2 marks)**

[S1999 P6 Q11]

7 **a** Give an example of two different irrational numbers q and r such that

$\frac{q}{r}$ is a rational number. **(2 marks)**

b Write down a number which is greater than 15 and less than 16 and which has a rational cube root. **(2 marks)**

[N1997 P5 Q14]

8 **a** $\sqrt{2\frac{1}{4}}$ $\quad \sqrt{8}$ $\quad \sqrt{2\frac{1}{2}}$ $\quad \sqrt{2\frac{3}{4}}$ $\quad \sqrt{3}$

Put a ring around any of the above numbers that are rational. **(1 mark)**

b Find the exact value of

$(\sqrt{7} + \sqrt{5})(\sqrt{7} - \sqrt{5})$ **(2 marks)**

c The number $(\sqrt{7} + \sqrt{5})$ is irrational.
Use your answer to part **b** to explain why the number $(\sqrt{7} - \sqrt{5})$ must also be irrational. **(1 mark)**

[N2001 P5 Q17]

9 Put a tick in the box underneath the rational numbers.

$\frac{\sqrt{12}}{\sqrt{3}}$	$\sqrt{10} - 3$	$\sqrt{9.5}$	$\sqrt{\frac{1}{36}}$	$\sqrt{12}$

(2 marks)

[S1997 P5 Q17]

3 Dealing with data

You need to know about:

- pie-charts
- scatter graphs
- lines of best fit
- correlation
- frequency polygons
- histograms with equal width groups
- histograms with unequal width groups
- frequency density

Pie chart

A **pie-chart** shows how something is divided up.
The angle of the sector represents the number of items in the sector.
You cannot read off accurate figures from a pie-chart.
It gives a quick visual representation of the data.

You will need to work out the angle for each sector.
To do this:
(1) Work out the total of the frequencies, n.
(2) Work out $360° \div n$.
This is the angle for each one of the data values.
(3) Multiply your answer to (2) by each frequency in turn to get the angle for each sector.

Scatter graph

A **scatter graph** is a diagram that is used to see if there is a connection between two sets of data.
One value goes on the x axis and the other goes on the y axis.
It does not matter which way round the data is plotted.

Line of best fit

If the points on a scatter graph lie roughly in a straight line then you can draw the **line of best fit**. This is the line that the points appear to be scattered around. When you draw a line of best fit, make sure that it is pointing in the same direction as the points appear to be following and that there are roughly the same number of points on either side.

You can use the line of best fit to read off estimates for missing values.
Always show the lines that you use as you go up and across or across and down.

Correlation

When there is a relationship between two sets of data, this is called **correlation**.

If you can draw a line of best fit on a scatter graph then there is correlation between the two sets of data being shown.
You have positive correlation if one data value increases as the other data value increases.
You have negative correlation if one data value decreases as the other data value increases.
Correlation is strong if the points are close to the line of best fit. Otherwise you can have weak correlation or no correlation.

Frequency polygon

A **frequency polygon** shows grouped data. It is drawn by plotting the frequency for each group at the mid-point of each group and then joining the points up with straight lines. The horizontal axis must be drawn as a graph scale. You must not label the groups like in a bar-chart.

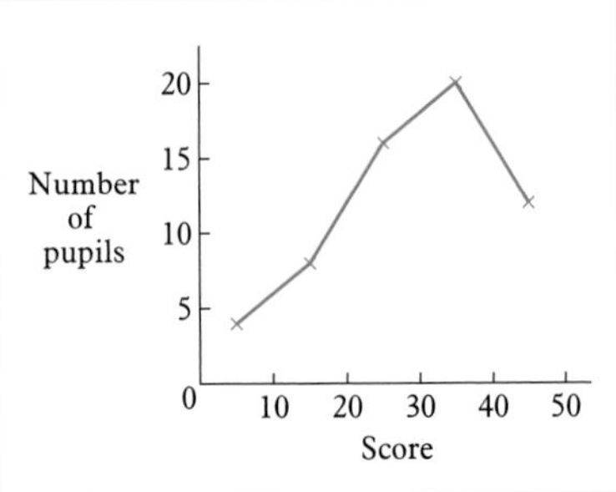

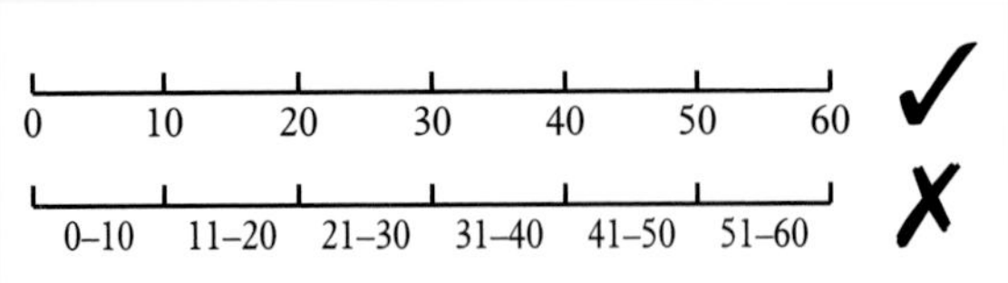

Histogram

A **histogram** is like a bar-chart but it can only be used for continuous data. The horizontal axis must be drawn as a graph scale as for a frequency polygon.
Histograms can be used with either equal width groups or unequal width groups. If you use equal width groups, you label the vertical axis frequency.

If the groups have unequal widths, you need to remember that it is the area of a histogram that represents the frequency and not the heights of the bars. You can choose a standard width and label the vertical axis *frequency per standard width*. So if your standard width is 10 marks, the label is *frequency per 10 marks*. Then you adjust the heights of the bars. If you have a width that is twice the standard width you halve the frequency to get the height and so on.

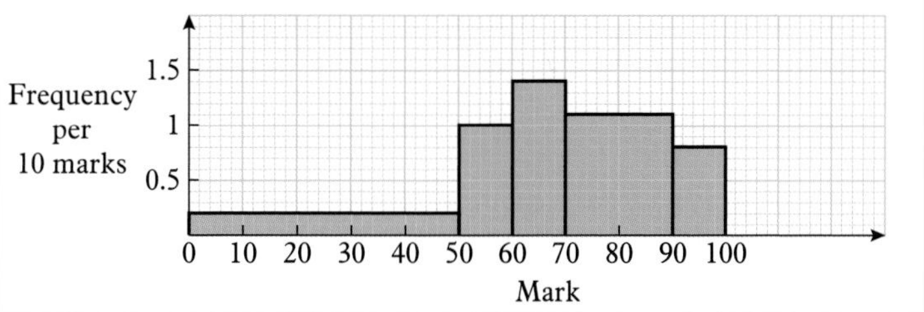

Frequency density

The **frequency density** is the frequency divided by the class width. To draw a histogram, you work out the frequency divided by the class width for all of the groups. This method has the advantage that it doesn't matter if there isn't an obvious standard width to use.
The final histogram will look exactly the same whichever method you use.

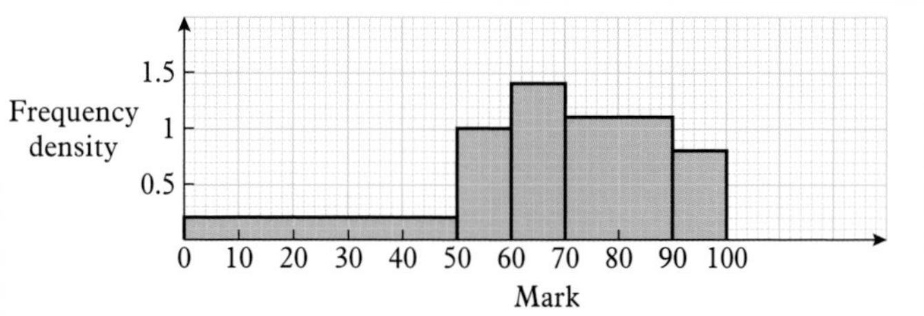

The frequency density is actually the frequency per unit width. You can work out the number of data values in each group by multiplying the height of the bar by the width.

If you have groups with gaps between them, you must adjust the class boundaries to get rid of the gaps. For example, if you have lengths to the nearest cm given in groups as 1–10, 11–20, etc. then you must use $0.5 \le \text{length} < 10.5$, $10.5 \le \text{length} < 20.5$, etc. This is because each length is only accurate to the nearest cm so it would round up to the next value if you pass halfway.
If you have ages given as 0–9, 10–19, etc. you must use $0 \le \text{age} < 10$, $10 \le \text{age} < 20$, etc. This is because age is not rounded up when you pass half a year.

1 This table shows the nutritional breakdown in a 50 ml serving of gravy granules.

	Protein	Carbohydrate	Fat	Fibre	Sodium
Weight (g)	0.1	2.2	0.6	0.3	0.2

a Draw a pie-chart to show this information.

Contained in the carbohydrate figure there is 0.5 g of sugars and with the fat figure there is 0.3 g of saturated fat.

b Draw a new pie-chart to show all of this information.

2 This table shows the trade-in value of a Ford Ka 2, with its age given in months.

Age in months, A	18	24	30	36	42	48
Value in £, V	4350	4050	3700	3400	3250	2950

Draw a scatter graph to show these values on graph paper.
Use a scale of 2 cm for 6 months on the age axis, starting from 0.
Use a scale of 2 cm for £500 on the value axis, going up to £6000.

a Describe the relationship between the value of the Ka 2 and its age.

b Draw a line of best fit on your scatter graph.

c Use your line of best fit to estimate the value of a Ka 2 that is 33 months old.

d The price of a new Ka 2 is £8000. Use your line of best fit to estimate the value of a new Ka 2.
Can you explain why the graph gives such a poor estimate of the new value? Write down your answer.

3 Mrs Jamieson organised an egg and spoon race for her Year 2 class. These are the times that the children took to finish the course.

Time in seconds, t	$30 \le t < 40$	$40 \le t < 50$	$50 \le t < 60$	$60 \le t < 200$
Number of children (frequency)	6	8	12	14

a Copy the table.

b Add a third row to your table. Label it class width. Fill it in.

c Add a fourth row to your table. Label it frequency density. Fill it in.

d Draw a histogram of these results.

e Why do you think that the last group is so wide?

4 Mr Zing and Mr Plod are dentists.
Miss Informed, the office manager, wants to compare the times that they spend with patients.
This is the data that she has collected for their last 40 patients.

Time spent in minutes t	Number of patients Mr Zing	Number of patients Mr Plod
$0 \le t < 5$	12	5
$5 \le t < 10$	10	10
$10 \le t < 15$	8	9
$15 \le t < 20$	6	6
$20 \le t < 25$	4	10

a Draw a frequency polygon for each dentist on the same axes.

b What deductions will Miss Informed draw from this data?

c How could she improve her survey?
Write down what you would recommend.

1 The times taken by students to complete a piece of homework is given in the table.

Times taken in minutes, t	$5 \leq t < 10$	$10 \leq t < 15$	$15 \leq t < 20$	$20 \leq t < 25$	$25 \leq t < 30$	$30 \leq t < 35$
Number of students	2	5	6	5	5	1

a What is the median class interval? **(1 mark)**

b Draw a frequency polygon for this information. **(2 marks)**

a The median class interval is the class interval in which the median lies.
The median is the middle value.
There are 24 students, so the middle ones are the 12th and 13th.
By the end of the second group there are $2 + 5 = 7$ students.
By the end of the third group there are $2 + 5 + 6 = 13$ students.
So the 12th and 13th students are in the 15–20 class.
The median class interval is $15 \leq t < 20$. **1 mark**

b

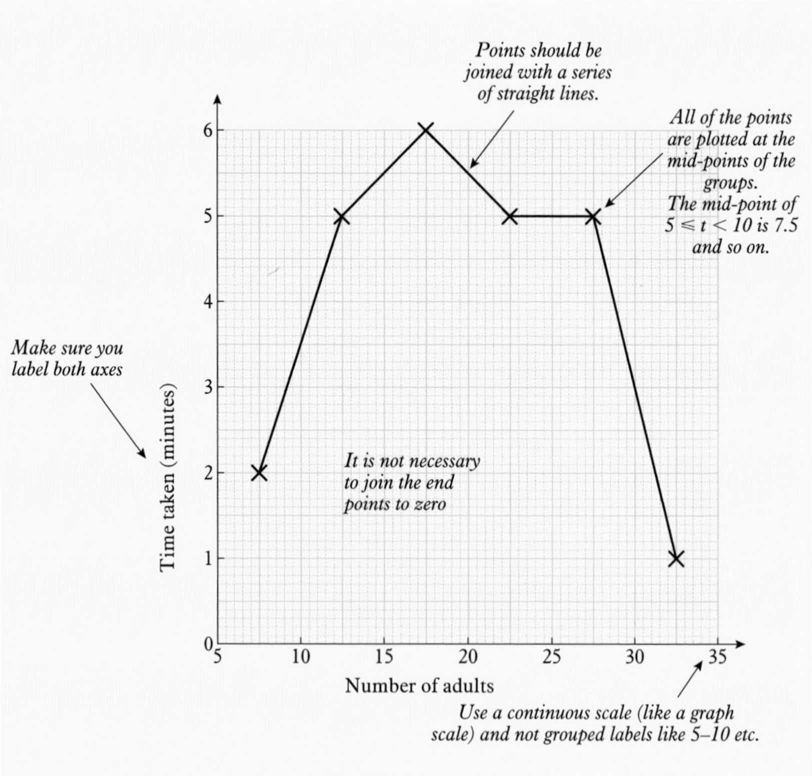
Points should be joined with a series of straight lines.
All of the points are plotted at the mid-points of the groups. The mid-point of 5 ≤ t < 10 is 7.5 and so on.
Make sure you label both axes
Time taken (minutes)
0
1
2
3
4
5
6
It is not necessary to join the end points to zero
5
10
15
20
25
30
35
Number of adults
Use a continuous scale (like a graph scale) and not grouped labels like 5–10 etc.

1 mark: points plotted at the correct heights and joined with straight-line segments
1 mark: points plotted at the mid-points of the classes

2 The incomplete table and histogram show information about some fish caught by fishermen.

Length of fish in cm, L	Frequency, f
$27 \leq L < 35$	70
$35 \leq L < 37$	
$37 \leq L < 39$	
$39 \leq L < 43$	36

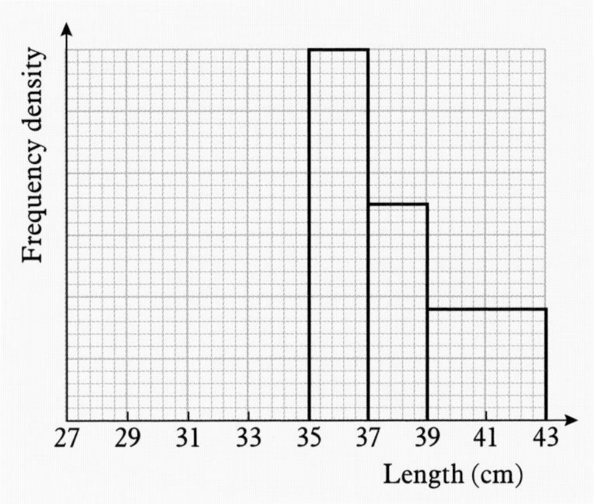

a Complete the table. **(2 marks)**

b Complete the histogram. **(2 marks)**

You have to find a class interval for which there is an entry in the table and a column drawn on the histogram: $39 \leq L < 43$ is the class that you need.
The important thing is to decide on the scale that has been used on the frequency density axis.
The class $39 \leq L < 43$ has a width of 4 and a frequency of 36, giving frequency density $= \frac{36}{4} = 9$.
So on the frequency axis, each small square is 1 unit.

a For $35 \leq L < 37$:
frequency density $= 30$, width $= 2$
frequency $=$ frequency density $\times$ width
$= 30 \times 2$
$= 60$ **1 mark**

Length of fish in cm, L	Frequency, f
$27 \leq L < 35$	70
$35 \leq L < 37$	60
$37 \leq L < 39$	36
$39 \leq L < 43$	36

For $37 \leq L < 39$:
frequency $=$ frequency density $\times$ width
$= 18 \times 2$
$= 36$ **1 mark**

b For $27 \leq L < 35$:
frequency $= 70$, width $= 8$
frequency density $= \frac{70}{8} = 8.75$ **1 mark**
So draw the rectangle shown to complete the histogram. **1 mark**

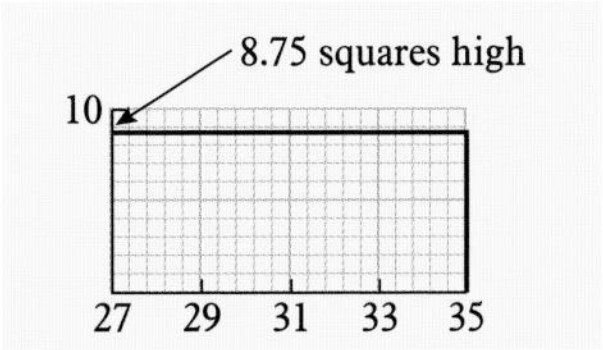

1 The table shows the hours of sunshine and the rainfall, in mm, in 10 towns during last summer.

Sunshine (hours)	650	455	560	430	620	400	640	375	520	620
Rainfall (mm)	10	20	15	29	24	28	14	30	25	20

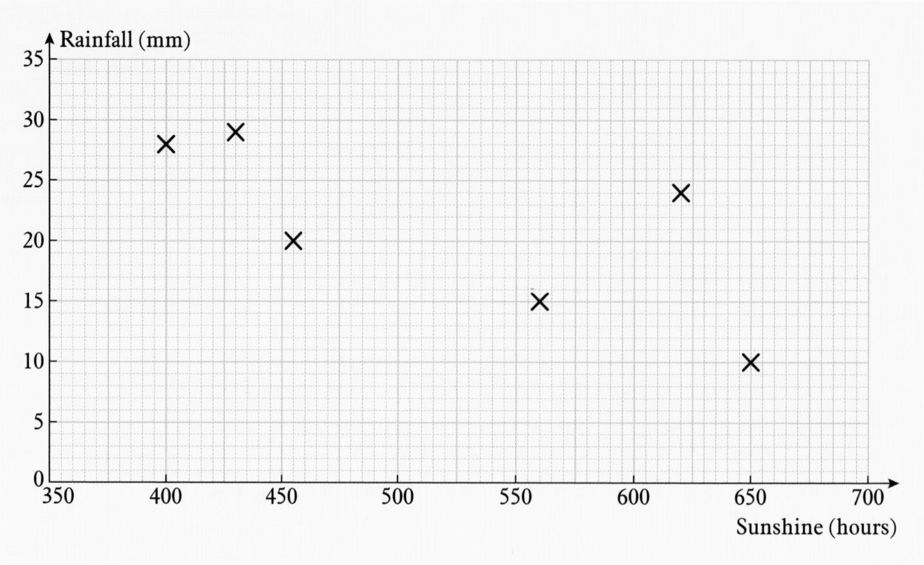

The points for the first six results in the table have been plotted in a scatter diagram.

a Plot the other four points to complete the scatter diagram. **(1 mark)**

b Describe the relationship between the hours of sunshine and the rainfall. **(1 mark)**

c Draw a line of best fit on your scatter diagram. **(1 mark)**

d Use your line of best fit to estimate
(1) the rainfall when there are 450 hours of sunshine, **(1 mark)**
(2) the amount of sunshine when there are 18 mm of rainfall. **(1 mark)**

[S2001 P5 Q1]

2 The scatter graph shows information about fourteen countries. For each country, it shows the birth rate and the life expectancy, in years.

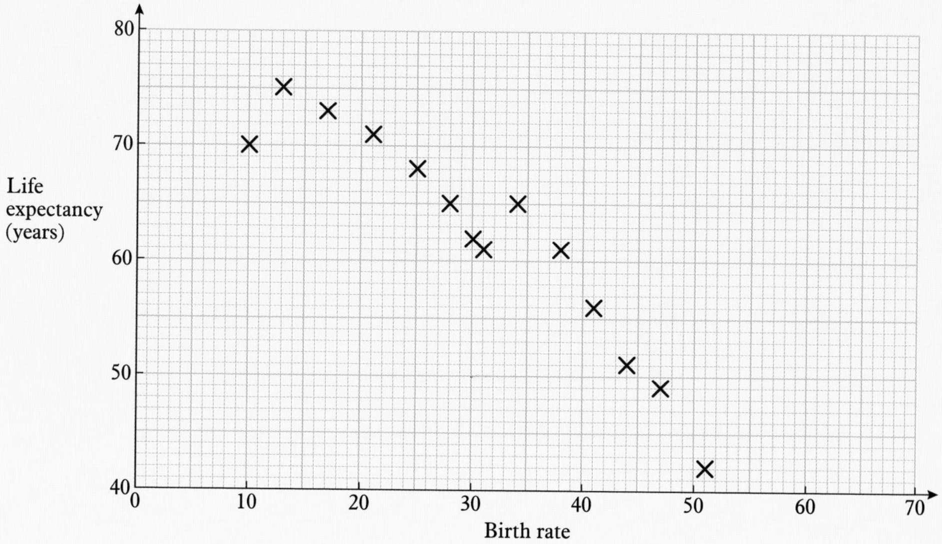

a Draw a line of best fit on the scatter graph. **(2 marks)**

The birth rate in a country is 42.

b Use your line of best fit to estimate the life expectancy in that country. **(1 mark)**

The life expectancy in a different country is 66 years.

c Use your line of best fit to estimate the birth rate in that country. **(1 mark)**

[S2001 P6 Q3]

3 The unfinished histogram and table give information about the heights, in centimetres, of the Year 11 students at Mathstown High School.

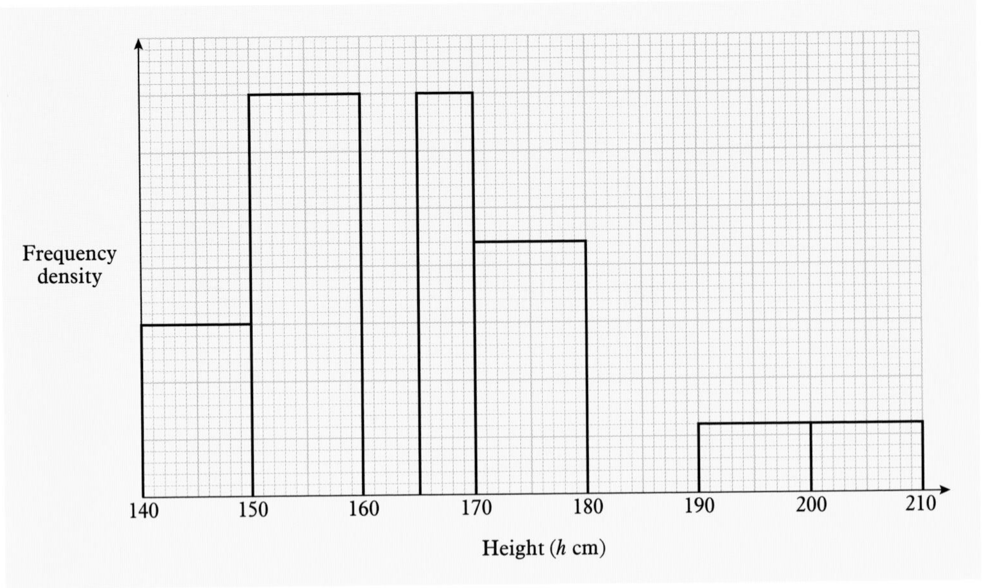

Height (h cm)	Frequency
$140 \leqslant h < 150$	15
$150 \leqslant h < 160$	
$160 \leqslant h < 165$	20
$165 \leqslant h < 170$	
$170 \leqslant h < 180$	
$180 \leqslant h < 190$	12
$190 \leqslant h < 210$	

a Use the histogram to complete the table. **(3 marks)**

b Use the table to complete the histogram. **(3 marks)**
[S2000 P6 Q11]

4 A travel company carried out a survey of the ages of its customers. The results of the survey are shown in the table.

Age group (years)	Percentage of customers in this age group
11–20	8
21–30	28
31–40	19
41–50	21
51–60	13
61–70	11

a On the grid below, draw a frequency polygon to show this information.

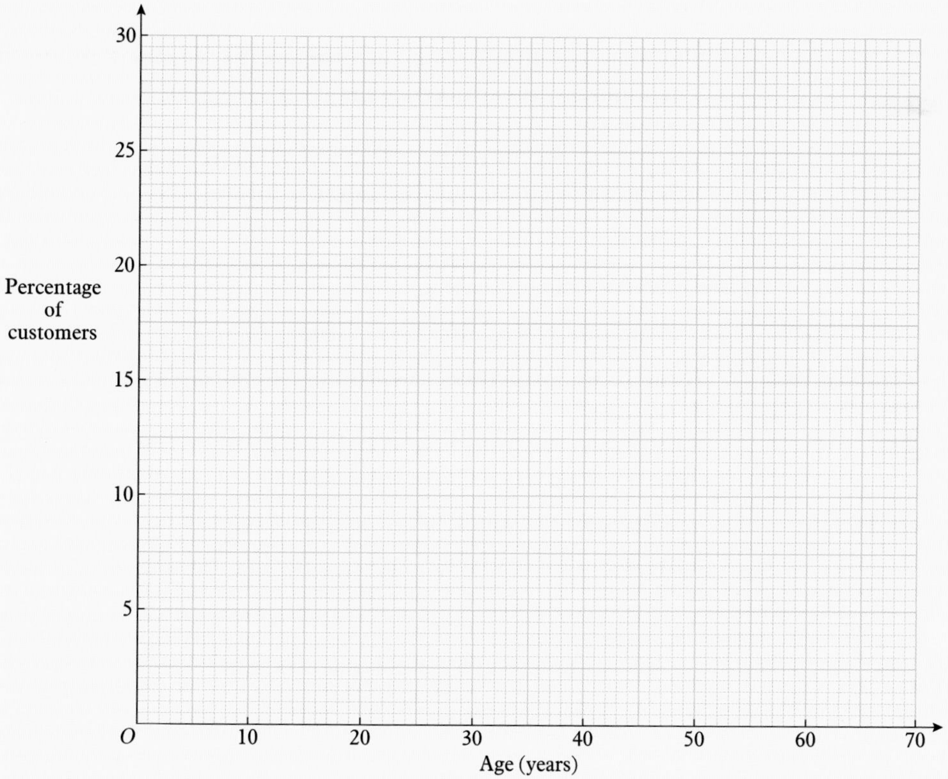

(2 marks)

b In which age group is the median age? **(2 marks)**

[S2000 P6 Q6]

4 Algebra: lining up

You need to know about:

- finding the formula for the *n*th term of sequences
- gradients of straight lines
- equations of straight lines
- solving linear equations
- solving simultaneous equations

Formulas that contain *n*

If the first difference is constant, the formula for the *n*th term of the sequence contains *n*.
The constant difference is the number that goes in front of *n*.
Then look for the difference between the *n* times table and the numbers in the sequence.
You can use the formula to find any term in the sequence.
Substitute the term number into the formula.

Add $5n - 1$	4,	9,	14,	19,	24,	…
	+5	+5	+5	+5		
	−1	−1	−1	−1	−1	
$5n$	5,	10,	15,	20,	25,	

Formulas that contain n^2

If the second difference is constant, the formula for the *n*th term contains n^2.
The number in front of n^2 is **half** the constant second difference.
Take the n^2 part away from the question and solve the rest as above.

Gradient

The **gradient** of a line tells you how steep the line is.

This line has positive gradient,

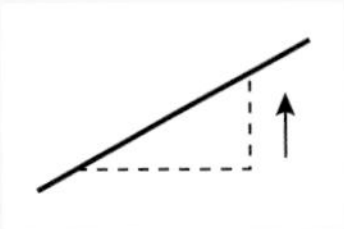

the vertical change is up.

This line has negative gradient,

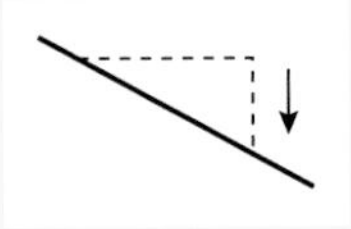

the vertical change is down.

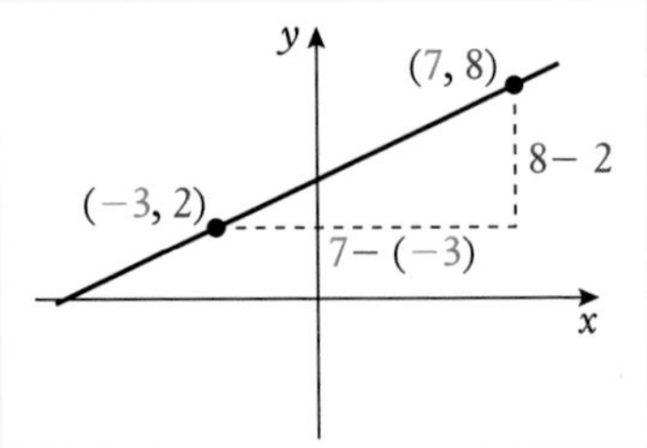

$$\text{Gradient} = \frac{\text{vertical change}}{\text{horizontal change}}$$

$$= \frac{8-2}{7-(-3)} = \frac{6}{10} = \frac{3}{5}$$

If you are not given the points to use, use two points as far away from each other as possible and with co-ordinates that are easy to read off.

Interpreting the equation of a straight line

A hire cost, C, is given by the formula $C = a + bn$ where n is the number of days.
The point where the line cuts the vertical axis is $(0, a)$. This is the fixed charge.
The gradient of the line gives the value of b. This is the daily rate.
For any straight line with equation $y = mx + c$, the gradient is m and the y intercept is at c.

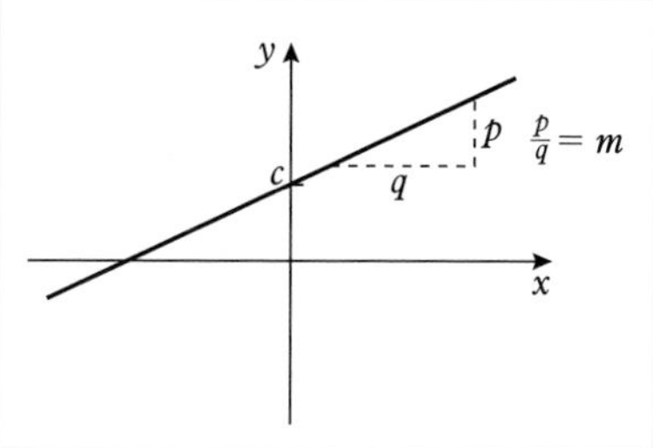

Parallel lines

Parallel lines have the same gradient.

Perpendicular lines

Perpendicular lines are at right angles to each other.
The product of the gradients of two perpendicular lines is -1.
The lines $y = 3x + 4$ and $y = -\frac{1}{3}x - 3$ are perpendicular as $3 \times -\frac{1}{3} = -1$
If a line has gradient m, a line that is perpendicular to it has gradient $-\frac{1}{m}$.

Solving equations

Solving an equation means finding the value of an unknown letter.

To solve	$11x - 20 = 6x + 15$
Take $6x$ from each side.	$11x - 6x - 20 = 6x - 6x + 15$
	$5x - 20 = 15$
Add 20 to each side.	$5x - 20 + 20 = 15 + 20$
	$5x = 35$
Divide both sides by 5	$x = 7$

Some equations have brackets in them.

To solve these, first multiply out the brackets.

Point of intersection

The point where two lines cross is called the **point of intersection** of the two lines.

Simultaneous equations

You can find points of intersection by drawing graphs. You can also use algebra and solve simultaneous equations. The solution of the simultaneous equations gives you the co-ordinates of the point of intersection.

You will need to have the same number of x or y. Then you **subtract** the two equations if the letter with the same number in front has the **same sign** and **add** the two equations if the letter with the same number in front has a **different sign** in each equation.

You sometimes have to multiply one or both of the equations before adding or subtracting.

Once you have found one of the letters, you substitute that value into either equation to find the missing letter. Then check that both values work in the other equation.

Solve $12x + 3y = 13$ (1)

$4x - y = 5$ (2)

Multiply (2) by 3:

$12x - 3y = 15$ (3)

Add (1) and (3) to get rid of y:

$14x = 28$

So $x = 2$

Put $x = 2$ into equation (1)

$2 \times 2 + 3y = 13$

$3y = 9$

$y = 3$

So the answer is $x = 2, y = 3$.

Use equation (2) to check your answer.

$4x - y = 4 \times 2 - 3 = 8 - 3 = 5$ ✓

The point (2, 3) is the point of intersection of the lines $2x + 3y = 13$ and $4x - y = 5$.

1 Find a formula for the nth term of each of these number sequences.

a 5, 12, 19, 26, 33, … **b** −4, −1, 2, 5, 8, …

2 The formula for the nth term of a sequence is $2n - 5$.
Which term has the value

a 45 **b** 39?

3 Find the formula for the nth term of each of these sequences.

a 1, 8, 17, 28, 41, … **b** 3, 4, 7, 12, 19, …

4 Write down the gradient of each of these lines.

a $y = 7x$ **b** $y = \frac{2}{3}x$ **c** $y = -2x$

5 Where will each of these lines cross the y axis?

a $y = 2x + 8$ **b** $y = x - 3$ **c** $y = 2x$

6 Which two of these lines are parallel?

$y = 5x + 2$ $y = 2x - 4$ $y = 2x + 2$ $y = x - 4$

7 **a** Find the gradient of the line shown.

b Write the equation of the line in the form $y = mx + c$.

c Use your equation to find the value of y when $x = -10$.

d What value of x gives a y value of 12?

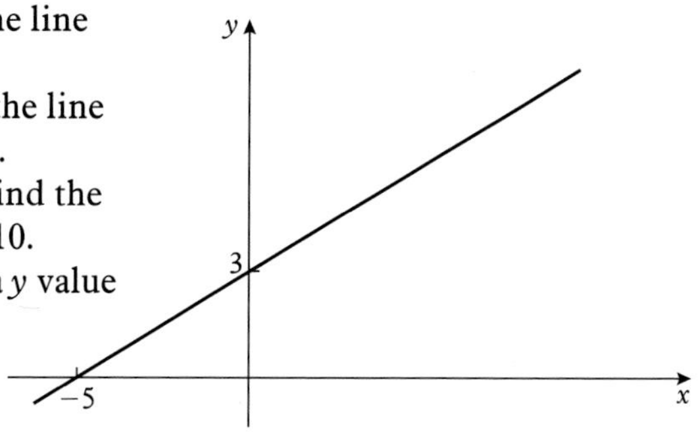

8 Solve these equations.

a $\frac{x}{2} = 4$ **d** $\frac{x}{6} + 1 = 8$

b $\frac{x}{3} = -4$ **e** $\frac{y}{2} - 4 = 3$

c $\frac{x}{2} + 3 = 14$ **f** $\frac{y}{6} - 6 = 12$

9 Solve these equations.

a $3x + 5 = 17 - x$ **c** $16 + 5x = 20 - 3x$

b $6 - 2x = 18 - 3x$ **d** $18 - 3x = 21 - x$

10 Solve these equations.

a $\frac{2x}{3} + 1 = 9$

b $\frac{3x}{5} + 2 = 5$

c $2(x + 3) = 11$

d $3(5x - 7) = 18$

11 A square's sides are $2x$ metres long.
A rectangle has a length of $3x$ metres and a width of 2 metres.

a Write down the perimeter of the square in terms of x.

b Write down the perimeter of the rectangle in terms of x.

The perimeters of the two shapes are equal.

c Write down an equation and solve it to find the perimeter of each shape.

12 Solve these pairs of simultaneous equations.

a $4x + 3y = 23$
$2x + 3y = 13$

b $2x - 3y = 12$
$x + y = 1$

c $3x - 4y = 5$
$5x + 3y = -11$

d $x - 2y - 24 = 0$
$2y = 8 - 3x$

13 **a** Write down the nth even number.

b Write down the mth even number.

c Use your answers to **a** and **b** to show algebraically that the product of any two even numbers is also even.

d Write down the pth odd number. This is the pth term of the sequence 1, 3, 5, . . .

e Use your answers to **a** and **d** to show algebraically that the product of an even number and an odd number is always even.

f Write down the qth odd number.

g Use your answers to **e** and **f** to show algebraically that the product of two odd numbers is always odd.

14 Jamie and Anthony enjoy their food.
Jamie buys 3 burgers and 2 fries and pays £3.55
Anthony buys 4 burgers and 3 fries and pays £4.95
Find the cost of a burger and the cost of fries.
You must use simultaneous equations.

1 A straight line has a gradient of 3 and passes through the point (1, 2).

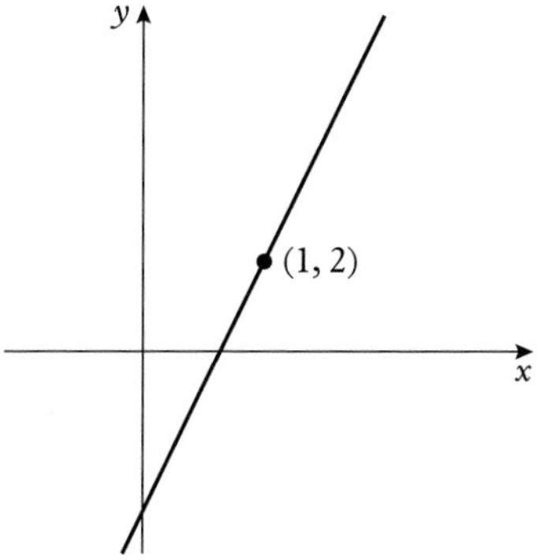

a Find the equation of this line. **(2 marks)**

b Write down the equation of the line that also passes through (1, 2) but is perpendicular to the first line. **(2 marks)**

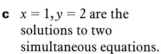

c $x = 1, y = 2$ are the solutions to two simultaneous equations. Write down two equations for which these are the solutions. **(2 marks)**

a *Use $y = mx + c$*

m is the gradient = 3 $y = 3x + c$ **1 mark**

(1, 2) tells you that when $x = 1, y = 2$.

Substitute these values in. $2 = (3 \times 1) + c$

$2 + 3 = c$, so $c = -1$

Remember to write down the equation in full. $y = 3x - 1$ **1 mark**

b *An equation that is perpendicular will have a gradient of $-\frac{1}{3}$.*

(1, 2) tells you that when $x = 1$, then $y = 2$. $y = -\frac{1}{3}x + c$

Substitute this in again. $2 = (-\frac{1}{3} \times 1) + c$ **1 mark**

$2 = -\frac{1}{3} + c$ so $c = 2\frac{1}{3}$

Write down the equation. $y = -\frac{1}{3}x + 2\frac{1}{3}$ **1 mark**

c *The two equations can be those already found, since these cross (intersect) at (1, 2).*

$y = 3x - 1$ **1 mark**

$y = -\frac{1}{3}x + 2\frac{1}{3}$ **1 mark**

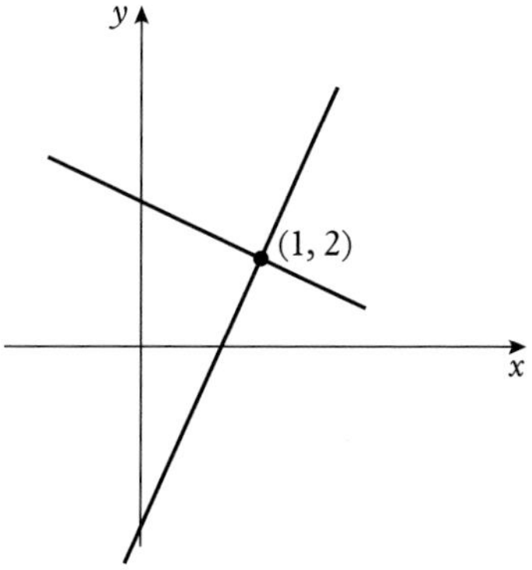

2 Solve $3x + 4 - (2x - 3) + 6 = 8(x + 4) + 10$. **(2 marks)**

Expand brackets $3x + 4 - 2x + 3 + 6 = 8x + 32 + 10$ **1 mark**
(Remember: $- \times - = +$*)*
Gather terms first. $x + 13 = 8x + 42$

Solve the equation. (-42) $x - 29 = 8x$
$(-x)$ $-29 = 7x$
Divide both sides by 7 to get x. $x = -\frac{29}{7}$
Do not do the division to get a decimal. $x = -4\frac{1}{7}$ **1 mark**
Change into a mixed fraction.

3 Solve $5x + 2y = 26$, $4x - 3y = 7$. **(4 marks)**

Write the equations under each other.
$5x + 2y = 26$ (1)
$4x - 3y = 7$ (2)
You need to have the same coefficients (numbers) in front of either the x or the y terms. To get this, you have a choice of operations:

either *(1)* $\times$ *4*
(2) $\times$ *5*

to get 20x in both

or *(1)* $\times$ *3*
(2) $\times$ *2*
to get 6y in both

Do it this way to get rid of y, the second letter.

Multiply every term in the equations:
(1) $\times$ 3 $15x + 6y = 78$
(2) $\times$ 2 $8x - 6y = 14$
The y terms have the same $23x = 92$
coefficient but different signs, $x = \frac{92}{23}$
so the equations should be added. **1 mark**
So $x = 92 \div 23 = 4$. $x = 4$ **1 mark**
Now you need to take this value of x and substitute it into one of the equations to help you find the other missing value. Choose $5x + 2y = 26$.

$$x = 4 \text{ so } (5 \times 4) + 2y = 26$$
$$20 + 2y = 26$$
$$2y = 6$$
$$y = 3$$ **1 mark**

State your solutions at the end of the problem. **1 mark**
So $x = 4$ and $y = 3$.

1 Here are the first six terms of a sequence.

4, 12, 24, 40, 60, 84.

Find in terms of n, an expression for the nth term of the sequence.

(2 marks)
[S2001 P5 Q13]

2 The diagrams show 3 shapes made with sticks.

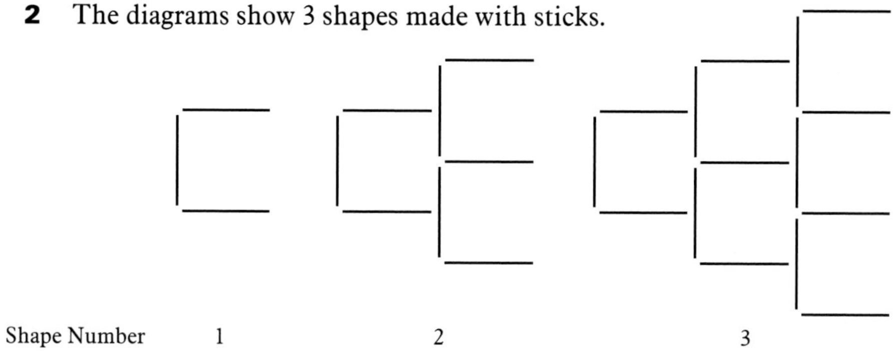

The table for the first 4 shapes is given below.

Shape Number	1	2	3	4
Number of sticks	3	8	15	24

Find an expression, in terms of n, for the number of sticks in Shape Number n.

(2 marks)
[N1999 P6 Q4]

3 Here are the first five terms of a sequence.

17, 14, 11, 8, 5.

Find, in terms of n, an expression for the nth term of the sequence.

(2 marks)
[N1998 P5 Q2]

4 **a** Factorise $p^2 - q^2$ **(1 mark)**

Here is a sequence of numbers

0, 3, 8, 15, 24, 35, 48, …

b Write down an expression for the nth term of this sequence.

(2 marks)

c Show algebraically that the product of any two consecutive terms of the sequence

0, 3, 8, 15, 24, 35, 48, . . .

can be written as the product of four consecutive integers. **(3 marks)**

[S2000 P5 Q15]

5 The diagram shows patterns made from square tiles.

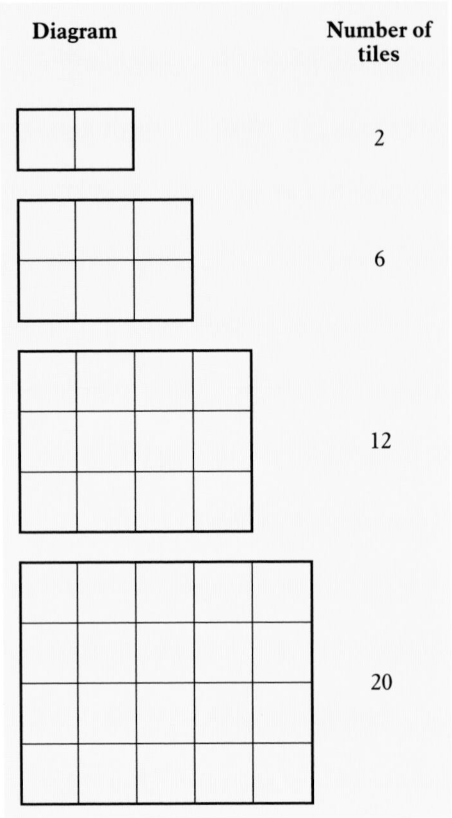

The numbers 2, 6, 12, 20, … form a number sequence.

a Write down an expression, in terms of n, for the nth number in the sequence. **(2 marks)**

b Work out the difference between the nth number and the $(n + 1)$th number.
Give your answer as simply as you can in terms of n. **(2 marks)**

6 **a** Factorise
$x^2 - 4y^2$. **(1 mark)**

b Solve the simultaneous equations
$x^2 - 4y^2 = 24$
$x + 2y = 6$. **(4 marks)**

[N1999 P5 Q20]

7 Solve the simultaneous equations
$2x + 5y = -1$
$6x - y = 5$ **(4 marks)**

[N2000 P6 Q9]

8 A straight line, **L**, has been drawn on the grid below.

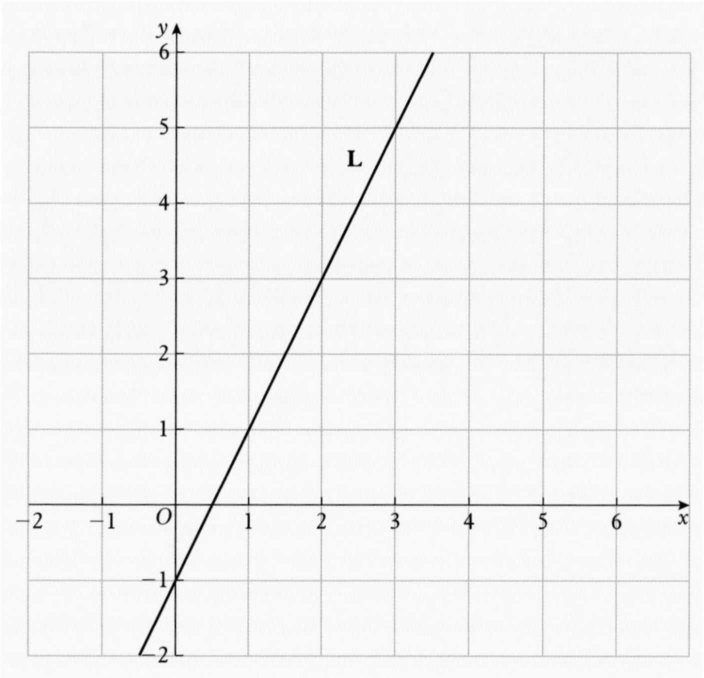

a Find an equation of the line **L**. **(2 marks)**
b Find an equation of the line through (1, 2) parallel to **L**. **(2 marks)**
[N1999 P6 Q2]

9 Solve the equation

$5(x + 2) = 3x + 7$ **(3 marks)**

[S2001 P6 Q2]

10 The perimeter of the pentagon is 200 cm.

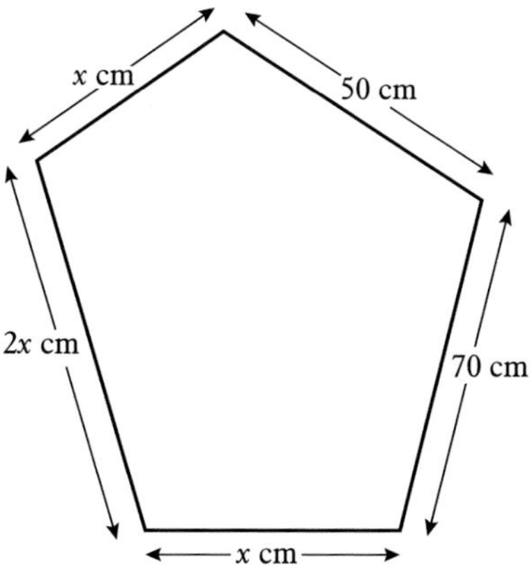

Diagram **NOT** accurately drawn

Work out the value of x. **(3 marks)**
[N1998 P6 Q1]

11

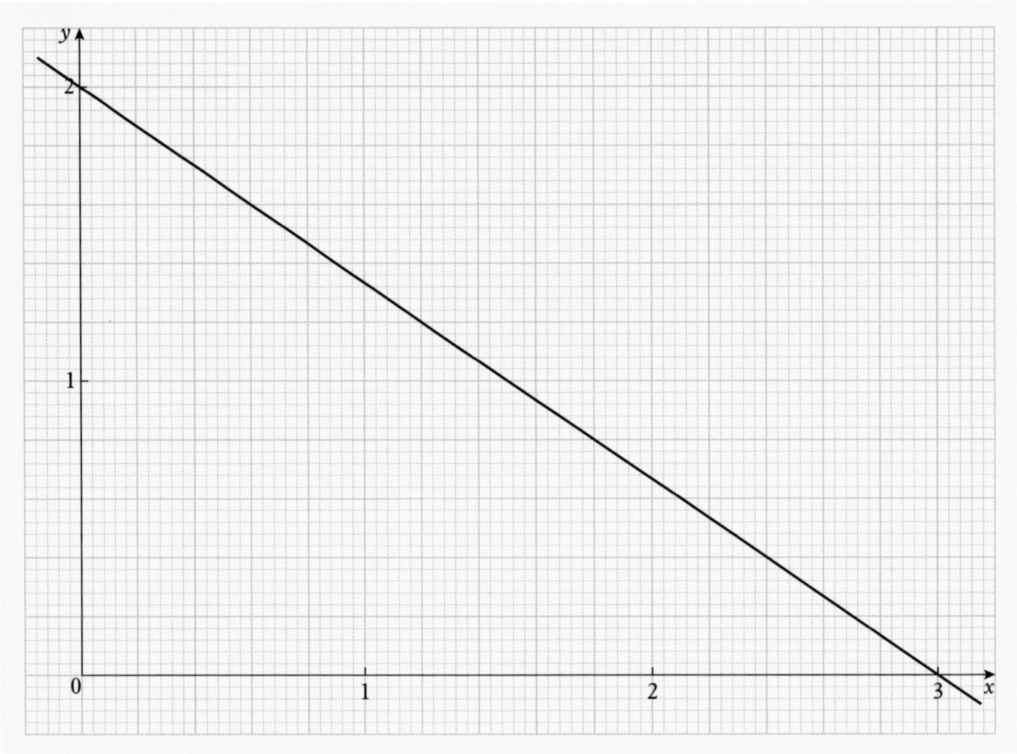

The line with equation $3y = -2x + 6$ has been drawn on the grid.

a Draw the graph of $y = 2x - 2$ on the same grid. **(2 marks)**

b Use the graphs to find the solution of the simultaneous equations

$3y = -2x + 6$
$y = 2x - 2$ **(2 marks)**

A line is drawn parallel to $3y = -2x + 6$ through the point (2, 1).

c Find the equation of this line. **(2 marks)**

[S1998 P6 Q2]

12 Solve the simultaneous equations

$4x + y = 4$
$2x + 3y = -3$ **(4 marks)**

[S2001 P6 Q9]

13 **a** Expand

$x(x + 4)$ **(1 mark)**

b Solve the simultaneous equations

$x + 8y = 5$
$3x - 4y = 8.$ **(4 marks)**

[N1999 P5 Q11]

5 Right-angled triangles

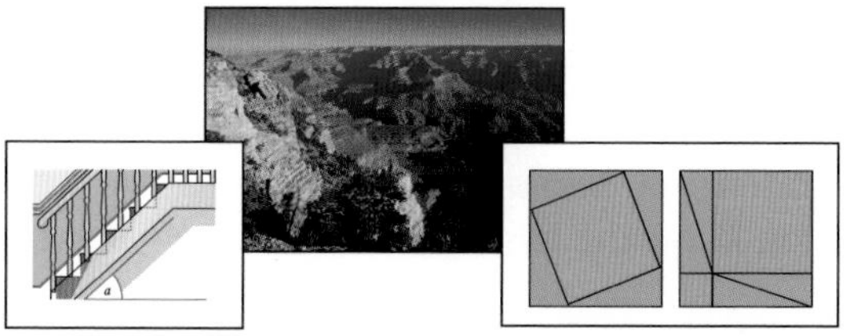

You need to know about:

- Pythagoras' theorem
- trigonometry
- finding an angle
- finding a side
- special angles

Pythagoras' theorem

$a^2 + b^2 = h^2$

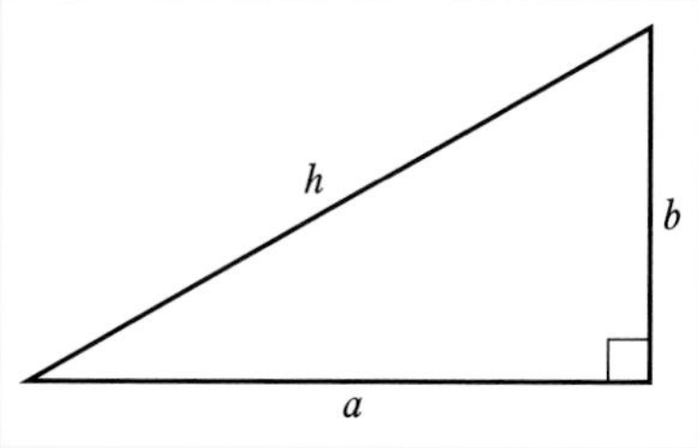

In a right-angled triangle, the two shorter sides squared and added is the same as the hypotenuse squared.

The hypotenuse is always opposite the right angle.

Example Find the side marked x in each of these triangles.

a

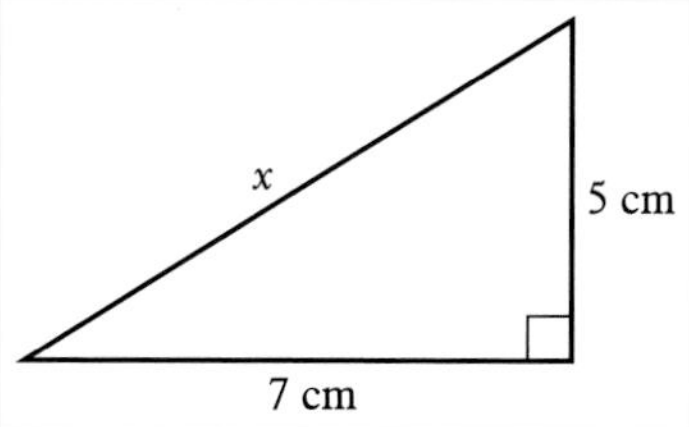

$$5^2 + 7^2 = x^2$$
$$25 + 49 = x^2$$
$$74 = x^2$$
$$x = 8.6 \text{ cm (1 dp)}$$

b

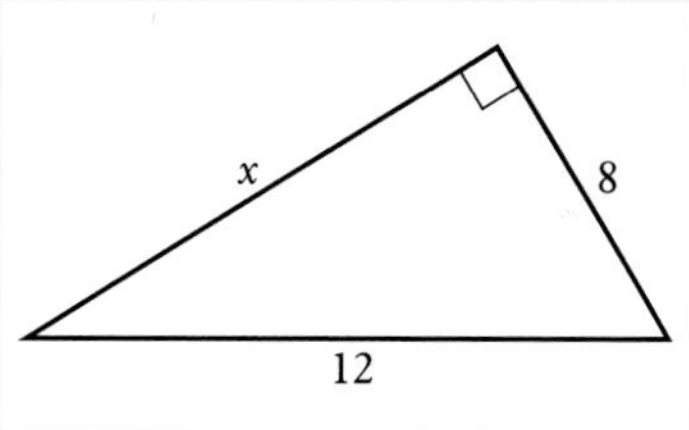

$$x^2 + 8^2 = 12^2$$
$$x^2 + 64 = 144$$
$$x^2 = 80$$
$$x = 8.9 \text{ cm (1 dp)}$$

Trigonometry

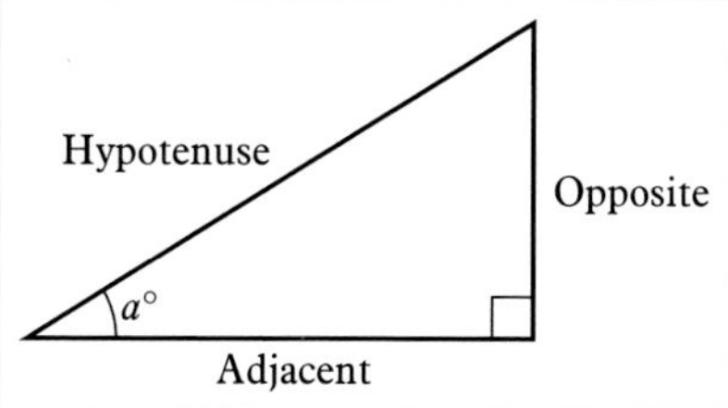

$$\text{Sin a}° = \frac{\text{Opposite}}{\text{Hypotenuse}}$$

$$\text{Cos a}° = \frac{\text{Adjacent}}{\text{Hypotenuse}}$$

$$\text{Tan a}° = \frac{\text{Opposite}}{\text{Adjacent}}$$

These formulas can be remembered using **SOH CAH TOA**.

Finding an angle

Example Find the angle marked $a°$ in this triangle.

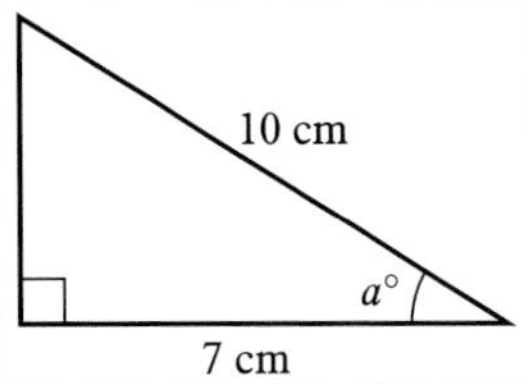

Write down SOH CAH TOA and cross out the sides that you know.
In this triangle, you know the hypotenuse and the adjacent.

SOH C~~A~~~~H~~ TO~~A~~

The cos formula has both parts crossed out

$$\cos a° = \frac{7}{10} \qquad a = 45.6° \text{ (1 dp)}$$

Finding a side

Example Find the side marked with a letter in each of these triangle.

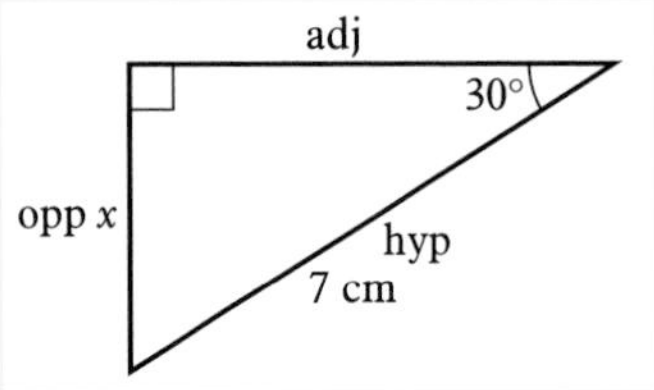

$$\sin 30° = \frac{x}{7}$$
$$7 \times \sin 30° = x$$
$$x = 3.5 \text{ cm}$$

S~~O~~~~H~~ CA~~H~~ TOA

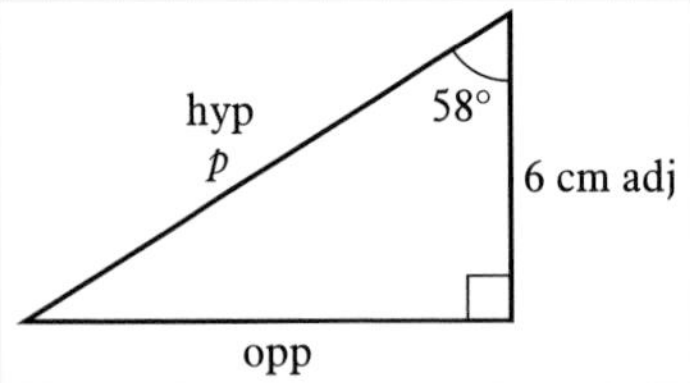

$$\cos 58° = \frac{6}{p}$$
$$p \times \cos 58° = 6$$
$$p = \frac{6}{\cos 58°}$$
$$p = 11.3 \text{ cm (1 dp)}$$

SOH C~~A~~~~H~~ TO~~A~~

Special angles

$$\sin 30° = \frac{1}{2} \qquad \cos 30° = \frac{\sqrt{3}}{2} \qquad \tan 30° = \frac{1}{\sqrt{3}}$$

$$\sin 45° = \frac{1}{\sqrt{2}} \qquad \cos 45° = \frac{1}{\sqrt{2}} \qquad \tan 45° = 1$$

$$\sin 60° = \frac{\sqrt{3}}{2} \qquad \cos 60° = \frac{1}{2} \qquad \tan 60° = \sqrt{3}$$

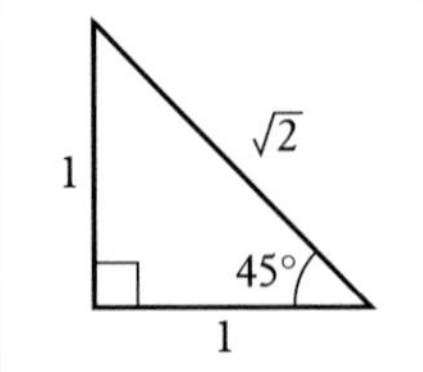

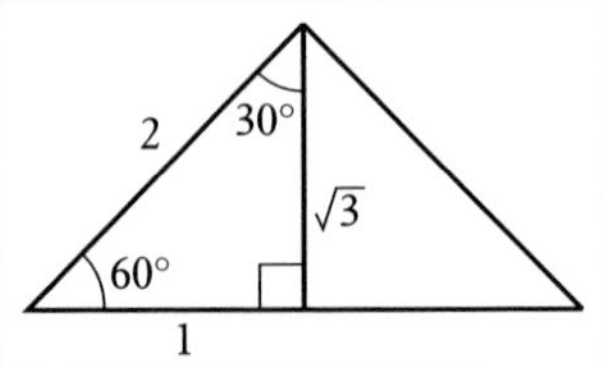

1 Find the length of the hypotenuse in each of these triangles.
Give each of your answers to 1 dp.

a

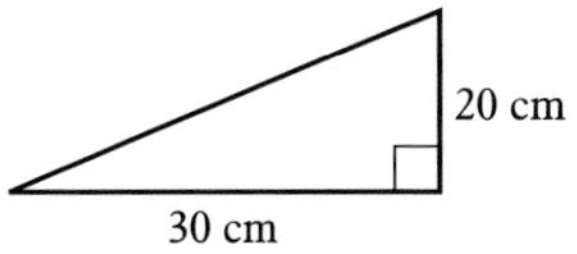

b

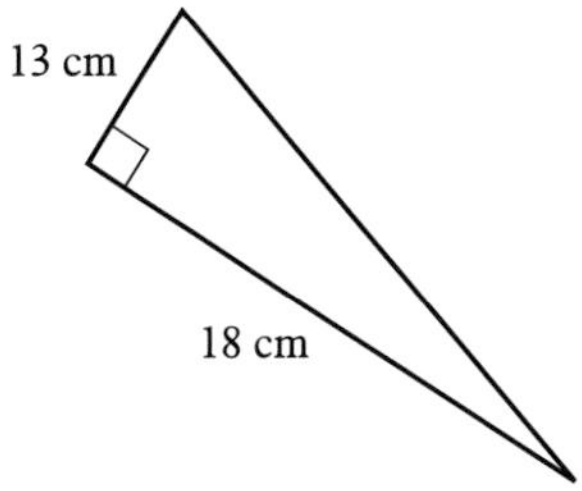

c

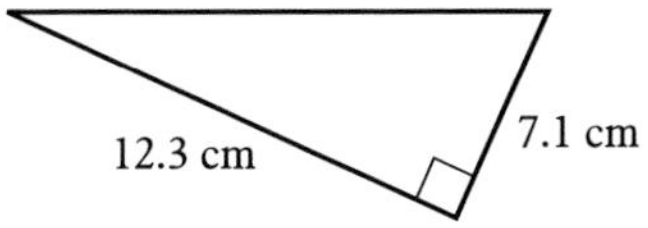

d

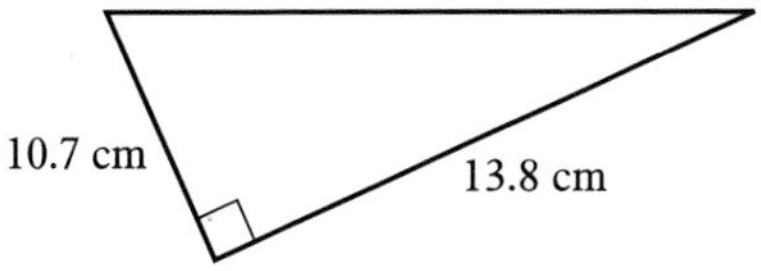

2 Find the length of the missing side in each of these triangles.
Give each of your answers to 1 dp.

a

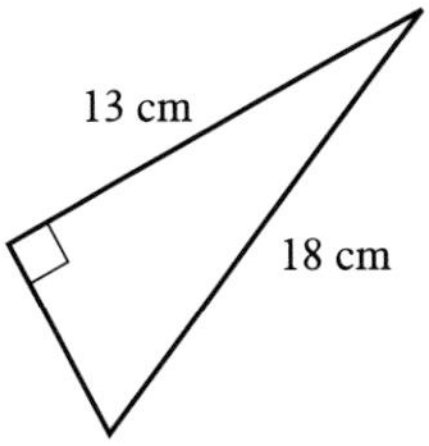

b

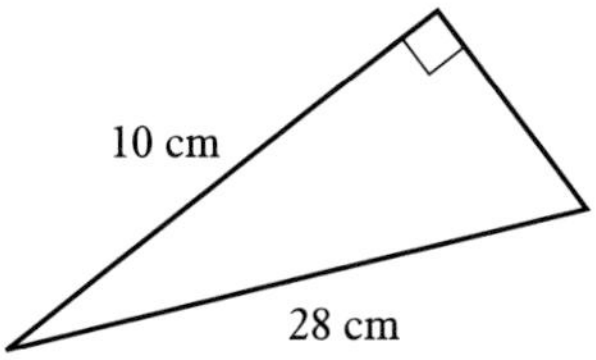

c

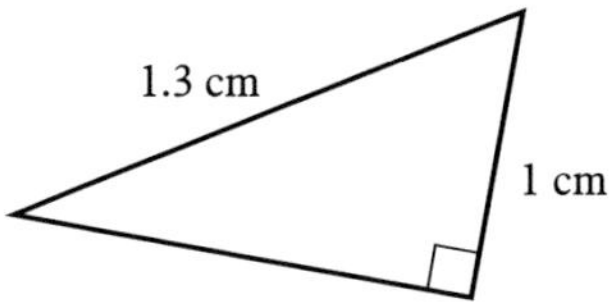

d

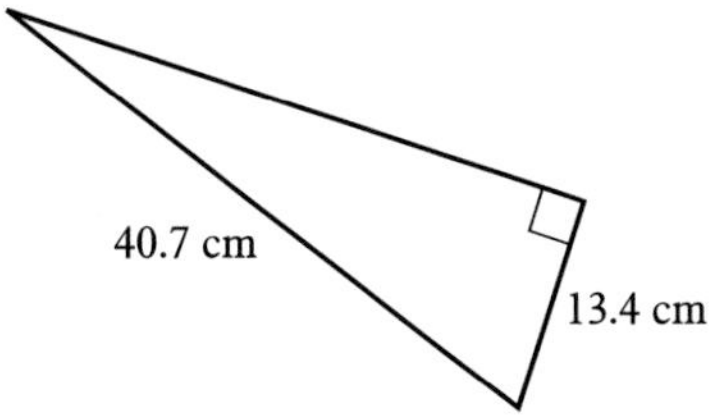

3 Find the length of the missing side in each of these triangles.
Leave your answers in surd form, simplified as much as possible.

a

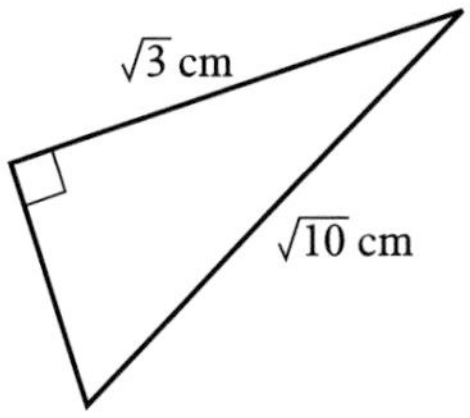

b

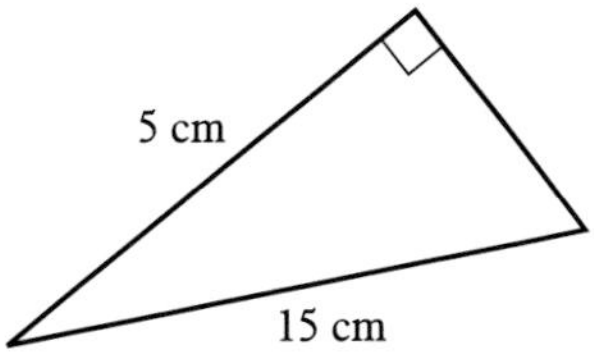

4 Find the height of an equilateral triangle of side 20 cm.

5 The diagram shows the badge of a secret society.
The circle has a radius of 3 cm.
The red chord is 4 cm long.
The purple lines meet at the centre of the circle.
Find the length of each of the blue lines.

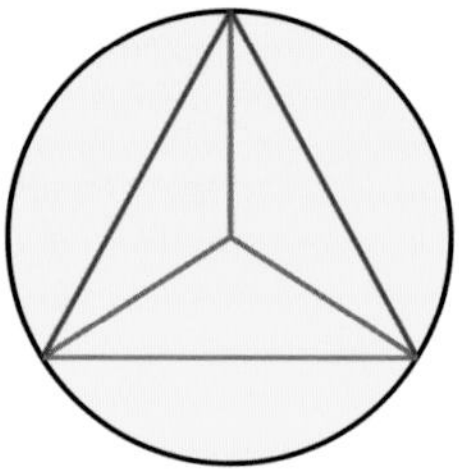

6 Use Pythagoras' theorem to find the value of x.

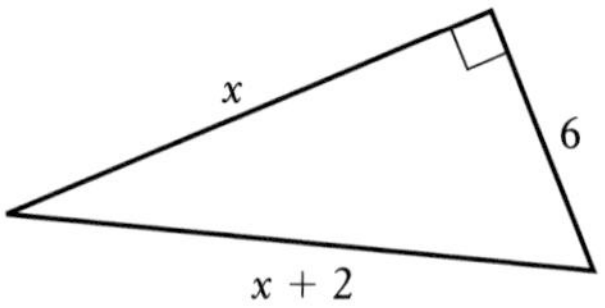

7 Find the labelled angles in these triangles.
Give your answers to the nearest degree.

a

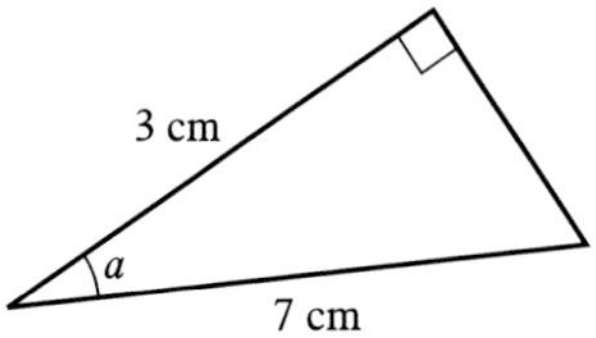

b

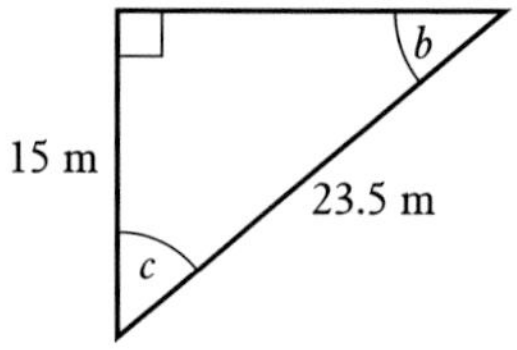

8 Find the labelled sides in these triangles.
Give each of your answers to 3 sf.

a

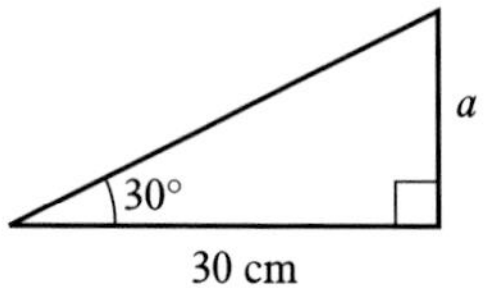

c

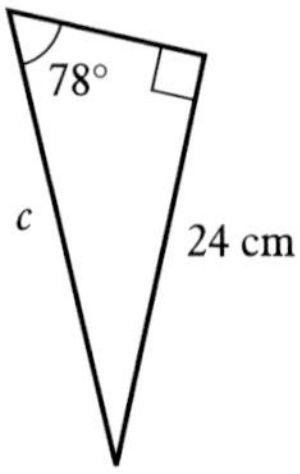

b

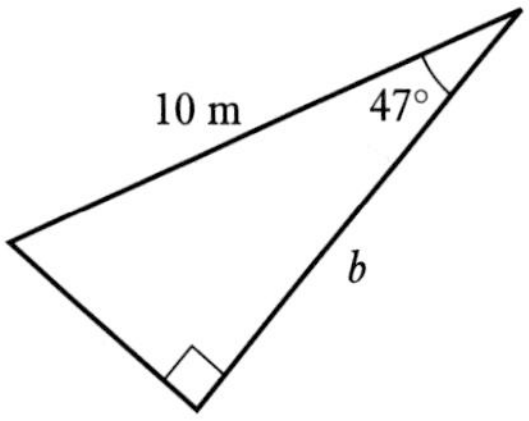

d

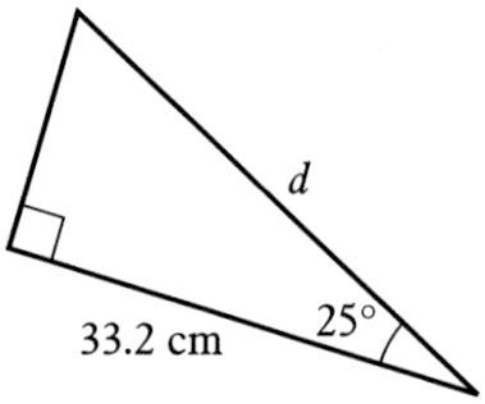

1 **a** Calculate angle x. **(3 marks)**
b Work out the length of BC. Give your answer correct to 3 significant figures. **(3 marks)**
c Work out the length of AD. Give your answer correct to 3 significant figures. **(3 marks)**

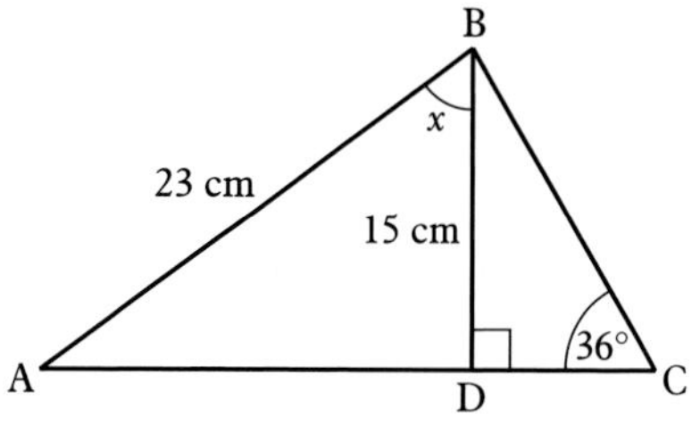

a *First identify the trigonometrical ratio to use.*
In this triangle you have the hypotenuse and the adjacent side.

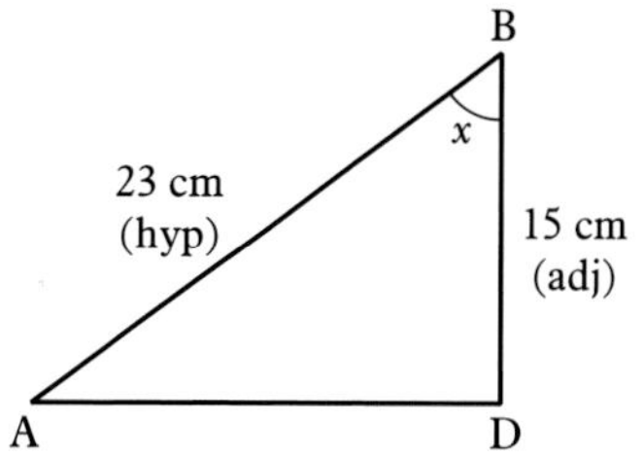

SOH   CAH   TOA → Cosine

Write down the cosine formula:

$$\cos x = \frac{\text{adj}}{\text{hyp}} = \frac{15}{23}$$ *Be sure to write down the full formula.* **1 mark**

So $\cos x = 0.65217$ *Write down both these values accurately from your calculator.* **1 mark**

$x = 49.29429 = 49.3°$ *Round appropriately.* **1 mark**

b *In this triangle you have the opposite side, and need to find the hypotenuse.*

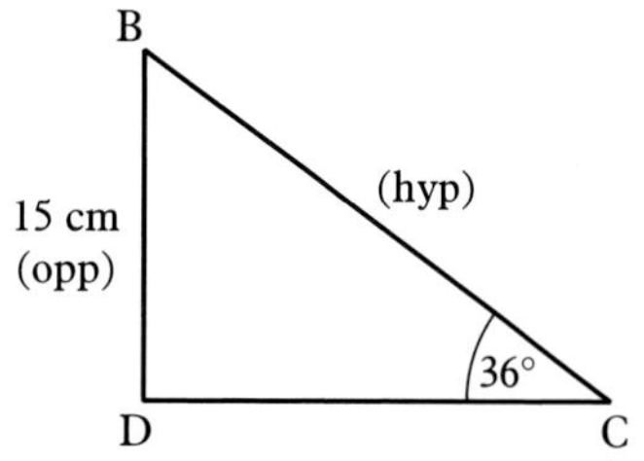

SOH   CAH   TOA

State the formula: $\sin x = \frac{\text{opp}}{\text{hyp}}$

Substitute figures: $\sin 36° = \frac{15}{\text{BC}}$ **1 mark**

Rearrange: So $\text{BC} = \frac{15}{\sin 36°}$ **1 mark**

$\text{BC} = 25.5195 = 25.5$ cm **1 mark**

c *It is possible to find AD using sine, since the angle x has been found in part* ***a****.*
Calculate with the accurate figures using your calculator.

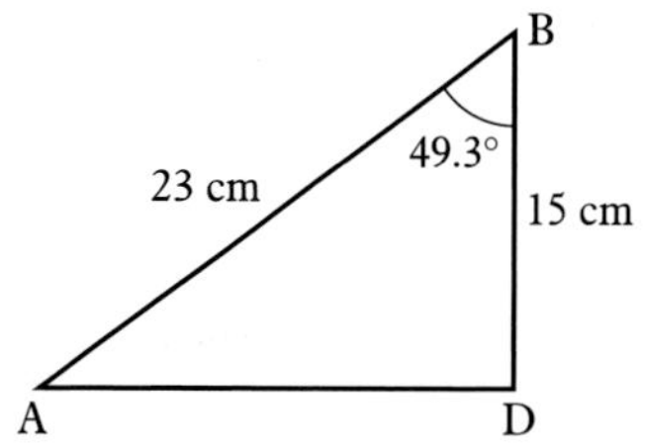

$\sin 49.3° = \frac{\text{AD}}{23}$ **1 mark**

$\text{AD} = 23 \times \sin 49.3°$ **1 mark**

$\text{AD} = 17.4$ cm **1 mark**

It is best, if possible, not to use an answer already found, but to use the values given in the question.

An alternative method is to use Pythagoras:

State Pythagoras' theorem: $AD = \sqrt{AB^2 - BD^2}$

Substitute values: $= \sqrt{23^2 - 15^2}$ **1 mark**

Show the first stage of the calculation (squaring the numbers): $= \sqrt{529 - 225}$ **1 mark**

$= \sqrt{304}$

$= 17.435595 = 17.4$ cm **1 mark**

2 Find the length of DC.
Give your answer correct to 3 significant figures. **(5 marks)**

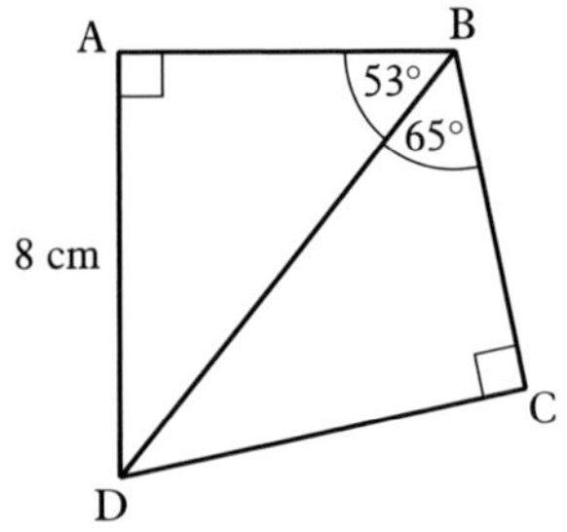

DC cannot be found directly: there is not enough information in triangle BCD.

First, use triangle ABD to find BD.

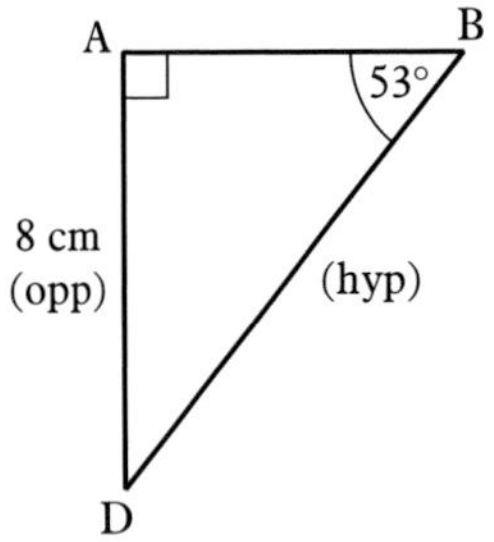

$$\sin 53° = \frac{8}{BD}$$ **1 mark**

$$BD = \frac{8}{\sin 53°} = 10.017\ldots = 10.0 \text{ cm}$$ **1 mark**

Keep the exact value of BD in the calculator memory to use later.

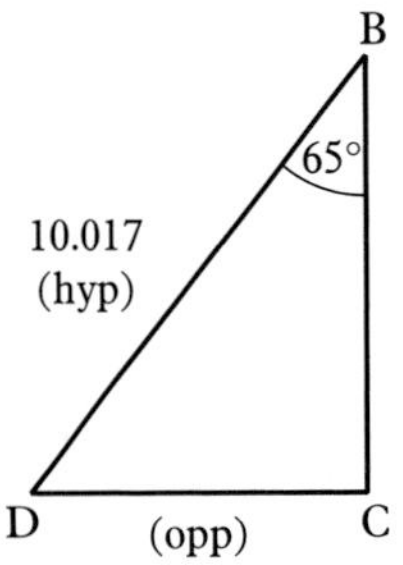

Now, in $\triangle BCD$, find DC.

$$\sin 65° = \frac{DC}{BD} = \frac{DC}{10.017\ldots}$$ **1 mark**

$DC = 10.017\ldots \times \sin 65°$ **1 mark**

$= 9.07856$

$DC = 9.08$ cm (3 sf) **1 mark**

3 sf is asked for in the question.

1 AB = 19.5 cm, AC = 19.5 cm and BC = 16.4 cm.
Angle ADB = 90°.
BDC is a straight line.

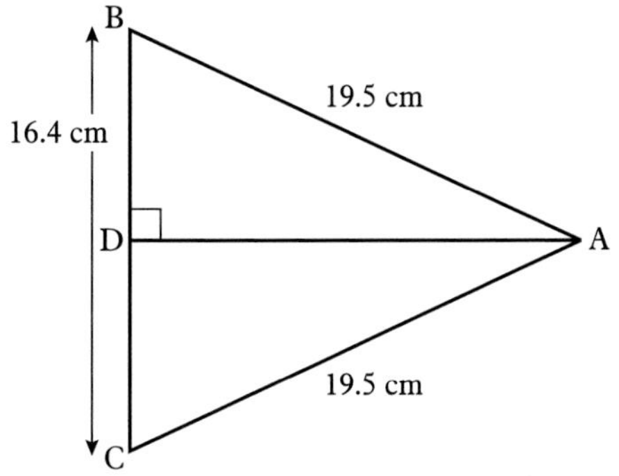

Diagram **NOT** accurately drawn

Calculate the length of AD.
Give your answer in centimetres, correct to 1 decimal place. **(4 marks)**
[N1999 P5 Q4]

2 ABC is a right-angled triangle.
AB = 4 cm, BC = 6 cm.

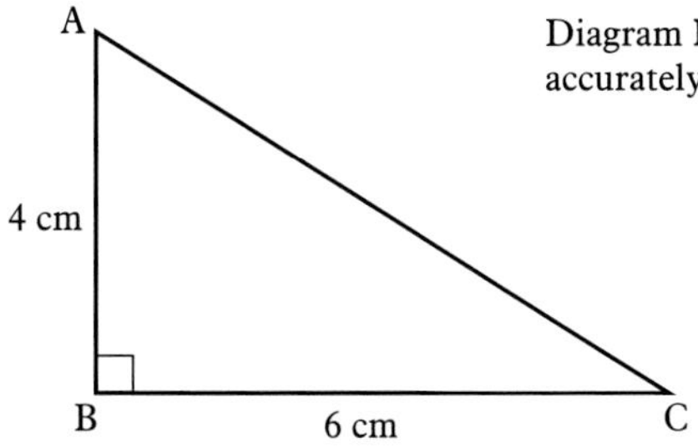

Diagram **NOT** accurately drawn

Calculate the length of AC.
Give your answer in centimetres, correct to 3 significant figures. **(3 marks)**
[S2001 P6 Q5]

3 A, B, C and D are four points on the circumference of a circle.
ABCD is a square with sides 20 cm long.

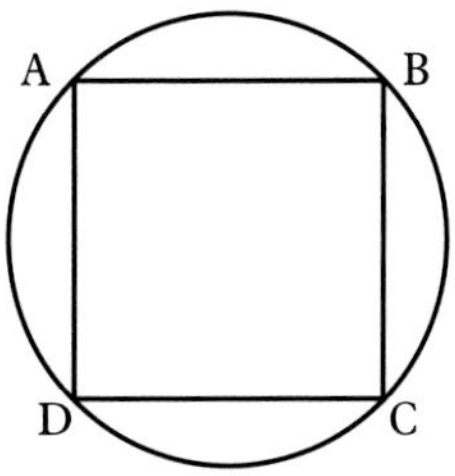

Diagram **NOT** accurately drawn

Work out the diameter of the circle.
Give your answer correct to 3 significant figures. **(4 marks)**
[N1998 P5 Q5]

4 AB and BC are two sides of a rectangle.
AB = 120 cm and BC = 148 cm.
D is a point on BC.
Angle BAD = 15°.

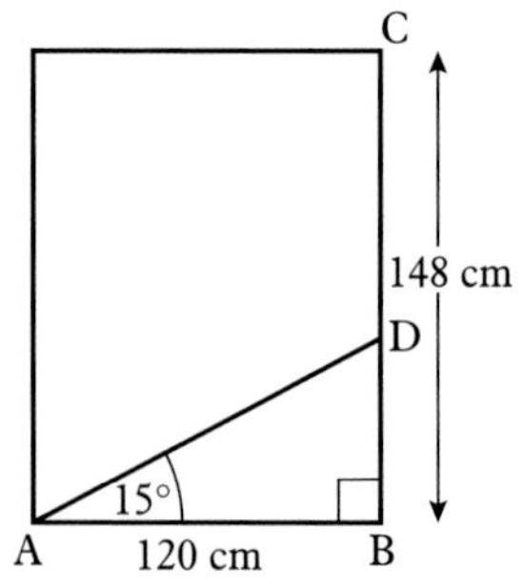

Diagram **NOT** accurately drawn

a Work out the length of CD.
Give your answer correct to the nearest centimetre. **(4 marks)**

148 is correct to 3 significant figures.
120 is correct to 3 significant figures.
15 is correct to the nearest whole number.

b Write down the three values which should be used to work out the lower bound for the length of CD. **(3 marks)**
[S2000 P6 Q9]

5 The diagram represents a rectangle which is 6 cm long.
A diagonal makes an angle of 23° with a 6 cm side.

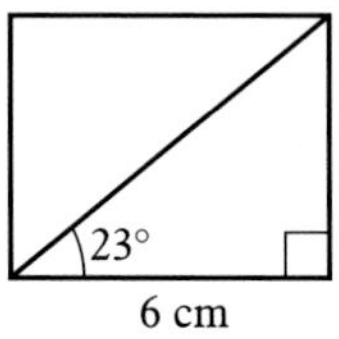

Diagram **NOT** accurately drawn

Calculate the length of a diagonal.
Give your answer correct to 3 significant figures. **(3 marks)**
[N1998 P5 Q14]

6 Angle Q = 90°.
Angle P = 32° and
PR = 2.6 metres.

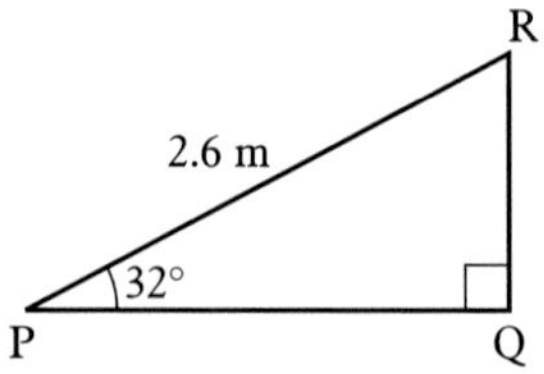

Diagram **NOT** accurately drawn

Calculate the length of QR.
Give your answer in metres, correct to 3 significant figures. **(3 marks)**
[N1999 P6 Q6]

7 The diagram represents the frame for part of a building.
BC and CD are equal in length.
BD and AE are horizontal.

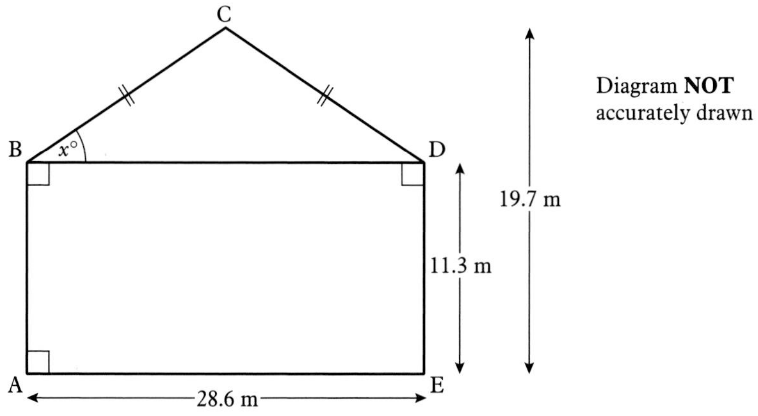

a Calculate the length BC.
Give your answer correct to 3 significant figures. **(3 marks)**

b Calculate the size of the angle marked $x°$.
Give your answer correct to 1 decimal place. **(3 marks)**

[N1998 P6 Q11]

8 ABCD is a quadrilateral.
Angle BDA = 90°, angle BCD = 90°, angle BAD = 40°.
BC = 6 cm, BD = 8 cm.

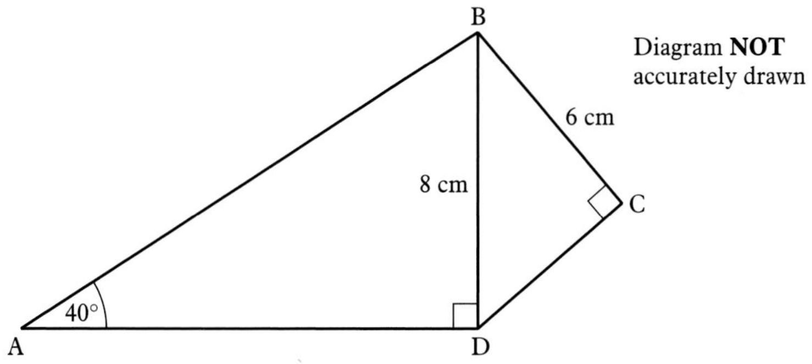

a Calculate the length of DC. Give your answer correct to 3 significant figures. **(3 marks)**

b Calculate the size of angle DBC.
Give your answer correct to 3 significant figures. **(3 marks)**

c Calculate the length of AB. Give your answer correct to 3 significant figures. **(3 marks)**

[N2000 P6 Q8]

9 Ballymena is due West of Larne.
Woodburn is 15 km due South of Larne.
Ballymena is 32 km from Woodburn.

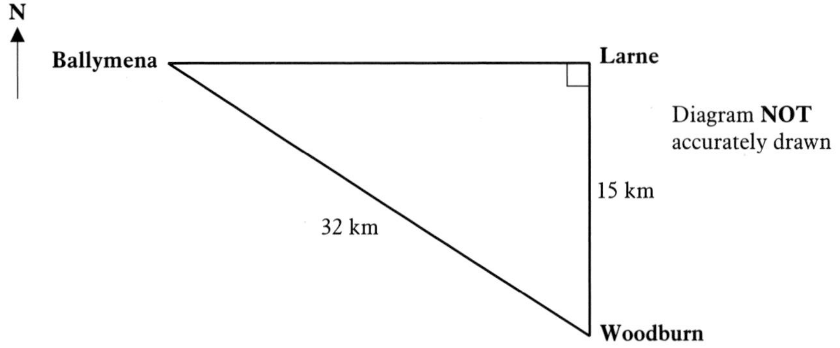

a Calculate the distance of Larne from Ballymena. Give your answer in kilometres correct to 1 decimal place. **(3 marks)**
b Calculate the bearing of Ballymena from Woodburn. **(4 marks)**
[S1998 P6 Q6]

10 A statue stands on a column.
In the diagram AB represents the statue and BC represents the column.
Angle ACD = 90°.
Angle BDA = 2.8°.
AD = 91.2 m and BD = 88.3 m.
ABC is a vertical straight line.

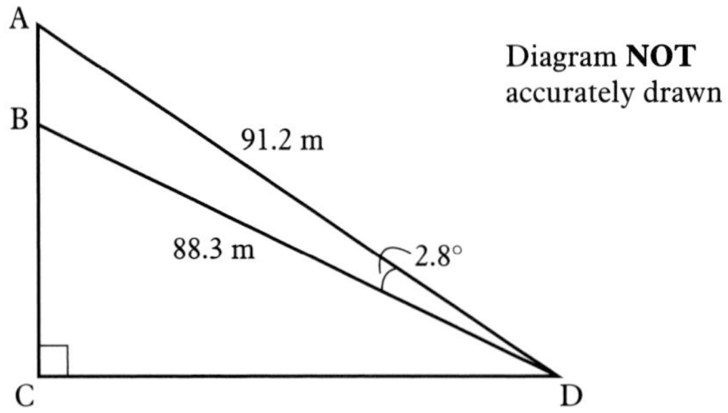

a Calculate the height, AB, of the statue. Give your answer, in metres, correct to 3 significant figures. **(3 marks)**
b Calculate the height, BC, of the column. Give your answer, in metres, correct to 3 significant figures. **(5 marks)**
[S1998 P6 Q13]

6 Ratio and proportion

You need to know about:

- using multipliers to work out percentage changes
- reversing percentage changes
- direct proportion
- equations of proportionality
- inverse proportion
- working with algebraic fractions

Multiplier

The **multiplier** for an increase of 6% is 1.06
(100% + 6% = 106% = 1.06)
To increase an amount by 6%, multiply by 1.06
For repeated percentage changes, multiply by all of the multipliers in turn.

Example

£5000 is invested at 6% per annum compound interest. Find the value of the investment after 5 years if no money is withdrawn.

$5000 \times 1.06 \times 1.06 \times 1.06 \times 1.06 \times 1.06$

$= 500 \times 1.06^5$

= £6691.13 to the nearest penny

The **multiplier** for a reduction of 6% is 0.94
(100% − 6% = 94% = 0.94)
To reduce an amount by 6% multiply by 0.94

Reverse prercentage charge

Sometimes you need to get back to the amount before an increase or decrease took place. To do this, you divide by the multiplier that would have been used to do the original increase or decrease.

Example

A CD player costs £135, including VAT at 17.5%.
What is the price before VAT is added?

To add VAT, you would multiply by 1.175 (17.5% increase).
To find the price before VAT was added you divide by 1.175

Price before VAT = 135 ÷ 1.175
= £114.89 to the nearest penny

Percentage profit and loss

$$\text{Percentage profit} = \frac{\text{Profit}}{\text{Cost price}} \times 100$$

$$\text{Percentage loss} = \frac{\text{Loss}}{\text{Cost price}} \times 100$$

Ratio

To divide £500 in the ratio 3 : 5 there are 3 + 5 = 8 shares altogether.
1 share = 500 ÷ 8 = £62.50
So 3 shares is 3 × 62.5 = £187.50 and
5 shares is 5 × 62.5 = £312.50

Direct proportion

Two quantities a and b are **directly proportional** if multiplying a by a number causes b to be multiplied by the same number.

Inverse proportion

Two quantities a and b are **inversely proportional** if multiplying a by a number causes b to be divided by the same number.

Equation of proportionality

If a and b are **directly proportional**, the **equation of proportionality** is $a = kb$ for some constant k, called the constant of proportionality.
If a and b are **inversely proportional**, the **equation of proportionality** is $a = \frac{k}{b}$ for some constant k.
When doing problems, set up the proportionality equation and find the value of k. Then use the full equation to find any subsequent values asked for.

Example

P is directly proportional to Q^2. $P = 12$ when $Q = 4$.
Find the value of Q when $P = 27$.

The equation of proportionality is $P=kQ^2$ and you need to find k.

When $P = 12$ and $Q = 4$,

$$12 = k \times 4^2,$$

$$k = \frac{12}{16}$$

$$= \frac{3}{4} \text{ so } P = \frac{3}{4}Q^2.$$

When $P = 27$, $\quad 27 = \frac{3}{4}Q^2,$

so $\quad Q^2 = \frac{27 \times 4}{3} = 36$

and $\quad Q = 6.$

Fractions

You can only **add** and **subtract** fractions that have the same denominator.

$$\frac{2}{x+1}+\frac{3}{x+1}=\frac{5}{x+1} \text{ and } \frac{9}{(a+b)^2}-\frac{4}{(a+b)^2}=\frac{5}{(a+b)^2}$$

If fractions do not have the same denominator you must change the denominators so that they are the same before you add or subtract.

So $\frac{5}{x}-\frac{3}{2x}$

$=\frac{10}{2x}-\frac{3}{2x}$

$=\frac{7}{2x}$

And $\frac{2}{x}+\frac{4}{x+1}$

$=\frac{2(x+1)}{x(x+1)}+\frac{4x}{x(x+1)}$

$=\frac{2x+2}{x(x+1)}+\frac{4x}{x(x+1)}$

$=\frac{6x+2}{x(x+1)}$

To **multiply** fractions, multiply the numerators together and the denominators together. If you can cancel first this will make sure that the answer you get is as simple as possible.

$$\frac{4x}{3}\times\frac{x+3}{x^2}=\frac{4\cancel{x}}{3}\times\frac{x+3}{x^{\cancel{2}}}=\frac{4(x+3)}{3x}$$ An x cancels.

At this last stage it is tempting to cancel 3 or x, but neither of these can be cancelled, as they are not factors of both the numerator and the denominator.

To **divide** two fractions, multiply the first fraction by the reciprocal of the second fraction.

$\frac{2x+2}{4}\div\frac{3x+3}{x}$

$=\frac{2\cancel{(x+1)}}{4}\times\frac{x}{3\cancel{(x+1)}}$

The $(x + 1)$ factors cancel.

$=\frac{2x}{12}$

$=\frac{x}{6}$

1 Write down the multiplier for each of these percentage increases.
a 1% **b** 65% **c** 12.5% **d** 12.2%

2 Write down the multiplier for each of these percentage reductions.
a 15% **b** 40% **c** 17.5% **d** 33.1%

3 Add VAT at 17.5% to each of these amounts.
a £10 **b** £124 **c** £7250 **d** £23 000

4 Each of these prices includes VAT at 17.5%.
Find the price before VAT was added.
a £188 **b** £129.25 **c** £185.65 **d** £3038.55

5 Samantha invests £5000 at 5% compound interest for 7 years.
a What is the value of the account at the end of this time, to the nearest penny?
b How much interest is gained?
c What is the overall percentage increase to 1 dp?

6 A car decreases in value by 12% each year for 5 years.
In 1990 the car cost £12 000.
a What was its value in 1995?
Give your answer to the nearest pound.
b What was the overall percentage decrease in its value?

7 T is proportional to v. $T = 20$ when $v = 5$.
a Find the equation of proportionality that connects T and v.
b Find the value of T when $v = 3$.
c Find the value of v when $T = 100$.

8 A is proportional to the square of r. When $A = 706.9$, $r = 15$.
Find the equation of proportionality that connects A and r.

9 y is proportional to h^4. $y = 32$ when $h = 2$.
a Find the equation of proportionality that connects y and h.
b Find the value of h when $y = 162$.

10 y is proportional to $\sqrt{x}$. $y = 24$ when $x = 9$.
a Find the equation of proportionality that connects y and x.
b Find the value of y when $x = 4$.
c Find the value of x when $y = 32$.

11 A is inversely proportional to B. $A = 20$ when $B = 2$.

a Find the equation of proportionality that connects A and B.

b Use your equation to find the value of A when $B = 10$.

12 V is inversely proportional to w^2. $V = 50$ when $w = 2$.

a Find the equation of proportionality that connects V and w.

b Find the value of V when $w = 4$.

c Find the values of w when $V = 8$.

13 A new car lost 20% of its value in its first year.
In the second year it lost 12% of its value at the end of the first year.
In the third year it lost 8% of its value at the end of the second year and in the fourth year it lost 7% of its value at the end of the third year.
What is the overall percentage loss in value of the car?
Give your answer to 1 dp.

14 Write the fractions $\frac{2}{9}$, $\frac{5}{18}$ and $\frac{7}{27}$ in order. Start with the smallest.

15 Simplify these expression.

a $\frac{2}{9} + \frac{5}{9}$ **b** $\frac{10}{p} - \frac{4}{p}$ **c** $\frac{9}{2r} + \frac{3}{2r} - \frac{1}{2r}$

16 Write each of these as a single fraction.

a $\frac{1}{y} + \frac{2}{y^2}$ **b** $9x - \frac{1}{x}$ **c** $5 - \frac{1}{x + 1}$

17 Simplify these expressions.

a $\frac{8}{x} \times \frac{x^2}{12}$ **b** $\frac{12pq}{5x} \times \frac{10x}{9p}$ **c** $\frac{2p + 6}{3} \div \frac{p + 3}{3p}$

18 Simplify these expressions.

a $\frac{3}{(x + 1)} - \frac{6}{(x + 1)(x - 2)}$ **b** $\frac{p + 2}{p} \times \frac{p}{2p + 4} \times \frac{q}{p + 4} \div \frac{q}{2p}$

19 Simplify these expressions.

a $\frac{2x}{9} + \frac{4x}{9}$ **b** $\frac{10}{3y} \times \frac{y^2}{5}$ **c** $\frac{5t}{8pr} \div \frac{15t}{16qr}$

1

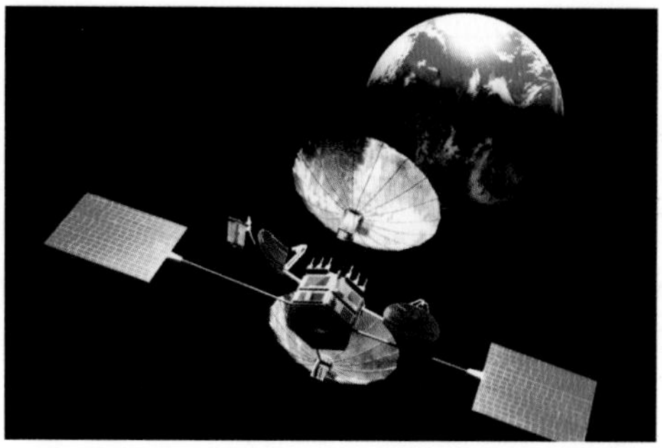

The speed at which a satellite needs to travel in orbit around the earth is inversely proportional to the square root of the distance of the satellite from the centre of the earth.
The satellite needs a speed of 7 km/s in an orbit 1630 km above the surface of the Earth. The radius of the Earth is 6370 km.
Find the speed needed to orbit 5150 km above the Earth. **(4 marks)**

Work out the distance from the centre of the Earth. $1630 + 6370 = 8000$

Start with the equation: $V = \frac{k}{\sqrt{d}}$

Substitute values to find k. So $7 = \frac{k}{\sqrt{8000}}$ **1 mark**

Leave it as $\frac{7}{\sqrt{8000}}$ rather than using a calculator here. $k = 7\sqrt{8000}$

Re-write the equation with k shown. So $V = \frac{7\sqrt{8000}}{\sqrt{d}}$ **1 mark**

When d is 5150: $V = \frac{7\sqrt{8000}}{\sqrt{5150}}$ **1 mark**

Press: 7 × √ 8 0 0 0 ÷ √ 5 1 5 0 =

$V = 8.72$ km/s (to 3 sf) **1 mark**

2 Find the interest earned when £200 is invested for 3 years at an interest rate of 9.5% per annum. The interest is compounded every year.
(4 marks)

You can work out the amounts for each year separately.

1st year: Interest $= \frac{£200 \times 9.5 \times 1}{100} = £19.00$ **1 mark**

New amount in the account = £200 + £19 = £219

2nd year: Interest $= \dfrac{£219 \times 9.5 \times 1}{100} = £20.81$ **Method: 1 mark**

New amount in the account $= £219 + £20.81 = £239.81$

3rd year: Interest $= \dfrac{£239.81 \times 9.5 \times 1}{100} = £22.78195$

Need to round to the nearest penny to make it sensible.

New amount in the account $= £239.81 + £22.78 = £262.59$

1 mark

The interest is the difference between the amount at the end and the amount at the beginning:

$£262.59 - £200 = £62.59$ interest **1 mark**

Alternatively, you can use the multiplier 1.095^3 to do all three years in one go.

$200 \times 1.095^3 = £262.59$ **3 marks**

Interest $= £62.59$ **1 mark**

3 Solve $\dfrac{3x - 7}{2} - \dfrac{x - 3}{5} = \dfrac{x}{3}$. **(5 marks)**

In order to solve the equation, first remove the denominators.
Multiply the whole equation by the LCM of 2, 5 and 3, which is 30.

Multiply each term by 30. Put brackets around the numerators to help you remember that you need to multiply all of each of them.

$$30 \times \frac{(3x - 7)}{2} - 30 \times \frac{(x - 3)}{5} = \frac{x}{3} \times 30$$

1 mark

Cancel each fraction

$$\overset{15}{\cancel{30}} \times \frac{(3x - 7)}{\cancel{2}} - \overset{6}{\cancel{30}} \times \frac{(x - 3)}{\cancel{5}} = \frac{x}{\cancel{3}} \times \overset{10}{\cancel{30}}$$

Multiply out the brackets.
$-6 \times -3 = +18$
Solve the equation.

$15(3x - 7) - 6(x - 3) = 10x$ **1 mark**

$45x - 105 - 6x + 18 = 10x$ **1 mark**

$39x - 87 = 10x$

$29x = 87$ **1 mark**

$x = \dfrac{87}{29}$

$x = 3$ **1 mark**

1 A shop is having a sale. Each day, prices are reduced by 20% of the price on the previous day.
Before the start of the sale, the price of a television is £450.
On the first day of the sale, the price is reduced by 20%.

a Work out the price of the television on
  (1) the first day of the sale,
  (2) the third day of the sale. **(5 marks)**

On the first day of the sale, the price of a cooker is £300.

b Work out the price of the cooker before the start of the sale. **(2 marks)**

[N1998 P5 Q11]

2 A clothes shop has a sale.
All the original prices are reduced by 24% to give the sale price.
The sale price of a jacket is £36.86.
Work out the original price of the jacket. **(3 marks)**

[N1999 P5 Q9]

3 The price of a new television is £423.
This price includes Value Added Tax (VAT) at $17\frac{1}{2}$%.

a Work out the cost of the television **before** VAT was added. **(3 marks)**

By the end of each year, the value of a television has fallen by 12% of its value at the start of that year.
The value of a television was £423 at the start of the first year.

b Work out the value of the television at the end of the **third** year.
Give your answer to the nearest penny. **(4 marks)**

[S2000 P6 Q8]

4 £500 is invested for 2 years at 6% per annum compound interest.

a Work out the total interest earned over the two years. **(3 marks)**

£250 is invested for 3 years at 7% per annum compound interest.

b By what single number must £250 be multiplied to obtain the total amount at the end of the 3 years? **(1 mark)**

[S1998 P5 Q4]

5 Tracey and Wayne share £7200 in the ratio 5 : 4
Work out how much each of them receives. **(3 marks)**

[N2000 P6 Q1]

6 A shop has a sale of jackets and shirts.
In the sale there are a total of 120 jackets and shirts.
The jackets and shirts are in the ratio 5 : 3.
a Work out the number of jackets. **(3 marks)**
b Calculate the percentage that are shirts. **(2 marks)**
[N1998 P6 Q2]

7 Jack shares £180 between his two children Ruth and Ben.
The ratio of Ruth's share to Ben's share is 5 : 4.
a Work out how much each child is given. **(3 marks)**
Ben then gives 10% of his share to Ruth.
b Work out the percentage of the £180 that Ruth now has. **(3 marks)**
[S1999 P5 Q1]

8

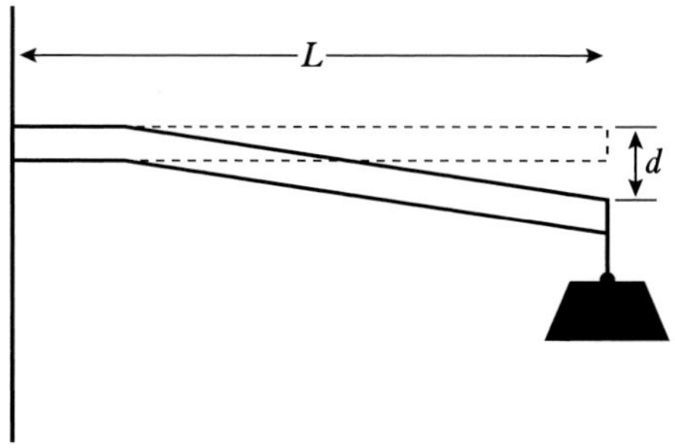

A weight is hung at the end of a beam of length L.
This causes the end of the beam to drop a distance d.
d is directly proportional to the cube of L.
$d = 20$ when $L = 150$.
a Find a formula for d in terms of L. **(3 marks)**
b Calculate the value of L when $d = 15$. **(2 marks)**
[S2000 P6 Q15]

9 y is directly proportional to the cube of x.
When $x = 2, y = 64$.
a Find an expression for y in terms of x. **(3 marks)**
Hence, or otherwise,
b **(1)** calculate the value of y when $x = \frac{1}{2}$,
(2) calculate the value of x when $y = 27$. **(3 marks)**
[S1999 P5 Q14]

10 y is inversely proportional to x^2.
$y = 3$ when $x = 4$.
a Write y in terms of x. **(3 marks)**
b Calculate the value of y when $x = 5$. **(5 marks)**
[N2000 P6 Q14]

11 y is directly proportional to x.
$y = 8$ when $x = 5$.
a Calculate the value of y when $x = 7$. **(2 marks)**

w is inversely proportional to u^2.
b Draw a sketch of the graph of w against u.
Use the axes given below.

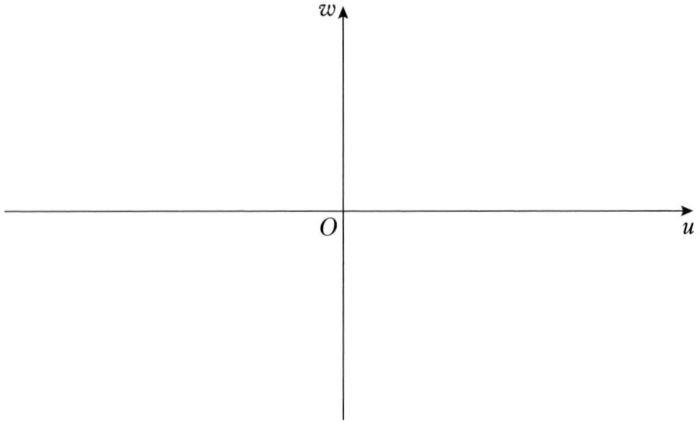

$w = 12$ when $u = 5$. **(2 marks)**
c Calculate the value of u when $w = 27$. **(3 marks)**
[N1999 P6 Q14]

12 Simplify

$$\frac{1}{2x+3} + \frac{1}{2x-1}$$

(3 marks)
[N1998 P6 Q19]

13 **a** Factorise
$x^2 + 3x + 2$. **(1 mark)**
b Write as a single fraction in its simplest form

$$\frac{3}{x+1} + \frac{3x}{x^2+3x+2}$$

(4 marks)
[S2000 P6 Q17]

14 Solve
a $3(x - 6) = 10 - 2x$ **(2 marks)**
b $\frac{y+1}{3} = \frac{1-y}{2}$ **(3 marks)**
[N1998 P5 Q8]

15 **a** Find the value of $\left(\frac{27}{125}\right)^{-\frac{1}{3}}$ **(2 marks)**

b Solve $\frac{2}{x} + \frac{3}{2x} = \frac{1}{3}$ **(2 marks)**

[N2000 P5 Q13]

16 **a** Show that the equation

$$\frac{2}{(x+1)} - \frac{1}{(x+2)} = \frac{1}{2}$$

can be written in the form

$x^2 + x - 4 = 0.$ **(4 marks)**

b Hence, or otherwise, find the values of x, correct to 2 decimal places, that satisfy the equation

$$\frac{2}{(x+1)} - \frac{1}{(x+2)} = \frac{1}{2}.$$ **(3 marks)**

[S1999 P6 Q14]

17 Joe put £5000 in a building society savings account.
Compound interest at 4.8% was added at the end of each year.

a Calculate the total amount of money in Joe's savings account at the end of 3 years. Give your answer to the nearest penny. **(4 marks)**

Sarah also put a sum of money in a building society savings account.
Compound interest at 5% was added at the end of each year.

b Work out the single number by which Sarah has to multiply her sum of money to find the total amount she will have after 3 years.

(2 marks)

[N1999 P6 Q5]

7 Probability diagrams

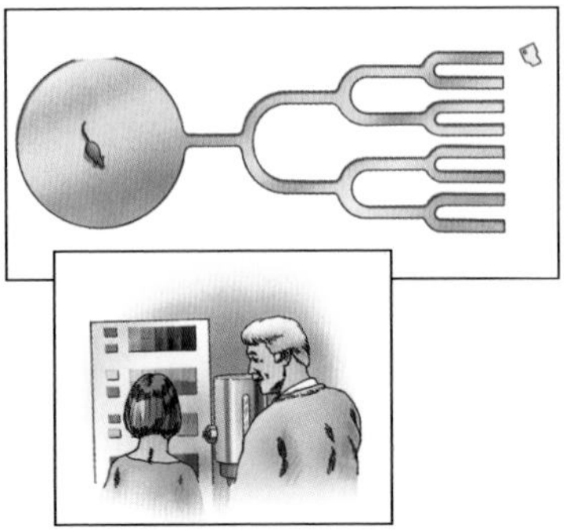

You need to know about:

- sample space diagrams
- independent events
- the probability of an event not happening
- mutually exclusive events
- tree diagrams
- the probability of at least one event happening

Sample space diagram

This **sample space diagram** shows all of the possible outcomes when a coin is tossed and a dice is rolled.

	1	2	3	4	5	6
H	H, 1	H, 2	H, 3	H, 4	H, 5	H, 6
T	T, 1	T, 2	T, 3	T, 4	T, 5	T, 6

In this diagram, all of the possible events are listed.
The probability of getting a tail and an even number $= \frac{3}{12} = \frac{1}{4}$

You may be interested in the total score when two dice are rolled. In this case your sample space diagram would look like this.

	1	2	3	4	5	6
1	2	3	4	5	6	7
2	3	4	5	6	7	8
3	4	5	6	7	8	9
4	5	6	7	8	9	10
5	6	7	8	9	10	11
6	7	8	9	10	11	12

In this diagram you do not show what actually happens, only the total score in each case.

The probability of getting a score of 6 $= \frac{5}{36}$

Independent

Two events are independent if the outcome of one has no effect on the outcome of the other.
If A and B are independent events then
$P(\text{A and B}) = P(\text{A}) \times P(\text{B})$.
In probability questions involving independent events, the words **AND** and **BOTH** mean that you should **MULTIPLY** the probabilities.

Probability of A not happening

For an event A, the **probability of A not happening** is written $P(A')$ and $P(A') = 1 - P(A)$.
$P(A')$ is sometimes written $P(\bar{A})$.

Mutually exclusive

Two events are **mutually exclusive** if they cannot happen at the same time. If A and B are mutually exclusive events then
$P(A \text{ or } B) = P(A) + P(B)$.
In probability questions involving mutually exclusive events, the word **OR** means that you should **ADD** the probabilities.

Tree diagram

A **tree diagram** allows you to show the outcomes for events that happen consecutively.
You **multiply** the probabilities as you go along the branches.
You **add** the probabilities if you need more than one route through the tree.
Some routes through the tree can stop before others.
The probabilities on branches of the tree can be different in the different sections of the tree.

This tree diagram shows the probabilities of getting red and blue balls, from a bag containing 6 red and 4 blue balls, when the first ball taken is replaced before the second is taken.

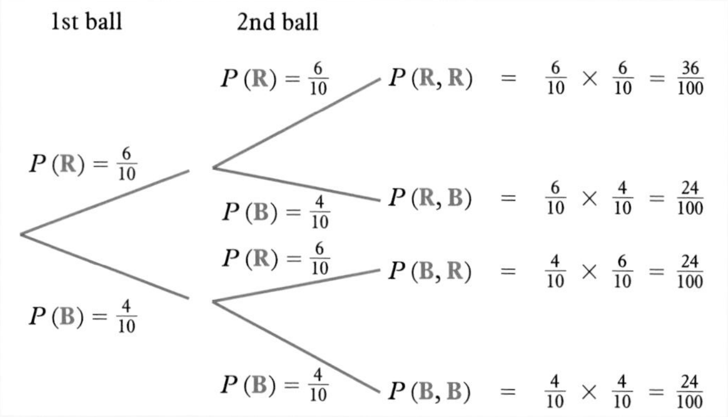

The probability of getting 2 red balls $= P(R, R) = \frac{36}{100} = \frac{9}{25}$

The probability of getting 1 ball of each colour
$= P(R, B) + P(B, R)$
$= \frac{24}{100} + \frac{24}{100} = \frac{48}{100} = \frac{12}{25}$

Probability of at least one

The **probability of at least one** of something happening is $1 -$ the probability of none happening.
This is because **none** and **at least one** are mutually exclusive and nothing else is possible.

1 **a** Draw a sample space diagram to show all the possible outcomes when a coin is tossed and an ordinary dice is rolled.

b Use your diagram to write down the probability of getting
(1) a head on the coin and a 6 on the dice
(2) a head on the coin and an even number on the dice
(3) a tail on the coin and a prime number on the dice
(4) a tail on the coin and a square number on the dice.

2 Events A, B and C are such that $P(\text{A}) = \frac{1}{5}$, $P(\text{B}) = \frac{3}{4}$ and $P(\text{C}) = \frac{1}{20}$.

a Show that $P(\text{C}') = P(\text{A}) + P(\text{B})$.

b Given that A and B are independent, find $P(\text{A and B})$.

c Given that $P(\text{A or B}) = \frac{17}{20}$, show that A and B are not mutually exclusive.

3 Two fair dice are rolled together. The score is the total on the two dice.

a Draw a sample space diagram to show all the possible scores.

b Find the probability of scoring 7.

c Find the probability of getting a double.

d Find the probability of getting a double or scoring 7.

e Find the probability of scoring 8.

f Find the probability of getting a double or scoring 8.

4 The diagram shows two spinners. The sections on each spinner are equal.

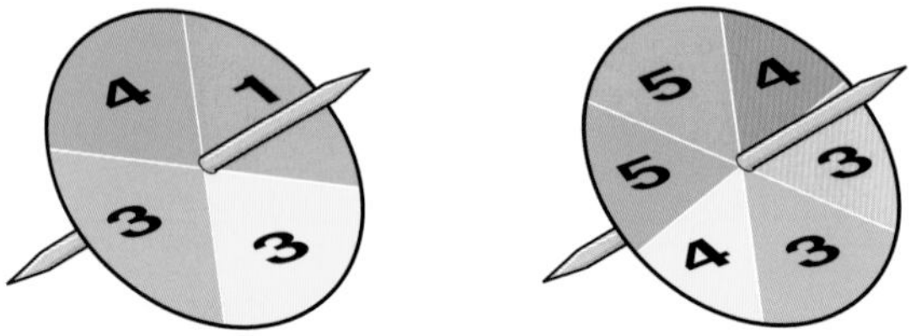

Both of the spinners are spun and the scores on the two spinners are added together.

a Copy this sample space diagram and complete it to show all the possible scores.

	3	3	4	4	5	5
1	4					
3					8	
3						
4						

b Find the probability that the score is at least 7.

5 A bag of sweets contains 7 toffees and 8 fruit creams.
Beth takes a sweet out and eats it.
Then she takes another sweet and eats it.

a Draw a tree diagram to show the possible outcomes.

b Find the probability that Beth

(1) eats 2 toffees

(2) eats a toffee followed by a fruit cream

(3) eats a toffee and a fruit cream

(4) eats at least one fruit cream.

6 Benny is building using blocks.
He has 12 red blocks and 10 green blocks.
He chooses his blocks at random and places them on top of each other to form a single stack.
Work out the probability that Benny will

a choose a red block first

b have 2 red blocks on the bottom of his stack

c have 1 block of each colour at the bottom of the stack

d use 3 red blocks as his first 3 blocks

e use at least 1 red block in his first 3 blocks.

7 Victoria is a high jumper.
In a competition, she has three attempts to achieve her personal best height.
If she clears the jump she does not take the remaining goes.
The probability that she clears the jump on the first attempt is 0.2.
If she fails, the probability of clearing the jump reduces to 0.1 on each subsequent try.

a Draw a tree diagram to show all the possible outcomes for the three attempts.

b Find the probability that Victoria is successful on her second attempt.

c Find the probability that Victoria is successful on her third attempt.

d Find the probability that Victoria achieves her personal best.

e Find the probability that Victoria fails to achieve her personal best.

1 In a game there are two fair spinners.
The first spinner is numbered 1, 2, 3, 4, 5, 6, 7, 8.
The second spinner is numbered 4, 5, 6, 7, 8.

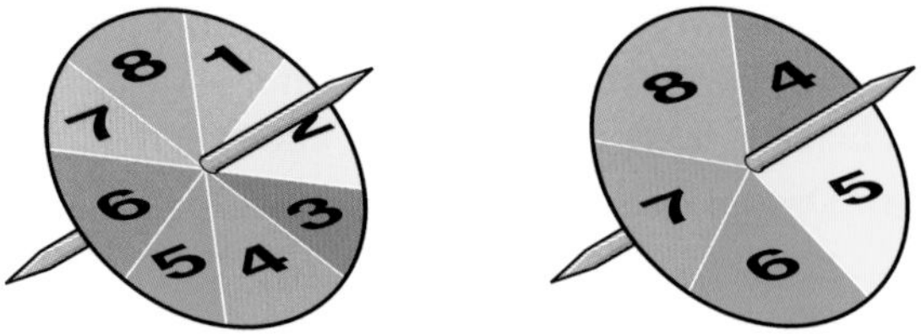

Both spinners are spun and the scores added together.

a Draw a sample space diagram to show all possible outcomes. **(2 marks)**

b Find the probability that the sum of the numbers on the two spinners will be
(1) 4 **(2)** 8 **(3)** 12 **(4)** **not** 15 **(4 marks)**

a *Draw a sample space diagram, adding the scores to give each value.*

+	1	2	3	4	5	6	7	8
4	5	6	7	8	9	10	11	12
5	6	7	8	9	10	11	12	13
6	7	8	9	10	11	12	13	14
7	8	9	10	11	12	13	14	15
8	9	10	11	12	13	14	15	16

2 marks

b **(1)** The number of 4s in the diagram is 0, so $P(4) = 0$. **1 mark**

(2) The number of 8s in the diagram is 4, so $P(8) = \frac{4}{40} = \frac{1}{10}$. **1 mark**

(3) The number of 12s in the diagram is 5, so $P(12) = \frac{5}{40} = \frac{1}{8}$. **1 mark**

(4) The number of 15s in the diagram is 2, so $P(15) = \frac{2}{40} = \frac{1}{20}$.

So $P(15') = \frac{19}{20}$. **1 mark**

2

Cherry	Apple	Banana	Orange	Strawberry
0.10	0.15	0.25	0.20	

The probabilities of obtaining fruits on a fruit machine are shown in the table.
Find the probability that the fruit selected will be

a strawberry
b cherry or apple
c banana or orange
d raspberry **(7 marks)**

a Since all five probabilities must add up to 1, the probability of strawberry is

$1 - (0.10 + 0.15 + 0.25 + 0.20)$ **1 mark**

$= 1 - 0.7 = 0.3$ **1 mark**

b The probability of cherry or apple is

$0.10 + 0.15$ **'OR' = ADD** **1 mark**

$= 0.25$ **1 mark**

c The probability of banana or orange is **'OR' = ADD**

$0.25 + 0.20$ **1 mark**

$= 0.45$ **1 mark**

d There is no raspberry in the table, so the probability is 0. **1 mark**

Write this as a '0' figure.

'None', 'nothing', $\frac{\text{'0'}}{10}$ will not be given a mark.

3 On an island, the probability that it will rain on any day is 0.2. The tree diagram below can be used to find the probability that it will rain on two consecutive days.

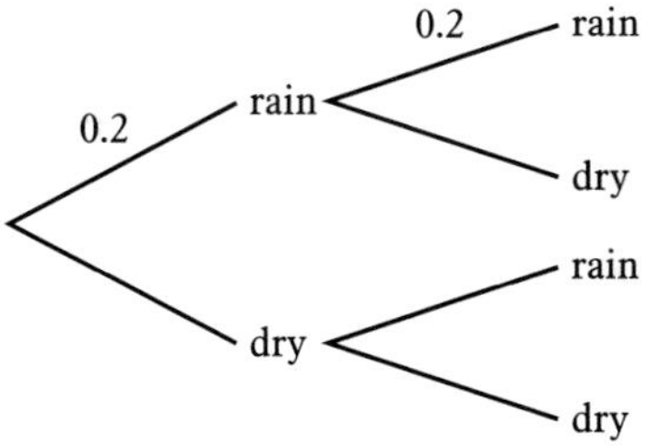

a Complete the tree diagram. **(2 marks)**

b Use the tree diagram to find the probability that

(1) it will rain on both days

(2) it will rain on just one of the days. **(5 marks)**

c *The probability of a rainy day is 0.2*
So the probability of a dry day is
$1 - 0.2 = 0.8$
Add these details to the tree diagram:

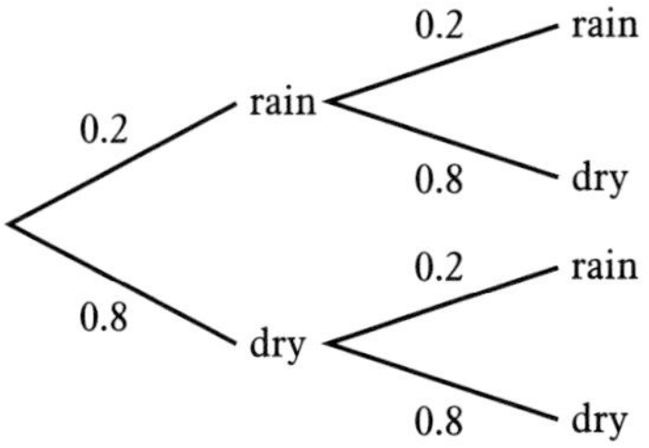

2 marks

d (1) *Follow the lines which show it will rain on both days:*

0.2×0.2 rain on both days is rain **1 mark**

$= 0.4$ **AND** rain, which means **MULTIPLY.** **1 mark**

(2) *Follow the lines which show it will rain on one day. There are two routes.*

rain **AND** dry	**OR**	dry **AND** rain
0.2×0.8	+	$0.8 \times 0.2 = 0.16 + 0.16 = 0.32$
1 mark	**1 mark**	**1 mark**

1 A lorry contains 232 boxes of crisps.
Each box has either plain crisps or cheese and onion flavour crisps.
The probability that a box selected at random holds plain crisps is $\frac{1}{3}$ of the probability that the box holds cheese and onion flavour crisps.

a Calculate the number of boxes of plain crisps. **(3 marks)**

Each box holds 48 packets of crisps.
One in every 8 packets of plain crisps has a prize in it. One in every 16 packets of cheese and onion flavour crisps has a prize in it.
A packet is to be selected at random from the lorry.

b Calculate the probability that the packet will have a prize in it **(3 marks)**

[S1998 P6 Q8]

2 Lauren and Yasmina each try to score a goal.
They each have one attempt.
The probability that Lauren will score a goal is 0.85.
The probability that Yasmina will score a goal is 0.6.

a Work out the probability that **both** Lauren **and** Yasmina will score a goal. **(2 marks)**

b Work out the probability that Lauren **will** score a goal **and** Yasmina **will not** score a goal. **(2 marks)**

[S1998 P5 Q1]

3 Tina has a biased dice.
When she rolls it, the probability that she will get a six is 0.09.
Tina is going to roll the biased dice **twice**.

Work out the probability that she will get

(1) **two** sixes, **(2)** **exactly one** six. **(5 marks)**

[S2000 P6 Q7]

4 Jack has two fair dice.
One of the dice has 6 faces numbered from 1 to 6.
The other dice has 4 faces numbered from 1 to 4.
Jack is going to throw the two dice.
He will add the scores on the two dice to get the total.

Work out the probability that he will get

(1) a total of 7, **(2)** a total of less than 5. **(4 marks)**

[S2000 P5 Q7]

5 The probability that a washing machine will break down in the first 5 years of use is 0.27.
The probability that a television will break down in the first 5 years of use is 0.17.
Mr Khan buys a washing machine and a television on the same day.
By using a tree diagram or otherwise, calculate the probability that, in the five years after that day

a both the washing machine and the television will break down, **(2 marks)**

b at least one of them will break down. **(4 marks)**

[S1998 P6 Q10]

6 Helen and Joan are going to take a swimming test.
The probability that Helen will pass the swimming test is 0.95.
the probability that Joan will pass the swimming test is 0.8.
The two events are independent.

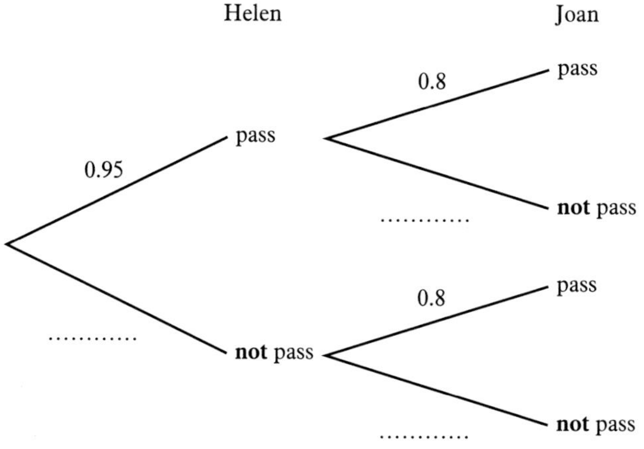

a Complete the probability tree diagram. **(2 marks)**

b Work out the probability that both Helen and Joan will pass the swimming test. **(2 marks)**

c Work out the probability that one of them will pass the swimming test and the other one will not pass the swimming test. **(3 marks)**

[N2000 P6 Q11]

7 Sharon has 12 computer discs. Five of the discs are red. Seven of the discs are black. She keeps all the discs in a box. Sharon removes one disc at random. She records its colour and replaces it in the box. Sharon removes a second disc at random, and again records its colour.

a Complete the tree diagram

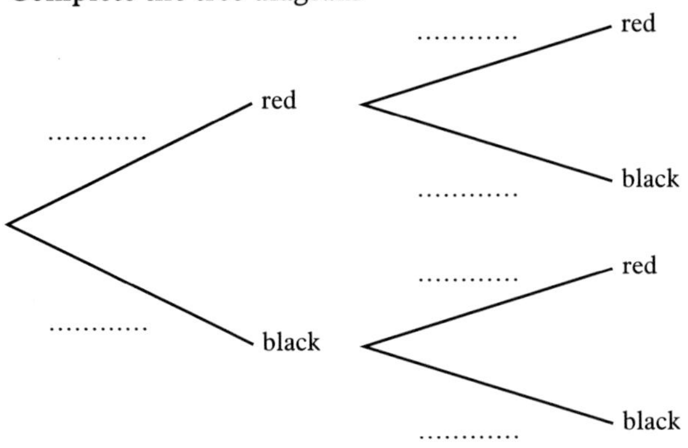

(2 marks)

b Calculate the probability that the two discs removed

(1) will both be red,

(2) will be different colours.

(5 marks)

[S1999 P6 Q7]

8 Jason has 10 cups. 6 of the cups are Star Battle cups. 4 of the cups are Top Pops cups. On Monday Jason picks at random one cup from the 10 cups. On Tuesday he also picks at random one cup from the same 10 cups.

a Complete the probability tree diagram

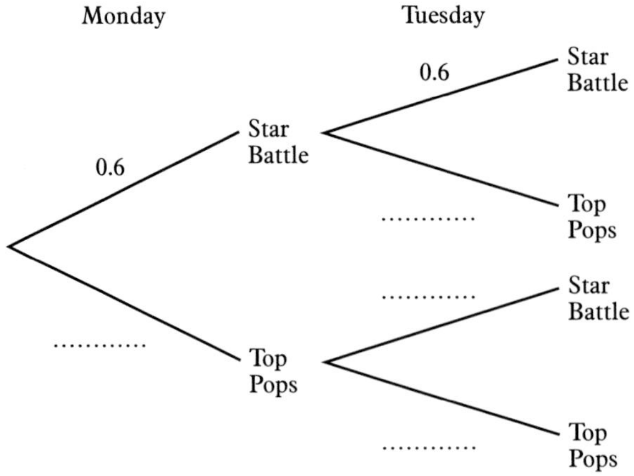

(2 marks)

b Work out the probability that Jason will pick a Star Battle cup on both Monday and Tuesday. **(2 marks)**

c Work out the probability that Jason will pick one of each type of cup. **(3 marks)**

[S2001 P5 Q11]

9 Nikki and Ramana both try to score a goal in netball.
The probability that Nikki will score a goal on her first try is 0.65.
The probability that Ramana will score a goal on her first try is 0.8.

(1) Work out the probability that Nikki and Ramana will both score a goal on their first tries.

(2) Work out the probability that neither Nikki nor Ramana will score a goal on their first tries. **(5 marks)**

[N1997 P5 Q13]

10 Tony carries out a survey about the words in a book.
He chooses a page at random.
He then counts the number of letters in each of the first hundred words on the page.
The table shows Tony's results.

Number of letters in a word	1	2	3	4	5	6	7	8
Frequency	6	9	31	24	16	9	4	1

A word is chosen at random from the hundred words.

a What is the probability that the word will have 5 letters? **(2 marks)**

The book has 25 000 words. **(2 marks)**

b Estimate the number of 5 letter words in the book. **(2 marks)**

The book has 125 pages with a total of 25 000 words.
The words on each of the first 75 pages are counted.
The mean is 192.

c Calculate the mean number of words per page for the remaining 50 pages. **(3 marks)**

[S1999 P5 Q6]

8 Mainly quadratics

You need to know about:

- changing the subject of a formula
- quadratic graphs
- factorising quadratic expressions
- solving quadratic equations
- completing the square
- solving simultaneous equations
- trial and improvement

Changing the subject

The subject of a formula is the letter that appears on its own on the left-hand side. To change the subject, reverse the operations and the order in which they have been done.

Example

Make t the subject of the formula $c = \sqrt{yt + p}$.
The t has been by **multiplied by** y, then p **has been added** and then the result has been **square rooted.**

Square both sides: $c^2 = yt + p$

Subtract p**:** $c^2 - p = yt$

Divide by y**:** $\dfrac{c^2 - p}{y} = t$

Always write the new subject letter on the left.

$$t = \frac{c^2 - p}{y}$$

If the subject letter appears more than once, get the terms together and factorise.

Example

Make t the subject of the formula.

$$ct = zt + yb$$

Get the terms with t in them together:

$$ct - zt = yb$$

Factorise out the t:

$$t(c - z) = yb$$

Divide by the bracket to leave t as the subject:

$$t = \frac{yb}{c - z}$$

Quadratic graphs

A **quadratic graph** will always be

 shaped if the coefficient of x^2 is positive

or

 shaped if the coefficient of x^2 is negative.

Factorising quadratic expressions

A **quadratic expression** is one of the form $ax^2 + bx + c$. Some quadratics will factorise into two brackets.

Example $2x^2 + 3x + 1$
$= (2x + 1)(x + 1)$

The number on its own is $+1$. The numbers in the brackets must have the same sign to give a product of $+1$. They must both be $+$ to give $+3x$.

Example $5x^2 - 22x - 15$
$= (5x + 3)(x - 5)$

The number on its own is -15. The numbers in the brackets must have different signs to give a product of -15.

Don't forget that a quadratic with no 'c' term factorises using common factors:

$2x^2 + 3x = x(2x + 3)$ $\quad$ $3x^2 - x = x(3x - 1)$ $\quad$ $9x^2 - 6x = 3x(3x - 2)$

Difference of two squares

$x^2 - a^2 = (x + a)(x - a)$

So, for example $25y^2 - 16z^2 = (5y)^2 - (4z)^2 = (5y + 4z)(5y - 4z)$

Solving a quadratic equation

To **solve a quadratic equation**, check if the quadratic factorises. If it does, factorise it and put each bracket equal to zero in turn to find the two solutions.

Examples

Solve $5x^2 - 22x - 15 = 0$

$(5x + 3)(x - 5) = 0$

So $5x + 3 = 0$ or $x - 5 = 0$

So $x = -\frac{3}{5}$ or $x = 5$

Solve $3x^2 + 4x = 0$

$x(3x + 4) = 0$

So $x = 0$ or $3x + 4 = 0$

So $x = 0$ or $x = -\frac{4}{3}$

Quadratic formula

If the quadratic does not factorise then use the quadratic formula.

If $ax^2 + bx + c = 0$ then $x = \dfrac{-b \pm \sqrt{b^2 - 4ac}}{2a}$.

This formula will be on your formula sheet, so you only need to remember how to use it.

Completing the square

This is another way of solving quadratic equations when they will not factorise. You need to change the quadratic $ax^2 + bx + c$ into an expression of the form $p(x + q)^2 + r$.
For $x^2 + 8x + 1$ you start with $(x + 4)^2$.
You use half the coefficient of x inside the bracket.
Now $(x + 4)^2 = x^2 + 8x + 16$, so $x^2 + 8x + 1 = (x + 4)^2 - 15$ because you only want $+1$ so you need to subtract 15.

For $2x^2 + 12x + 1$ you start by taking the 2 out.
$2x^2 + 12x + 1 = 2(x^2 + 6) + 1$.
You need to complete the square on $(x^2 + 6x)$ and you need to use $(x + 3)^2$ for this.
Again use half the coefficient of x inside the bracket.

$$\begin{aligned} \text{Now} \qquad 2(x + 3)^2 &= 2(x^2 + 6x + 9) \\ &= 2x^2 + 12x + 18 \\ \text{So} \qquad 2x^2 + 12x + 1 &= 2(x + 3)^2 - 17. \end{aligned}$$

Simultaneous equations

You can be asked to solve **simultaneous equations** when one of the equations is linear and one is quadratic.

In particular, you may have to deal with an equation of the form $x^2 + y^2 = r^2$ (a circle centre (0, 0) and radius r). To do this, make x or y the subject of the linear equation and substitute into the quadratic to get a quadratic equation to solve. Then put your answers back into the linear equation to find the simultaneous solutions. The solutions to these simultaneous equations are the points of intersection of the line and the circle.

Trial and improvement

For more difficult equations, you can solve them by trial and improvement. To find a solution to 1 dp get a pair of consecutive 1 dp numbers, one of which gives a value that is too big and one a value too small. Then check the 2 dp mid-value of these 1 dp numbers. If this value is too big then the answer is the 1 dp value that gave the small value and vice versa.

1 Make the red letter the subject of each of these formulas

a $t = \dfrac{a + g}{7}$

b $p = \dfrac{4b + 7r}{2}$

c $k = \dfrac{pc + 2t}{y}$

d $yk = \dfrac{ud + g}{4}$

e $w = \sqrt{5e - l}$

f $d = f^2 - 5$

g $t = \sqrt{\dfrac{g}{5}}$

h $h = \dfrac{4h + j}{5}$

i $i = \dfrac{gi + 6j}{p}$

j $aj = \dfrac{4h + bj}{5}$

k $u = \dfrac{4h - 3k}{k}$

l $a = \dfrac{4h + bl}{5l}$

m $m = \sqrt{5m^2 - l}$

n $an = \sqrt{6n^2 - 4g}$

2 **a** Copy this table. Fill it in.

x	-4	-3	-2	-1	0	1	2	3	4
x^2									
$+3x$									
-1									
y									

b Draw an x axis from -4 to $+4$ and a y axis from -5 to 30.

c Draw the graph of $y = x^2 + 3x - 1$.

3 Ken is playing cricket. He hits the ball trying for a six. The height of the ball above the ground, h metres, t seconds after he hits it is given by the formula below.

$$h = 1.3 + t - 0.25t^2$$

a Copy this table. Fill it in.

t	0	1	2	3	4	5
h						

b Draw a graph to show the height of the ball. Use a scale of 2 cm for each second on the t axis and 2 cm for 1 m on the h axis.

c The ball passes over the boundary rope at a height of 0.8 m. Use your graph to estimate the time that it takes for the ball to cross the rope.

4 Multiply out each of these pairs of brackets.

a $(x + 1)(x - 6)$ **b** $(2x + 1)(5x - 2)$

5 Factorise these quadratic expressions.

a $x^2 + 9x + 18$ **c** $3x^2 - 10x + 3$

b $3x^2 + 4x - 4$ **d** $6x^2 - x - 12$

6 Solve these quadratic equations.

a $x^2 - 8x + 12 = 0$ **c** $3x^2 - 4x - 4 = 0$

b $2x^2 + 3x + 1 = 0$ **d** $6x^2 - 17x + 5 = 0$

7 Solve these quadratic equations.
Give your answers to 2 dp.

a $x^2 + 9x + 12 = 0$ **c** $3x^2 - 2x - 4 = 0$

b $2x^2 + 3x - 1 = 0$ **d** $4x^2 - 10x + 3 = 0$

8 **a** Write down an expression for the area of this rectangle, which has width x and length $2x + 8$.

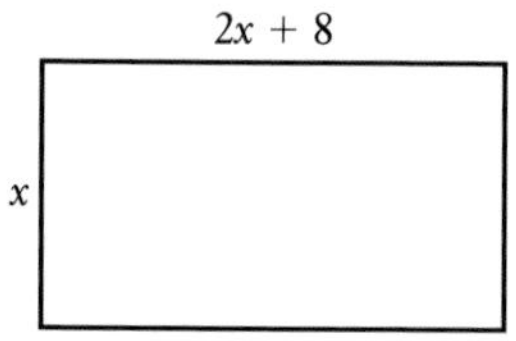

Another rectangle is drawn.
Its length is half the length of the rectangle above.

b Write down an expression for the length of the new rectangle.

The width of the new rectangle is 5 units more than the width of the rectangle above.

c Write down an expression for the width of the new rectangle.

d Write down an expression for the area of the new rectangle.

e The areas of the two rectangles are the same.
Find the value of x.

9 **a** Write $x^2 + 4x - 1$ in the form $(x + a)^2 + b$.

b Write $3x^2 - 6x - 4$ in the form $p(x + q)^2 + r$.

10 Solve this pair of simultaneous equations.

$$x + y = 2$$
$$x^2 + y^2 = 25$$

11 Solve each of these equations by trial and improvement.
Give all of your answers to 1 dp.

a $x^3 + 15 = 200$ **c** $2x^3 + x = 81$

b $x^2 + \frac{3}{x} = 1$ **d** $5x^2 - \frac{1}{x} = 2$

1 The formula for the volume of a hemisphere is $V = \dfrac{2\pi r^3}{3}$.

a Calculate the volume of a hemisphere of radius 5.47 cm.
Give your answer correct to 3 significant figures. **(2 marks)**

b Make r the subject of the formula. **(3 marks)**

a Volume $= \dfrac{2}{3} \times \pi \times 5.47^3$

Although this is an easy calculation on a calculator, it is also easy to make mistakes.

Always put $\frac{2}{3}$ in as 2 ÷ 3 or 2 **$a^b/_c$** *3. Rounding to a few dp (e.g. 0.666) will give an inaccurate answer.*

Always use the π value on your calculator. Don't use a rounded value.

Make sure 5.47^3 is put in as 5.47 **x^y** *3 or 5.47 × 5.47 × 5.47 Don't use rounded values.*

$$\frac{2}{3} \quad \times \quad \pi \quad \times \quad 5.47^3$$ **1 mark**

2 **$a^b/_c$** **3** **×** **π** **×** **5** **.** **4** **7** **x^y** **3** **=**

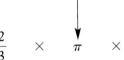

Always write down an accurate answer ***before*** *you give a rounded answer.*

$= 342.610\,262\,8$

$= 343\text{ cm}^3$ (3 sf) **1 mark**

b $V = \dfrac{2\pi r^3}{3}$

Start by getting r on the left: $\dfrac{2\pi r^3}{3} = V$

In making r the subject you want to end up with 'r ='

Multiply both sides by 3 to get rid of the denominator: $\not{3} \times \dfrac{2\pi r^3}{\not{3}} = V \times 3$ **1 mark**

$2\pi r^3 = 3V$

Divide both sides by 2π to get r^3: $\dfrac{\not{2}\not{\pi} r^3}{\not{2}\not{\pi}} = \dfrac{3V}{2\pi}$

$r^3 = \dfrac{3V}{2\pi}$ **1 mark**

To find r, cube root both sides: $r = \sqrt[3]{\dfrac{3V}{2\pi}}$ **1 mark**

2 Use a trial and improvement method to find the solution to the equation $x^3 + x = 20$.
Give your answer correct to one decimal place. **(4 marks)**

Try $x = 2$: $2^3 + 2 = 10$ too small
Try $x = 3$: $3^3 + 2 = 31$ too big

A solution lies between 2 and 3. **1 mark**

Try $x = 2.5$: $2.5^3 + 2.5 = 18.125$ too small
Try $x = 2.6$: $2.6^3 + 2.6 = 20.176$ too big

Always show what you have worked out, and the answers to the calculations.

A solution lies between 2.5 and 2.6. **1 mark**

Now use $x = 2.55$: $2.55^3 + 2.55 = 19.13$ too small **1 mark**

The solution lies between 2.55 and 2.6.

The solution is the value of x that has been used, not the resulting answer.

So the solution, to one decimal place, is 2.6. **1 mark**

3 Solve $x^2 + y^2 = 9$
$3x + y = 4$

Give your answers correct to 2 decimal places. **(5 marks)**

The first step is to rearrange the linear equation to make x or y the subject.

So $3x + y = 4$ *becomes* $y = 4 - 3x$

Substitute the expression for y into the quadratic equation:

Substitute in:

$$x^2 + y^2 = 9$$

$$x^2 + (4 - 3x)^2 = 9$$ **1 mark**

$$x^2 + (4 - 3x)(4 - 3x) = 9$$

Multiply out the brackets: $$x^2 + 16 - 12x - 12x + 9x^2 = 9$$

Gather terms: $$10x^2 - 24x + 7 = 0$$ **1 mark**

Solve this quadratic by using the formula with $a = 10$, $b = -24$ *and* $c = 7$.

$$x = \frac{-b \pm \sqrt{b^2 - 4ac}}{2a} = \frac{24 \pm \sqrt{576 - 280}}{20} = \frac{24 \pm \sqrt{296}}{20}$$ **1 mark**

$$x = \frac{24 + 17.2\ldots}{20} \quad \text{or} \quad x = \frac{24 - 17.2\ldots}{20}$$

$x = 2.06$ or $x = 0.34$ **1 mark**

Now find the corresponding values of y. Use $y = 4 - 3x$

$y = -2.18$ or $y = 2.98$ **1 mark**

The solutions are $x = 2.06, y = -2.18$ and $x = 0.34, y = 2.98$.

1 $C = 180R + 2000$

The formula gives the capacity, C litres, of a tank needed to supply water to R hotel rooms.

Make R the subject of the formula $C = 180R + 2000$. **(2 marks)**
[S1999 P5 Q5]

2 $t = \frac{8(p+q)}{pq}$

$p = 2.71$
$q = -3.97$

a Calculate the value of t.
Give your answer to a suitable degree of accuracy. **(2 marks)**

b Make q the subject of the formula

$$t = \frac{8(p+q)}{pq}$$

(4 marks)
[S2000 P6 Q12]

3 The diagram shows a solid.

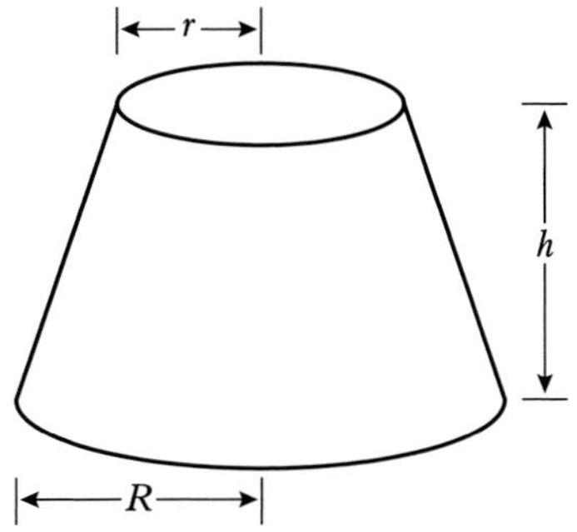

The volume, V, of the solid is given by the formula

$$V = \frac{\pi h}{3}(R^2 + Rr + r^2)$$

a $h = 6.8$, $R = 9.7$ and $r = 5.3$
Calculate the value of V.
Give your answer correct to 3 significant figures. **(3 marks)**

b Make h the subject of the formula. **(2 marks)**

c $V = 200$, $h = 10$ and $R = 2r$.
Calculate the value of R.
Give your answer correct to 3 significant figures. **(3 marks)**
[N1998 P5 Q13]

4 The diagram shows a rectangle with length $3x + 2$ and width $2x$.
All measurements are given in centimetres.
The perimeter of the rectangle is P centimetres.
The area of the rectangle is A square centimetres.

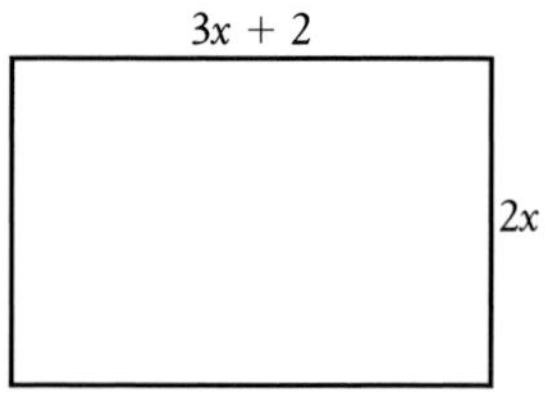

a Write down an expression in its simplest form, in terms of x, for
(1) P, **(2)** A. **(3 marks)**

$P = 44$.
b Work out the value of A. **(3 marks)**
[S1999 P6 Q2]

5 Expand and simplify
(1) $(2x - 3)(x + 4)$
(2) $(x^2 + y^2)^2$
(3) $(x + y)^2 - (x - y)^2$ **(6 marks)**
[S1998 P5 Q15]

6 **a** Expand and simplify
$(2x + 1)(x - 3)$ **(2 marks)**
b Simplify
$3a^3 \times 2ab^2$ **(1 mark)**
c Factorise completely
$2t^2 + 4t$ **(1 mark)**
d Factorise completely
$9x^2 - 4$ **(2 marks)**
[N1998 P6 Q8]

7 Solve the equation
$(2x - 3)^2 = 100$ **(3 marks)**
[N2000 P5 Q20]

8 **a** Expand and simplify
$(2x - 5)(x + 3)$ **(2 marks)**
b **(1)** Factorise
$x^2 + 6x - 7$
(2) Solve the equation
$x^2 + 6x - 7 = 0$ **(3 marks)**
[S1999 P6 Q8]

9 **(1)** Factorise
$x^2 - 6x + 8$
(2) Solve the equation
$x^2 - 6x + 8 = 0$ **(3 marks)**
[S2001 P6 Q11]

10 The length of a rectangle is $(x + 4)$ cm.
The width is $(x - 3)$ cm.
The area of the rectangle is 78 cm^2.

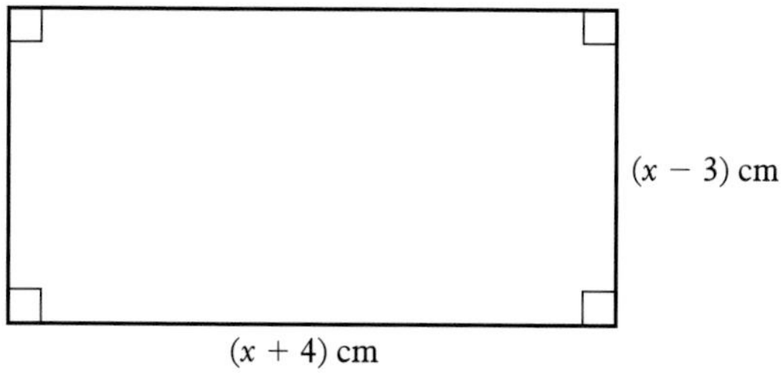

Diagram **NOT** accurately drawn

a Use this information to write down an equation in terms of x. **(2 marks)**

b **(1)** Show that your equation in part **a** can be written as
$x^2 + x - 90 = 0$

(2) Find the values of x which are the solutions of the equation
$x^2 + x - 90 = 0$

(3) Write down the length and the width of the rectangle. **(6 marks)**

[N2000 P5 Q11]

11 Use the method of trial and improvement to solve the equation

$x^3 - 2x = 37$

Give your answer correct to two decimal places.
You must show **ALL** your working. **(4 marks)**

[N1998 P6 Q9]

12 A solution of the equation

$x^3 - 9x = 5$

is between 3 and 4.
Use the method of trial and improvement to find this solution.
Give your answer correct to 2 decimal places.
You must show **all** your working. **(4 marks)**

[S1999 P5 Q3]

13 Using trial and improvement, or otherwise, solve the equation

$t^3 + t = 17$

Show **ALL** your working and give your answer correct to 2 decimal places. **(4 marks)**

[S1998 P5 Q2]

14 The equation

$$x^3 - 5x = 38$$

has a solution between 3 and 4.
Use a trial and improvement method to find this solution.
Give your answer correct to 1 decimal place.
You must show **ALL** your working. **(4 marks)**
[S2000 P6 Q2]

15 Use the method of trial and improvement to solve the equation

$$x^3 + x = 26$$

Give your answer correct to 1 decimal place. You must show **all** your working. **(4 marks)**
[S2001 P6 Q6]

16

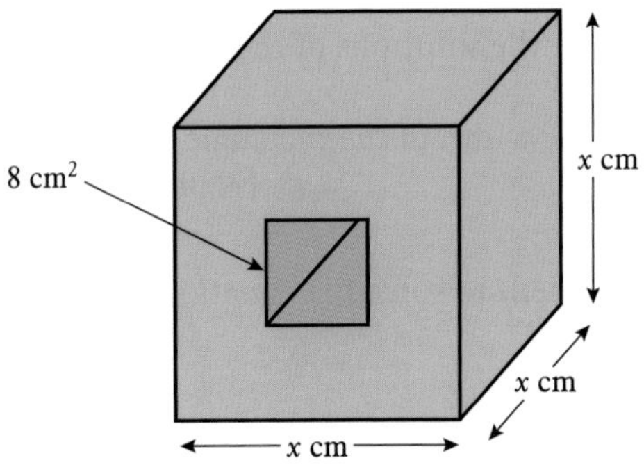

Diagram **NOT** accurately drawn

The diagram shows a prism
The prism is made from a cube of side x cm.
A hole of uniform cross-sectional area 8 cm^2 is cut through the cube.
Find, in terms of x, an expression for the volume of the prism. **(4 marks)**
[S2001 P5]

17 **a** Expand and simplify
$4(x + 3) + 3(2x - 3)$ **(2 marks)**
b Expand and simplify
$(2x - y)(3x + 4y)$ **(3 marks)**
[N2000 P6]

18 Use the method of trial and improvement to find the positive solution of

$$x^3 + x = 37$$

Give your answer correct to 1 decimal place. **(4 marks)**
[S1997 P6]

9 Vectors

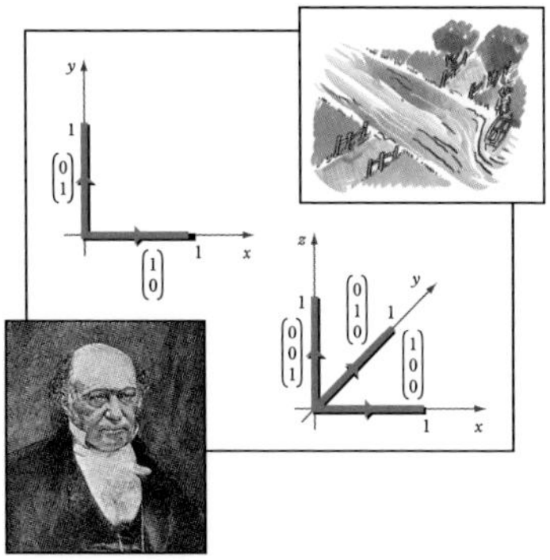

You need to know about:

- vector notation
- adding and subtracting vectors
- multiplying a vector by a scalar
- the magnitude of a vector
- position vectors
- collinear vectors
- resultant vectors

Vector quantity

A **vector quantity** has both size and direction.
You can show a vector using a line with an arrow to show the direction. The length of the line represents the size of the vector. You use a small underlined letter to represent the vector, although books will use a bold letter. So this vector is written here as **a** which you would write as $\underset{\sim}{a}$.

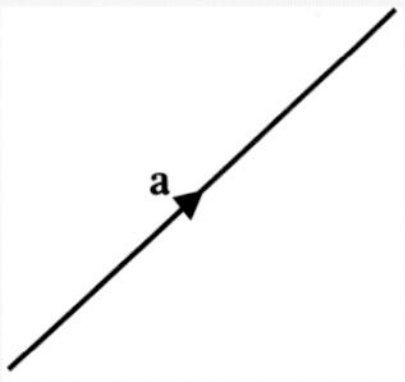

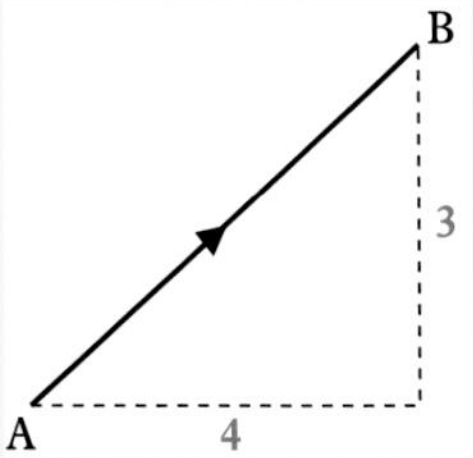

This vector passes from A to B. The notation for this is $\overrightarrow{AB}$.
This vector is 4 units to the right and 3 units up.

This can be written as the column vector $\begin{pmatrix}4\\3\end{pmatrix}$.

Adding vectors

To **add vectors** start the second vector from the end of the first. The sum of the two vectors is from the start of the first to the end of the second.

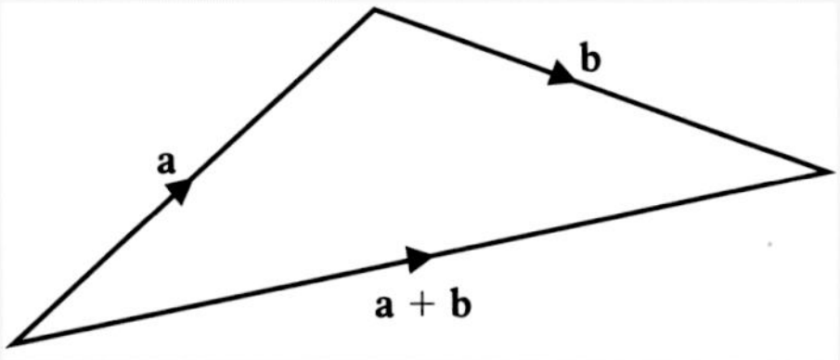

For column vectors, add the corresponding parts together.

$$\begin{pmatrix}4\\3\end{pmatrix}+\begin{pmatrix}5\\-2\end{pmatrix}=\begin{pmatrix}4+5\\3+(-2)\end{pmatrix}=\begin{pmatrix}9\\1\end{pmatrix}$$

Subtracting vectors

To **subtract vectors** add the negative of the second vector to the first. The negative of **b** is the same length as **b** in the opposite direction.

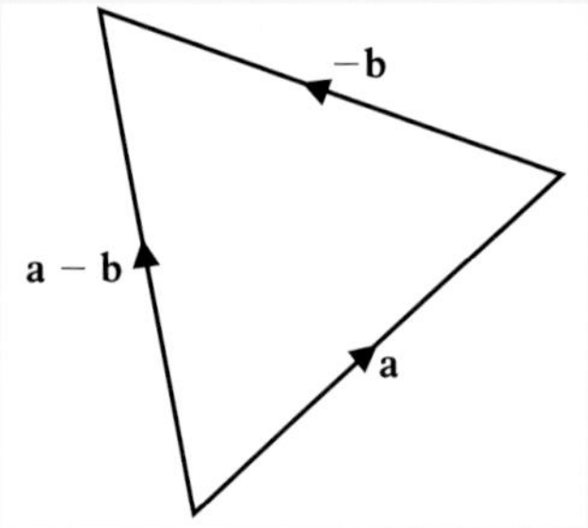

For column vectors, subtract the corresponding parts.

$$\begin{pmatrix}4\\3\end{pmatrix} - \begin{pmatrix}5\\-2\end{pmatrix} = \begin{pmatrix}4-5\\3-(-2)\end{pmatrix} = \begin{pmatrix}-1\\5\end{pmatrix}$$

On a vector diagram, any route that starts and finishes at the same points as another route gives the same vector. This allows you to solve geometrical problems where you need to use different routes round a diagram and use the equivalent vectors.

Multiplying a vector by a scalar

Multiplying a vector by a scalar multiplies the length of the vector. It does not change the direction.

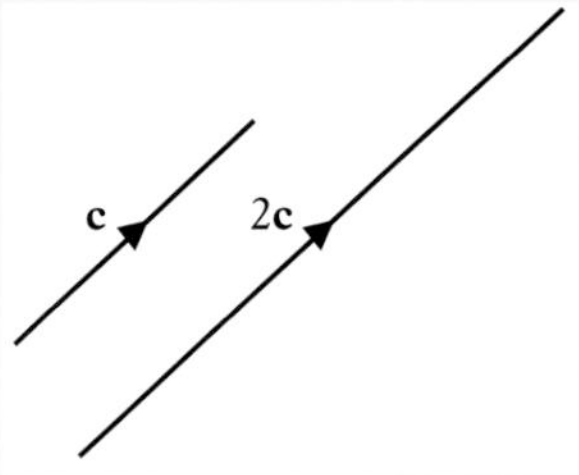

For column vectors, multiply each part by the scalar.

$$2 \times \begin{pmatrix}3\\2\end{pmatrix} = \begin{pmatrix}2\times3\\2\times4\end{pmatrix} = \begin{pmatrix}6\\8\end{pmatrix}$$

Magnitude of a vector

The **magnitude of a vector** is the size of the vector.

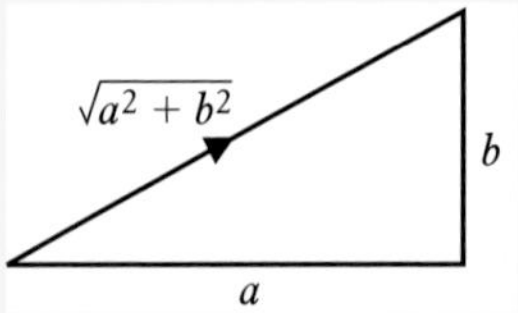

The magnitude of the vector $\begin{pmatrix}a\\b\end{pmatrix}$ is $\sqrt{a^2 + b^2}$.

Position vector

The **position vector** of a point is the vector from the origin to the point. So the position vector of A is $\overrightarrow{OA}$.

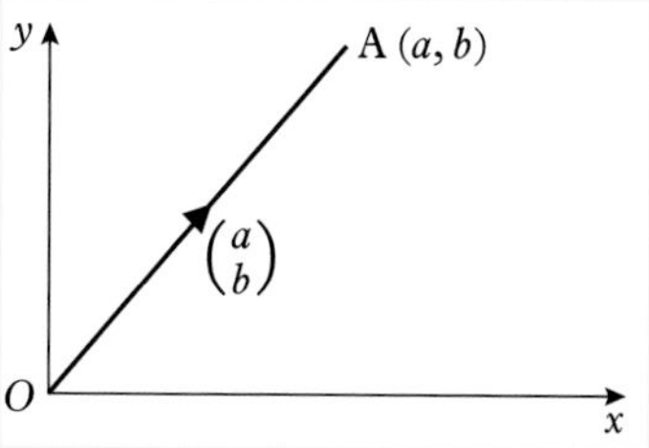

If A is the point (a, b) the position vector of A is $\begin{pmatrix} a \\ b \end{pmatrix}$.
Be careful in exam questions to give the co-ordinates of A as (a, b) and the position vector of A as $\begin{pmatrix} a \\ b \end{pmatrix}$.

Mid-point

If A has position vector **a** and B has position vector **b**, then the position vector of M, the **mid-point** of AB, is $\frac{1}{2}(\mathbf{a} + \mathbf{b})$.

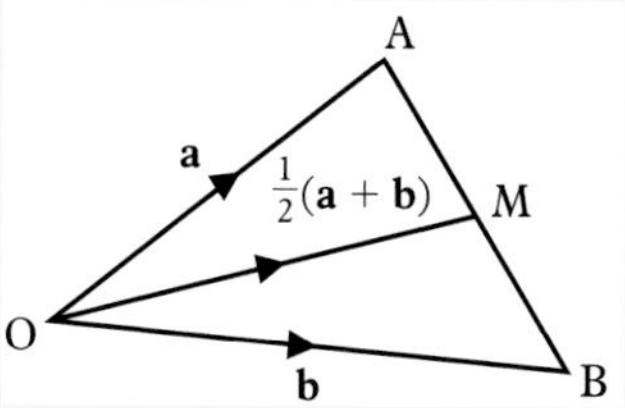

Parallel vectors

Two vectors are **parallel** if one is a multiple of the other.

Collinear

Points that are **collinear** lie on the same straight line. To show that 3 points, A, B and C are collinear, show that the vectors $\overrightarrow{AB}$ and $\overrightarrow{AC}$ are parallel. Then because A is a common point, the points must be collinear.

Resultant vector

When you add two or more vectors together, the vector that you get is called the **resultant vector**.

Component

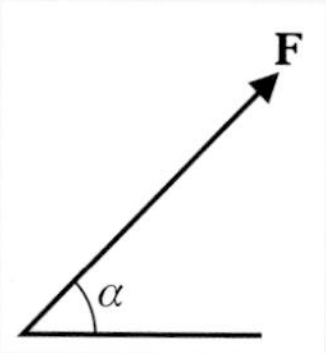

A **component** of a vector is the amount of the vector that acts in a given direction.
Components are often very useful when you are looking at forces.
If you have a force F, acting at an angle α above the horizontal, you can split the force into components. The horizontal component is F$\cos\alpha$ and the vertical component is F$\sin\alpha$.
It is usually easiest to find a resultant vector by working with horizontal and vertical components.

1 Copy these. Fill in the gaps.

a $\begin{pmatrix} 8 \\ \dots \end{pmatrix} - \begin{pmatrix} \dots \\ -1 \end{pmatrix} = \begin{pmatrix} -1 \\ 5 \end{pmatrix}$

b $\begin{pmatrix} 3 \\ \dots \end{pmatrix} + \begin{pmatrix} \dots \\ -8 \end{pmatrix} = \begin{pmatrix} 9 \\ 2 \end{pmatrix}$

c $\begin{pmatrix} \dots \\ -5 \end{pmatrix} - \begin{pmatrix} -6 \\ -2 \end{pmatrix} = \begin{pmatrix} -1 \\ \dots \end{pmatrix}$

d $\overrightarrow{AB} + \overrightarrow{\dots F} = \overrightarrow{AF}$

e $\overrightarrow{P\dots} - \overrightarrow{QB} = \overrightarrow{PQ}$

f $\overrightarrow{GH} + \overrightarrow{BA} - \overrightarrow{BH} = \dots$

2 Write $\overrightarrow{AB}$ in terms of **p**, **q** and **r**.

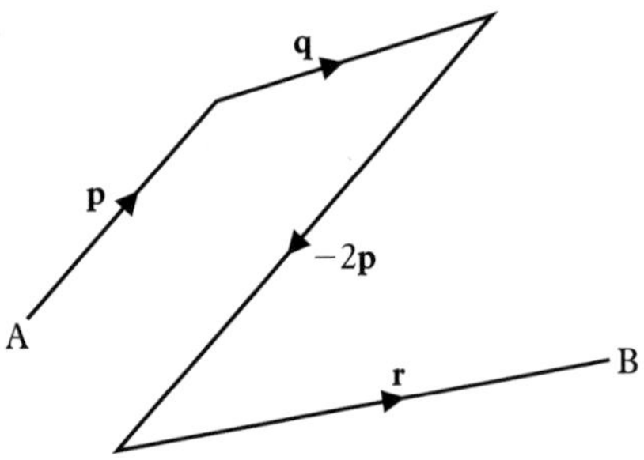

3 $\mathbf{p} = \begin{pmatrix} 7 \\ 2 \end{pmatrix}$ $\quad \mathbf{q} = \begin{pmatrix} 6 \\ -1 \end{pmatrix}$ $\quad \mathbf{r} = \begin{pmatrix} 10 \\ -3 \end{pmatrix}$

Find:

a $\mathbf{p} + \mathbf{q}$

b $\mathbf{q} + \mathbf{r}$

c $\mathbf{p} - \mathbf{q}$

d $\mathbf{p} - \mathbf{q} + \mathbf{r}$

e $\mathbf{p} - \mathbf{q} - \mathbf{r}$

f $\mathbf{r} - \mathbf{q} - \mathbf{p}$

4 $\mathbf{s} = \begin{pmatrix} 8 \\ -1 \end{pmatrix}$ $\quad \mathbf{t} = \begin{pmatrix} -6 \\ 4 \end{pmatrix}$

Write these as column vectors.

a $3\mathbf{s}$

b $-2\mathbf{t}$

c $\frac{1}{2}\mathbf{s}$

d $-\frac{1}{2}\mathbf{t}$

e $2\mathbf{s} - 3\mathbf{t}$

f $\frac{1}{2}(\mathbf{s} - \mathbf{t})$

5 Find the magnitude of each of these vectors.

a $\begin{pmatrix} -3 \\ 4 \end{pmatrix}$

b $\begin{pmatrix} -5 \\ -12 \end{pmatrix}$

c $\begin{pmatrix} 8 \\ -15 \end{pmatrix}$

d $\begin{pmatrix} 3 \\ -5 \end{pmatrix}$

e $\begin{pmatrix} 7 \\ -1 \end{pmatrix}$

f $\begin{pmatrix} -5 \\ -5 \end{pmatrix}$

6 A is the point (5, −2) and B is the point (7, 2).

a Write down the position vectors of A and B.

b Find $\overrightarrow{AB}$.

c Find the position vector of the mid-point of A and B.

7 $\overrightarrow{KL} = \mathbf{a} + 2\mathbf{b}$ $\overrightarrow{KM} = 2\mathbf{a} + 4\mathbf{b}$

Show that K, L and M are collinear.

8 $\overrightarrow{OR} = \mathbf{a} + \mathbf{b}$, $\overrightarrow{OS} = 7\mathbf{a} + 3\mathbf{b}$ and $\overrightarrow{OT} = 10\mathbf{a} + 4\mathbf{b}$.

a Show that R, S and T are collinear.

b Show that S divides RT in the ratio 2 : 1.

9 ABCDEF is a regular hexagon.
$\overrightarrow{AF} = \mathbf{a}$ and $\overrightarrow{FE} = \mathbf{b}$ as shown.
O is the centre of the hexagon.

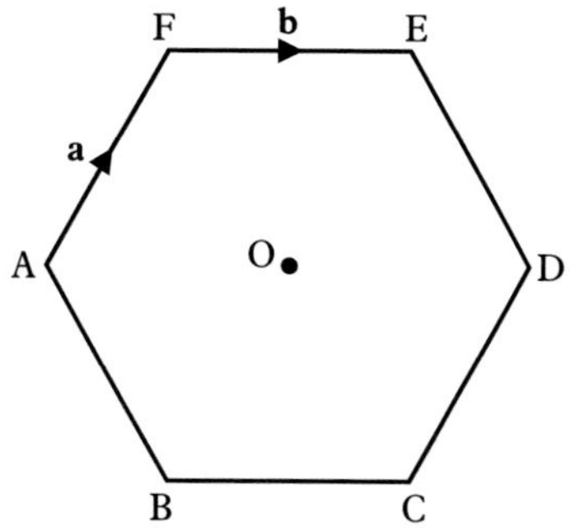

a Write down, in terms of **a** and **b**,

(1) $\overrightarrow{CD}$ **(3)** $\overrightarrow{AO}$

(2) $\overrightarrow{BC}$ **(4)** $\overrightarrow{AD}$.

b Work these out.

(1) $\overrightarrow{ED}$ **(2)** $\overrightarrow{BA}$

10 Aimee is going to row across a river. The speed of the river is 3 m/s.
She tries to row straight across at 0.5 m/s.
The width of the river is 50 metres.

a At what speed does she travel?

b What direction does she travel across the river?

c How long does the journey take her?

11 Two tugs are pulling a cruise ship.
Tug A is pulling with a force of 20 000 N on a bearing of 045°.
Tug B is pulling with a force of 30 000 N on a bearing of 150°.

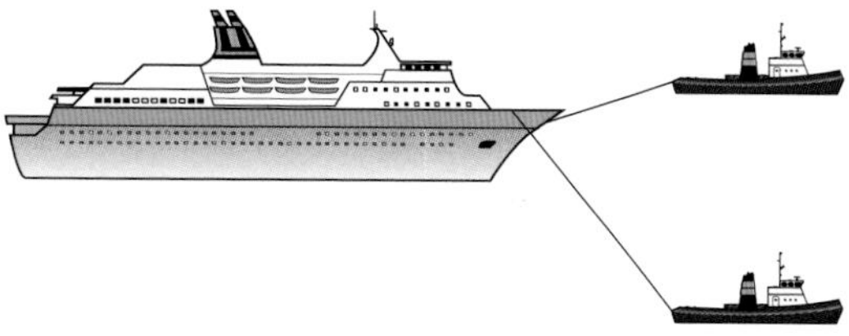

a Find the overall easterly component force on the ship.

b Find the overall southerly component force on the ship.

c Find the magnitude and direction of the resultant force on the ship.

1

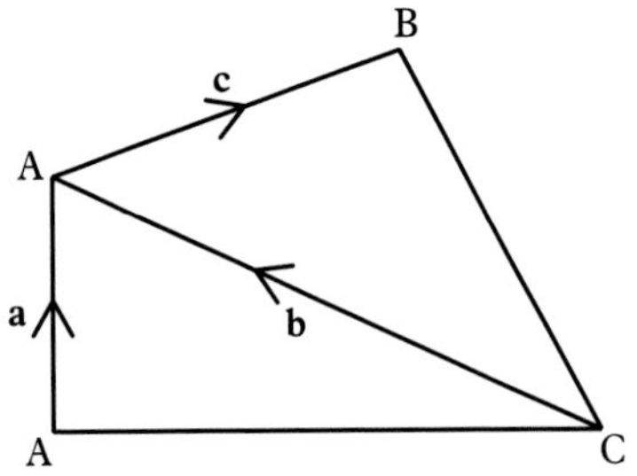

Using the information in the diagram, find, in terms of **a**, **b** and **c**:

a $\overrightarrow{DC}$ **b** $\overrightarrow{BC}$ **c** $\overrightarrow{DB}$ **(3 marks)**

For each vector, look for an alternative 'route' for which the vectors are given.

a $\overrightarrow{DC} = \overrightarrow{DA} + \overrightarrow{AC} = \mathbf{a} + (-\overrightarrow{CA}) = \mathbf{a} - \mathbf{b}$ **1 mark**

b $\overrightarrow{BC} = \overrightarrow{BA} + \overrightarrow{AC} = -\mathbf{c} - \mathbf{b}$ **1 mark**

c $\overrightarrow{DB} = \overrightarrow{DA} + \overrightarrow{AB} = \mathbf{a} + \mathbf{c}$ **1 mark**

2 The mid-points of the sides AB and AC, of triangle ABC, are D and E respectively.

Prove that

a DE is parallel to BC

b DE = $\frac{1}{2}$BC.

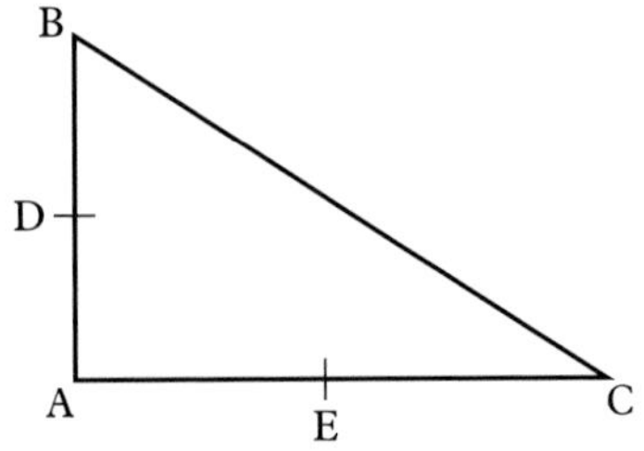

(3 marks)

*Start by writing $\overrightarrow{AD}$ and $\overrightarrow{DB}$ as **a**, and $\overrightarrow{AE}$ and $\overrightarrow{EC}$ as **b**. Show this on your diagram.*

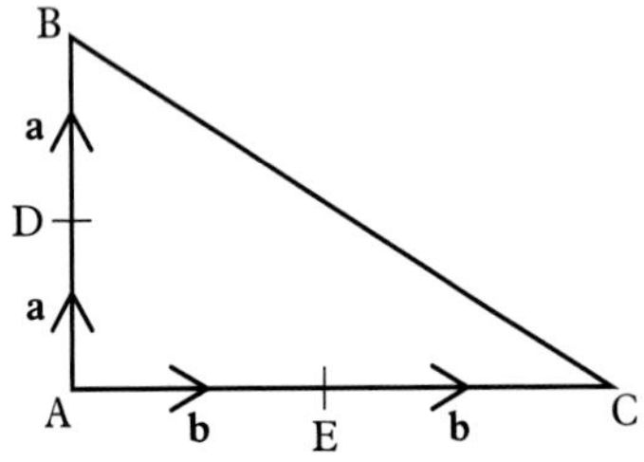

$\overrightarrow{BC} = \overrightarrow{BA} + \overrightarrow{AC} = -2\mathbf{a} + 2\mathbf{b} = 2(-\mathbf{a} + \mathbf{b})$ **1 mark**

$\overrightarrow{DE} = -\mathbf{a} + \mathbf{b}$

Compare these two answers.

So $\overrightarrow{BC} = 2\overrightarrow{DE}$, since $2(-\mathbf{a} + \mathbf{b}) = 2 \times (-\mathbf{a} + \mathbf{b})$.

a BC is parallel to DE since the same vector $(-\mathbf{a} + \mathbf{b})$ appears in both expressions. **1 mark**

b BC = 2DE so DE = $\frac{1}{2}$BC **1 mark**

3 A boat takes an hour to sail from a harbour, H, to a point, Y, which is 20 km from H on a bearing of 045°.
The current is 4 km/h due east.
Find the speed and direction that the boat takes.
Give the direction as a bearing. **(3 marks)**

Choose a scale to do a scale drawing.
Use 1 cm : 4 km/h.
Start by drawing the line HY 5 cm long at 45°. This line is shown in red.
Then show the current by drawing the line XY 1 cm long.
Now HX is the route the boat tries to sail. **1 mark**

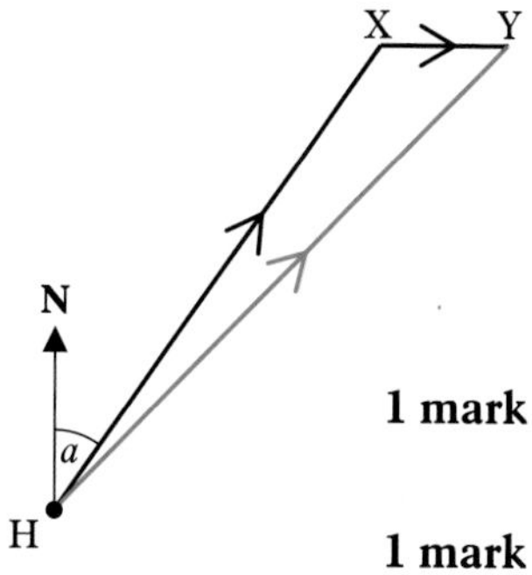

HX is 4.4 cm long, so the boat travels at
$4.4 \times 4 = 18$ km/h (to nearest whole number) **1 mark**

The bearing is equal to the angle a, which is 36°.
As a bearing, this is 036°. **1 mark**

4 Given that $\mathbf{a} = \begin{pmatrix} 10 \\ 6 \end{pmatrix}$, $\mathbf{b} = \begin{pmatrix} 4 \\ 0 \end{pmatrix}$ and $\mathbf{c} = \begin{pmatrix} -8 \\ 4 \end{pmatrix}$

write as a column vector **a** $\frac{1}{2}\mathbf{a}$ **b** $5\mathbf{b} + 4\mathbf{c}$ **c** $\mathbf{a} + \mathbf{b}$ **(3 marks)**

a $\frac{1}{2}\mathbf{a} = \frac{1}{2}\begin{pmatrix} 10 \\ 6 \end{pmatrix} = \begin{pmatrix} 5 \\ 3 \end{pmatrix}$ **1 mark**

b $5\mathbf{b} + 4\mathbf{c} = 5\begin{pmatrix} 4 \\ 0 \end{pmatrix} + 4\begin{pmatrix} -8 \\ 4 \end{pmatrix} = \begin{pmatrix} 20 \\ 0 \end{pmatrix} + \begin{pmatrix} -32 \\ 10 \end{pmatrix}$

$= \begin{pmatrix} 20 - 32 \\ 0 + 10 \end{pmatrix} = \begin{pmatrix} -12 \\ 10 \end{pmatrix}$ **1 mark**

c $\mathbf{a} + \mathbf{b} = \begin{pmatrix} 10 \\ 6 \end{pmatrix} + \begin{pmatrix} 4 \\ 0 \end{pmatrix} = \begin{pmatrix} 14 \\ 6 \end{pmatrix}$ **1 mark**

1 $\mathbf{p} = \begin{pmatrix} 2 \\ 1 \end{pmatrix}$ and $\mathbf{q} = \begin{pmatrix} 1 \\ -2 \end{pmatrix}$.

a Write down as a column vector

i $2\mathbf{p} + \mathbf{q}$, **ii** $\mathbf{p} - 2\mathbf{q}$. **(2 marks)**

A is the point (15, 15). O is the point (0, 0).

The vector $\overrightarrow{OA}$ can be written in the form $c\mathbf{p} + d\mathbf{q}$, where c and d are scalars.

b Using part **a**, or otherwise, find the values of c and d. **(3 marks)**

[N1999 P6 Q12]

2 A is the point (0, 4).

$\overrightarrow{AB} = \begin{pmatrix} 3 \\ 2 \end{pmatrix}$

a Find the co-ordinates of B. **(1 mark)**

C is the point (3, 4). BD is a diagonal of the parallelogram ABCD.

b Express $\overrightarrow{BD}$ as a column vector. **(3 marks)**

$\overrightarrow{CE} = \begin{pmatrix} 1 \\ -3 \end{pmatrix}$

c Calculate the length of AE. **(3 marks)**

[S1999 P5 Q13]

3 A is the point (2, 3) and B is the point (−2, 0).

a Find $\overrightarrow{AB}$ as a column vector. **(1 mark)**

C is the point such that $\overrightarrow{BC} = \begin{pmatrix} 4 \\ 9 \end{pmatrix}$

b Write down the co-ordinates of the point C. **(1 mark)**

X is the mid-point of AB. O is the origin.

c Find $\overrightarrow{OX}$ as a column vector. **(2 marks)**

[S1998 P6 Q12]

4 A is the point (2, 3) and B is the point (−2, 0).

a **i** Write $\overrightarrow{AB}$ as a column vector.

ii Find the length of the vector $\overrightarrow{AB}$. **(4 marks)**

D is the point such that

$\overrightarrow{BD}$ is parallel to $\begin{pmatrix} 0 \\ 1 \end{pmatrix}$ and

the length of $\overrightarrow{AD}$ = the length of $\overrightarrow{AB}$.

O is the point (0, 0).

b Find $\overrightarrow{OD}$ as a column vector. **(2 marks)**

C is the point such that ABCD is a rhombus.
AC is a diagonal of the rhombus.
c Find the coordinates of C. **(2 marks)**
[S2001 P5 Q16]

5

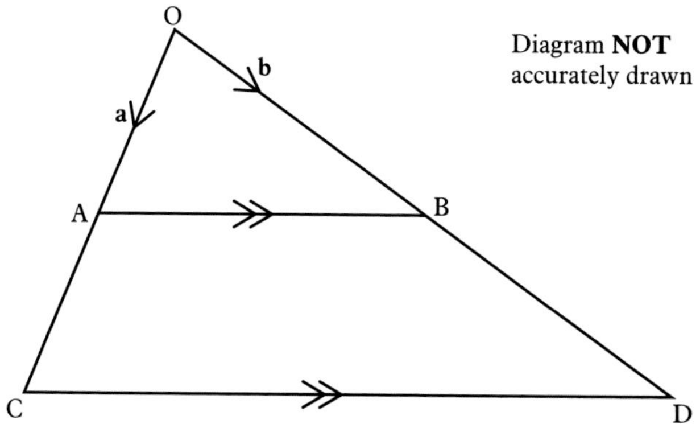

The diagram shows two triangles OAB and OCD.

OAC and OBD are straight lines.
AB is parallel to CD.

$\overrightarrow{OA} = \mathbf{a}$ and $\overrightarrow{OB} = \mathbf{b}$.

The point A cuts the line OC in the ratio OA : OC = 2 : 3.

Express $\overrightarrow{CD}$ in terms of **a** and **b**. **(3 marks)**
[N2001 P6 Q21]

6

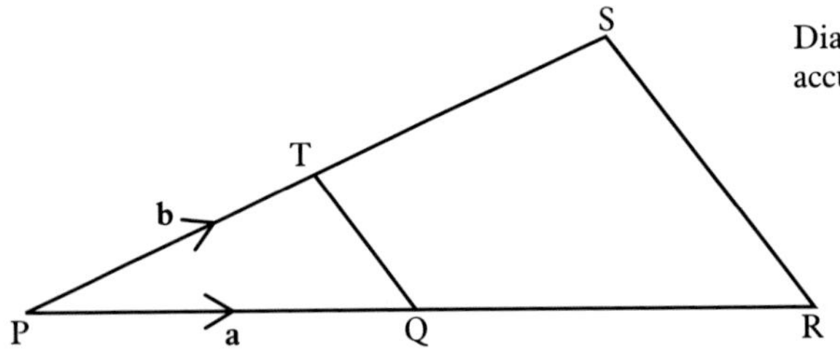

Diagram **NOT** accurately drawn

Q is the mid-point of the side PR and T is the mid-point of the side PS of triangle PRS.

$\overrightarrow{PQ} = \mathbf{a}$, $\overrightarrow{PT} = \mathbf{b}$

a Write down, in terms of **a** and **b**, the vectors

(1) $\overrightarrow{QT}$ **(2)** $\overrightarrow{PR}$ **(3)** $\overrightarrow{RS}$ **(3 marks)**

b Write down one geometrical fact about QT and RS which could be deduced from your answers to part **a**. **(1 mark)**

[S1997 P5 Q16]

7

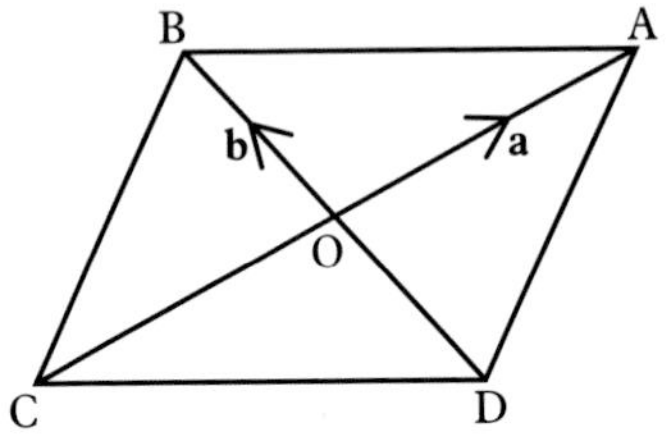

Diagram **NOT** accurately drawn

ABCD is a parallelogram. The diagonals of the parallelogram intersect at O.

$\overrightarrow{OA} = \mathbf{a}$ and $\overrightarrow{OB} = \mathbf{b}$

a Write an expression, in terms of **a** and **b**, for

(1) $\overrightarrow{CA}$, **(2)** $\overrightarrow{BA}$ **(3)** $\overrightarrow{BC}$ **(3 marks)**

X is the point such that $\overrightarrow{OX} = 2\mathbf{a} - \mathbf{b}$

b **(1)** Write down an expression, in terms of **a** and **b**, for $\overrightarrow{AX}$.

(2) Explain why B, A and X lie on the same straight line.

(3 marks)

[S1997 P5 Q17]

8

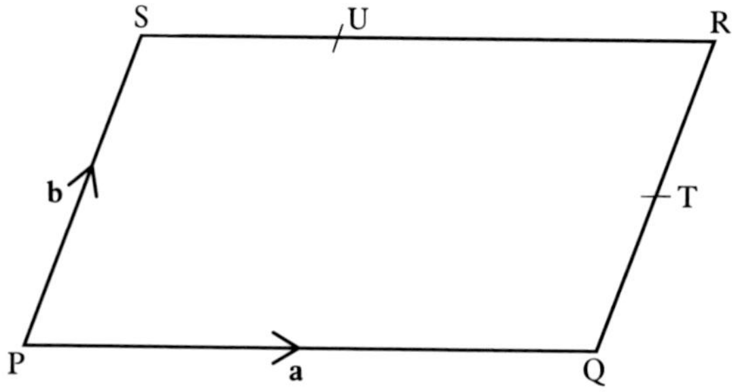

Diagram **NOT** accurately drawn

PQRS is a parallelogram.
T is the mid-point of QR.
U is the point on SR for which SU : UR = 1 : 2

$\overrightarrow{PQ} = \mathbf{a}$ and $\overrightarrow{PS} = \mathbf{b}$.

Write down, in terms of **a** and **b**, expressions for

(1) $\overrightarrow{PT}$, **(2)** $\overrightarrow{TU}$ **(2 marks)**

[N2000 P5 Q21]

9

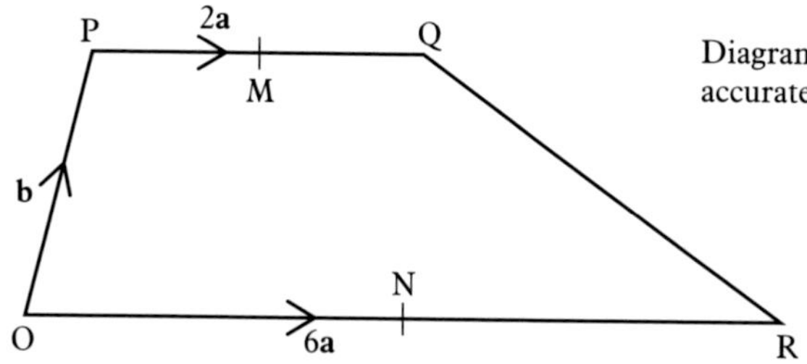

Diagram **NOT** accurately drawn

OPQR is a trapezium. PQ is parallel to OR.
$\overrightarrow{OP} = \mathbf{b}$, $\overrightarrow{PQ} = 2\mathbf{a}$, $\overrightarrow{OR} = 6\mathbf{a}$.

M is the mid-point of PQ.
N is the mid-point of OR.

a Find, in terms of **a** and **b**, the vectors

(1) $\overrightarrow{OM}$, **(2)** $\overrightarrow{MN}$. **(2 marks)**

X is the mid-point of MN.

b Find, in terms of **a** and **b**, the vector $\overrightarrow{OX}$. **(2 marks)**

The lines OX and PQ are extended to meet at the point Y.

c Find, in terms of **a** and **b**, the vector $\overrightarrow{NY}$. **(3 marks)**

[S2000 P5 Q20]

10 Angles

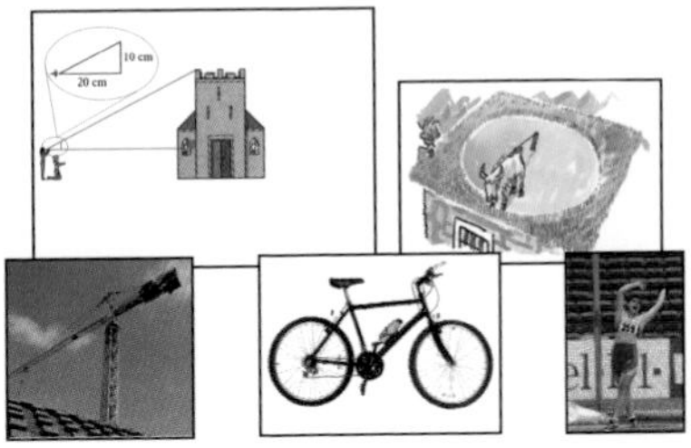

You need to know about:

- angle facts for triangles and parallel lines
- interior and exterior angles of polygons
- the rules for showing that two triangles are similar
- the rules for showing that two triangles are congruent
- loci
- circle theorems
- proofs

Alternate, corresponding and opposite angles

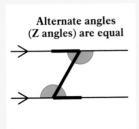

Alternate angles (Z angles) are equal

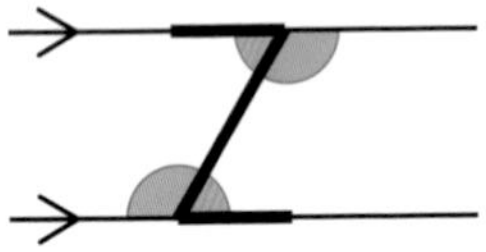

Corresponding angles (F angles) are equal

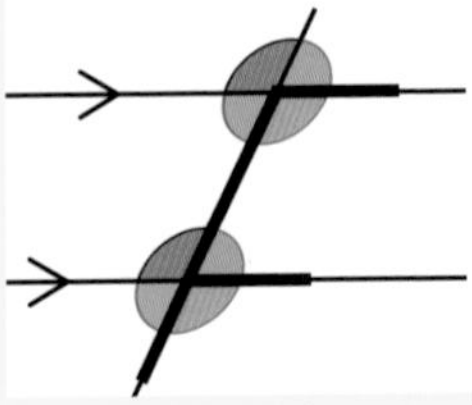

Opposite angles (X angles) are equal

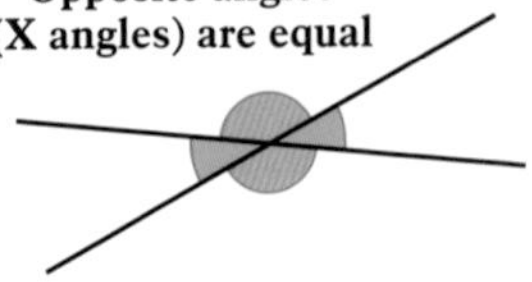

Exterior and interior angles

The exterior angles of any polygon add up to 360°

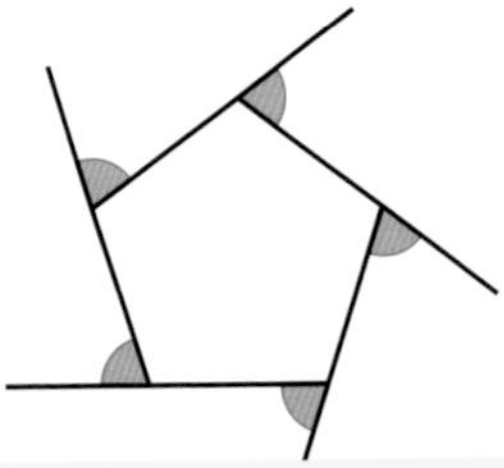

The interior angles of an n sided polygon add up to $(n - 2) \times 180°$.

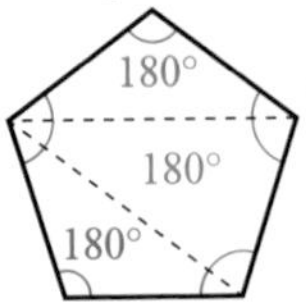

There are $n - 2$ triangles in an n-sided shape

In any triangle, the exterior angle at a vertex is equal to the sum of the opposite two interior angles. $e = a + b$

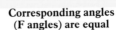

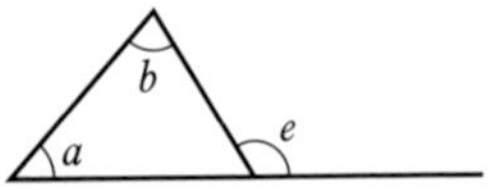

You need to know how to prove this result. You split angle e using a line parallel to the line between angles a and b and use the parallel line results to see that $q = a$ and $p = b$. So $p + q = a + b$ and hence the result.

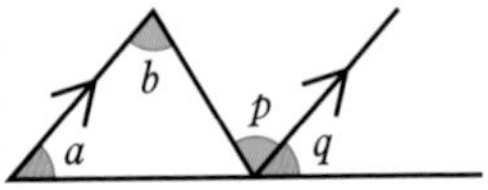

This is also the method that you should use to prove the angles in a triangle add up to 180°. If the third angle of the triangle is c, then $c + p + q = 180°$ since they are angles on a straight line. But since $q = a$ and $p = b$ it follows that $c + a + b = 180°$ and hence the result.

Similar triangles

Similar triangles have all 3 pairs of angles equal.
The 3 pairs of sides are in the same ratio.

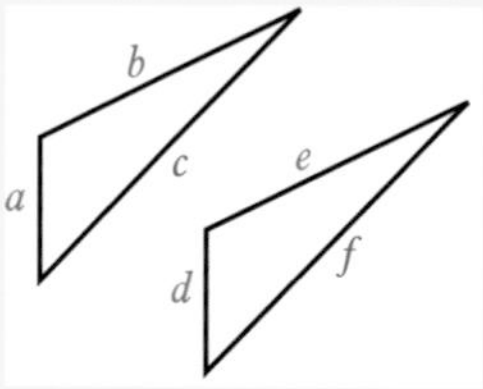

So if these triangles are similar $\frac{d}{a} = \frac{e}{b} = \frac{f}{c}$.

This allows you to find missing lengths in questions involving similar triangles.

Congruent

Two objects are **congruent** if they are identical to each other. They must have exactly the same shape *and* size.
To prove that two triangles are **congruent**, you must show that one of these rules is true.

Rule 1 All three pairs of sides are equal. You can remember this as SSS.

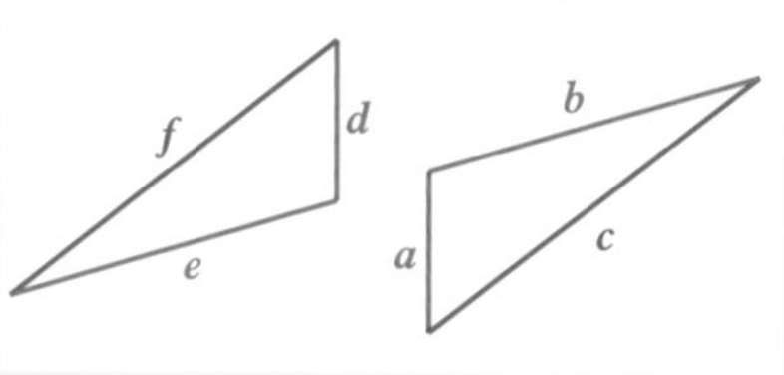

Here $a = d \quad b = e \quad c = f$

Rule 2 Right angle, hypotenuse and side are equal. Remember this as RHS.

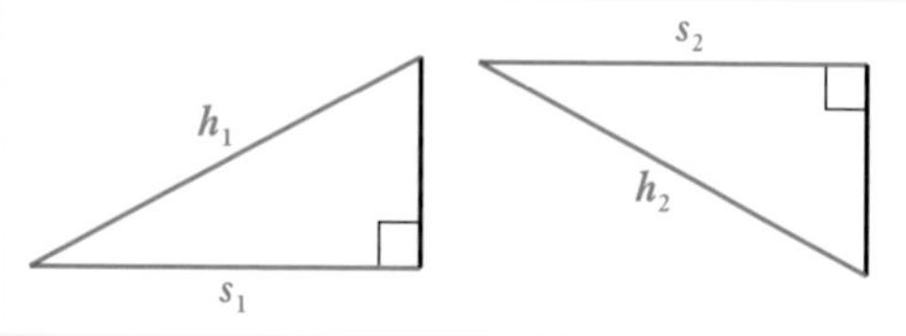

Here $h_1 = h_2 \quad s_1 = s_2$

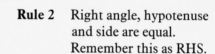

Rule 3 Two pairs of corresponding sides are equal and the angles *between* each pair of sides are also equal. Remember this as SAS.

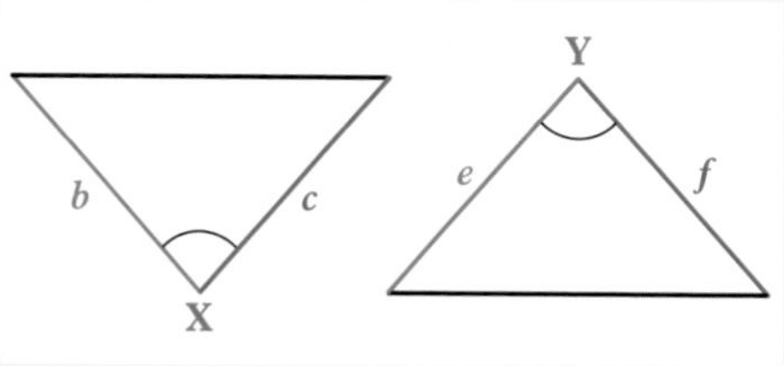

Here $b = e \quad c = f$ and $\angle X = \angle Y$

Rule 4 Two pairs of angles are equal and a pair of corresponding sides is also equal. Remember this as **AA corr. S**.

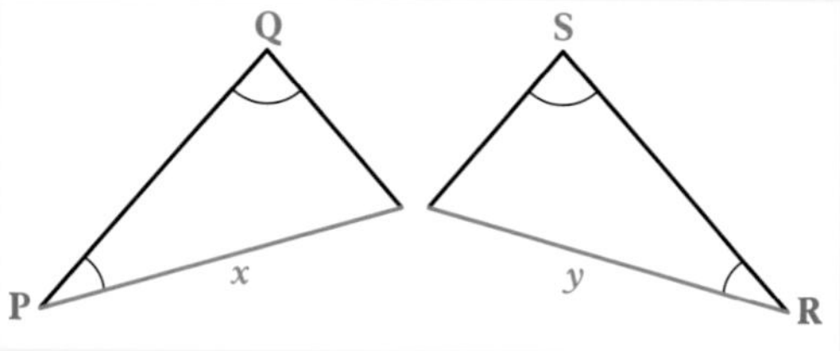

Here $x = y \quad \angle P = \angle R$ and $\angle Q = \angle S$

Locus

The **locus** of an object that is moving according to a rule is the path of the object. You can describe a locus in words or with a diagram.

Tangent

A **tangent** to a circle is a straight line that touches the circle at one point.

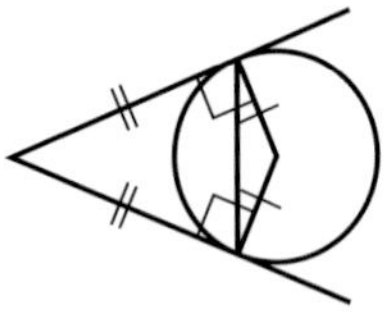

There are two tangents to a circle from a point outside the circle.
A tangent to a circle is perpendicular to the radius of the circle drawn to the point of contact.

Cyclic quadrilateral

A **cyclic quadrilateral** is one in which all the vertices of the quadrilateral lie on a circle.

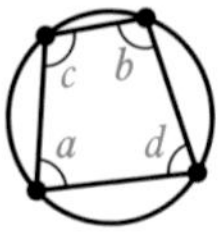

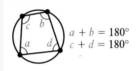

Opposite angles in a cyclic quadrilateral add up to 180°.

Circle theorems

All angles in the same segment are equal.

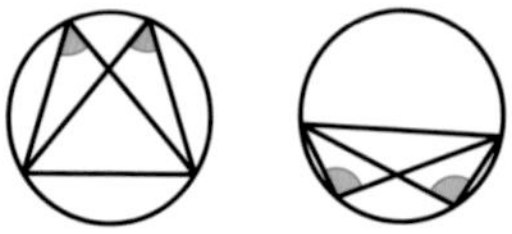

The angle in a semi-circle is 90°.

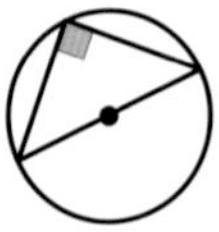

The angle at the centre is twice the angle at the circumference.

The angle between a tangent and a chord drawn to the point of contact is equal to the angle in the alternate segment.

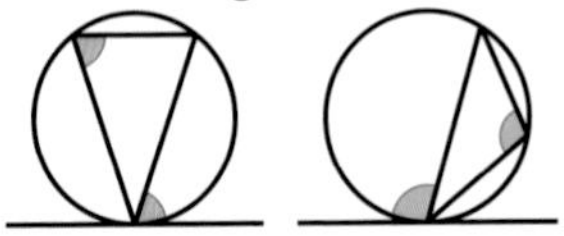

1 For each part, write down the angles marked with letters.
Give a reason for each answer.

a

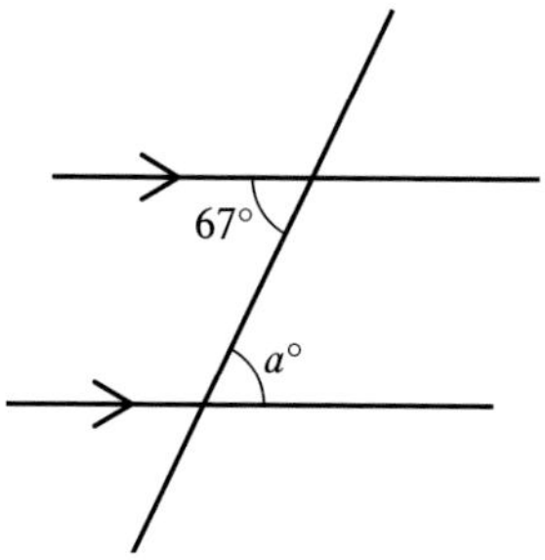

b

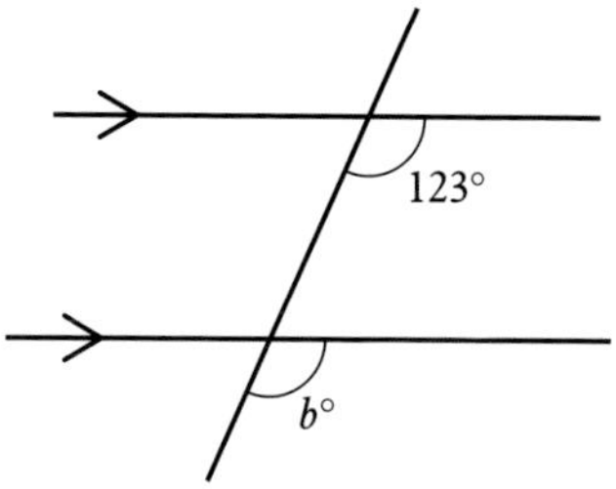

c

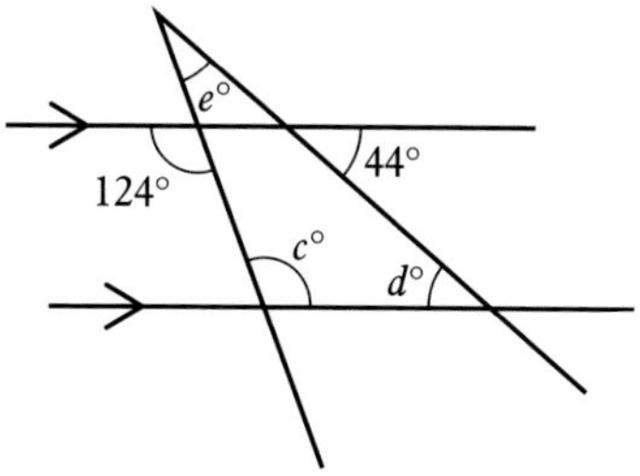

d

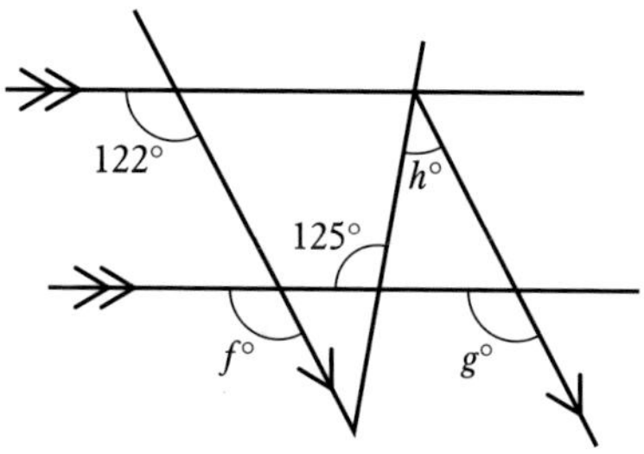

2 A decagon has 10 sides.

a Find the size of an exterior angle of a regular decagon.

b Work out the sum of the interior angles of a decagon.

3 A regular polygon has an interior angle of 160°.
How many sides does the polygon have?

4 Triangles ABC and XYZ are similar.

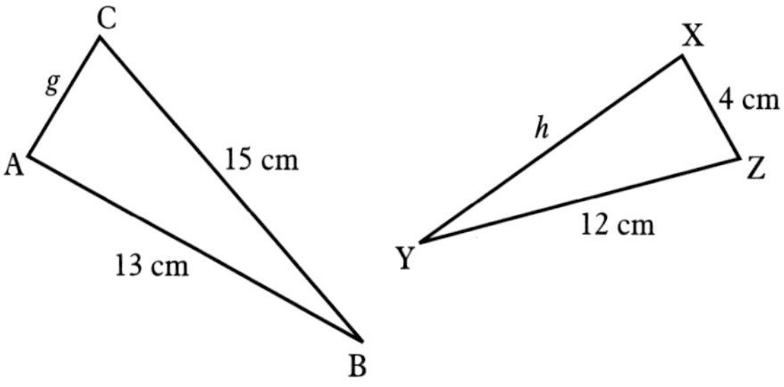

a Use colours to show equal angles.

b Work out the lengths marked with letters.

5 P, Q, R and S are the mid-points of the sides of the parallelogram ABCD as shown.
Prove that triangle PQS is congruent to triangle RSQ.
You must produce a full explanation and state the rule that you use to prove congruence.

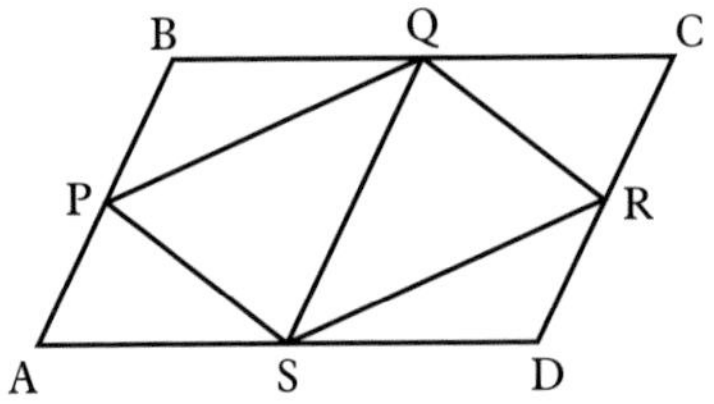

6 Joey sails his boat so that he always stays the same distance from the two lighthouses A and B.

Construct accurately the route on which Joey sails.

7 For each part, write down the angles marked with letters.
Give a reason for each answer.

a

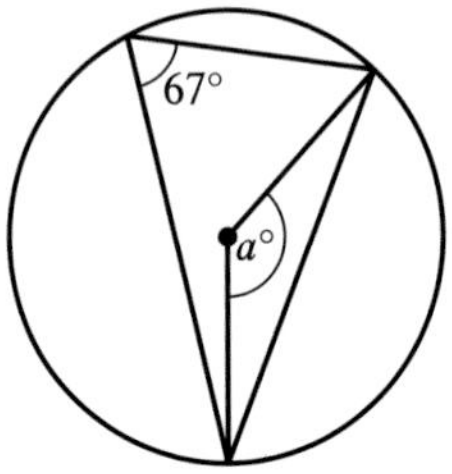

c

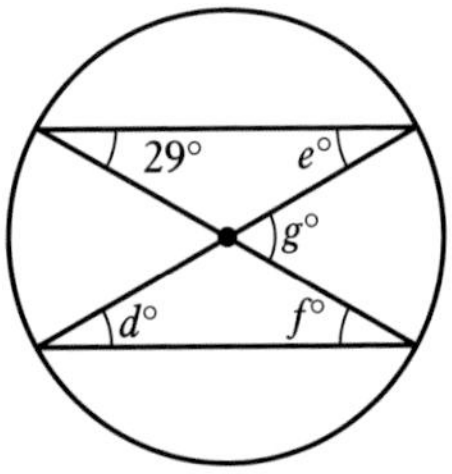

b

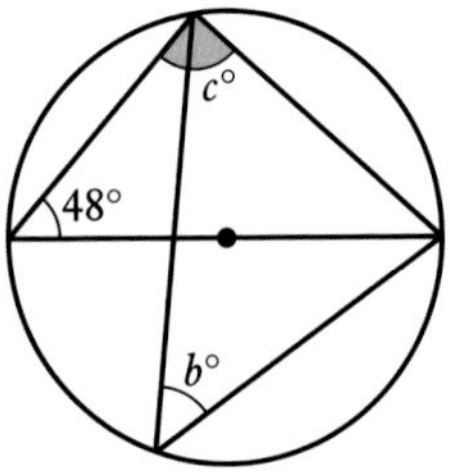

d

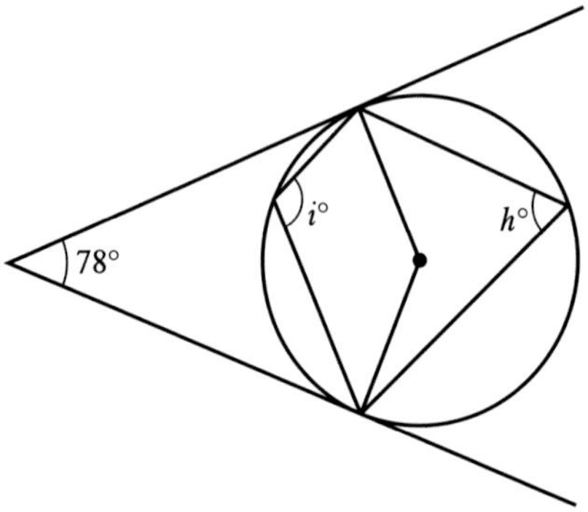

8 **a** AB is a diameter of the circle centre O.
ST is a tangent to the circle that touches the circle at A.
Prove that $a = p$.

b The proof in **a** is a special case of a circle theorem.
State the circle theorem for which part **a** is a special case.

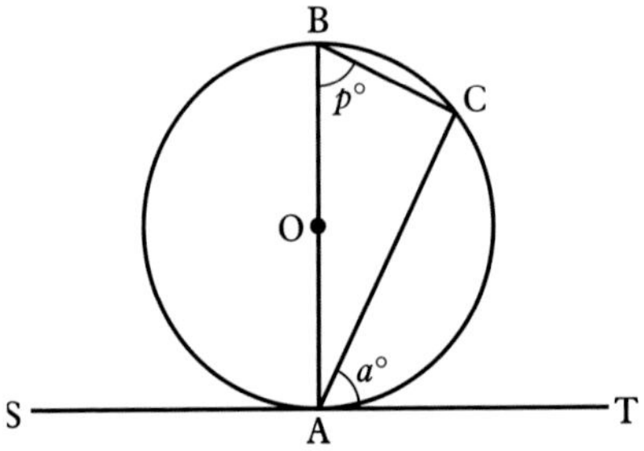

1 The diagram shows four connected regular pentagons.

a Calculate the interior angle of each pentagon. **(2 marks)**

b How many sides has the regular polygon that would fit exactly in the gap at vertex A? **(3 marks)**

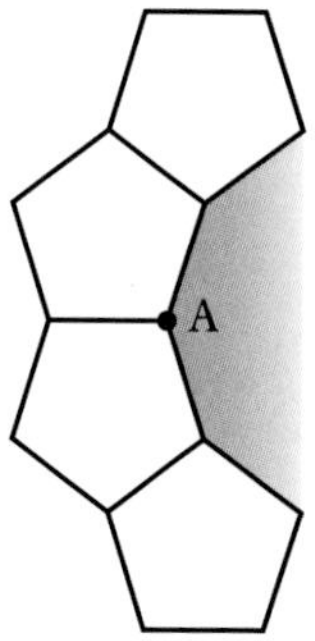

You need to remember that the total of the exterior angles of any polygon is 360°.

a Exterior angle of a pentagon $= 360° \div 5 = 72°$ **1 mark**

Interior angle $= 180° -$ exterior angle $= 180° - 72° = 108°$ **1 mark**

b The interior angle of the polygon shaded, that would fit at A, is

$360° - (2 \times 108°) = 144°$

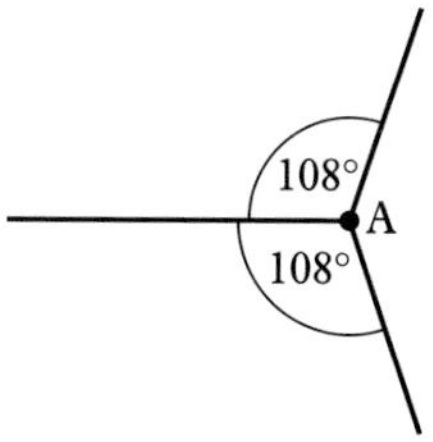

1 mark

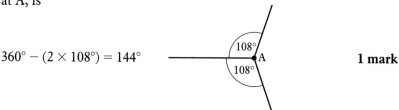

So the exterior angle of this polygon is $180° - 144° = 36°$. **1 mark**

Number of sides of shaded polygon $= \dfrac{360}{36} = 10$ **1 mark**

2 Copy the diagram. Inside the rectangle draw:

a the locus of points 1 cm from point E **(1 mark)**

b the locus of points 1 cm from the line AD **(1 mark)**

c the locus of points that are the same distance from AD and DC **(1 mark)**

d the locus of points that are the same distance from B and C. **(1 mark)**

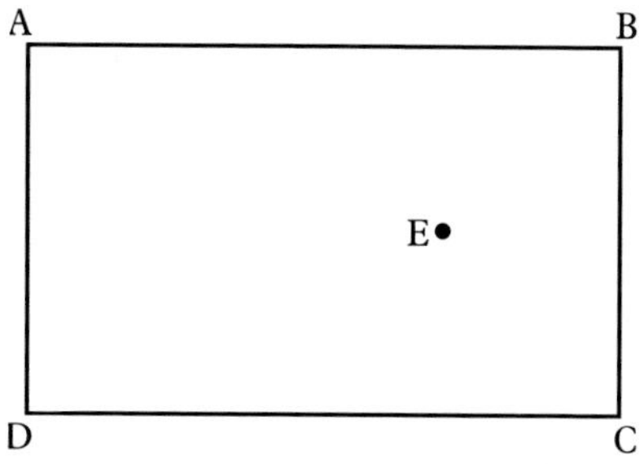

These are the loci you need to draw:

a a circle of radius 1 cm, centre E **1 mark**

b a line parallel to AD, and 1 cm from it **1 mark**

c the bisector of the angle ADC. Show your construction lines. **1 mark**

d the perpendicular bisector of BC. Show your construction lines. **1 mark**

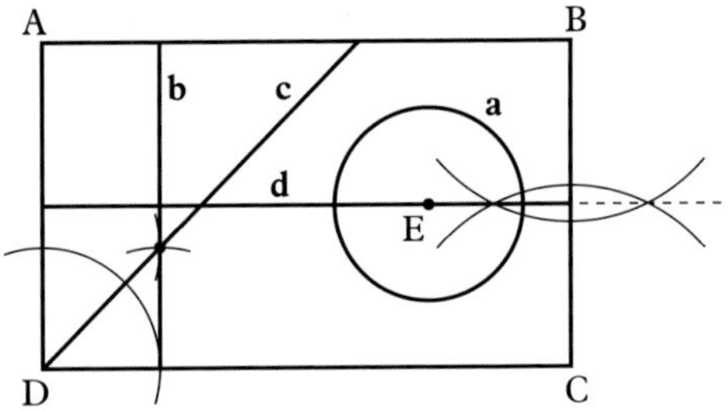

3 AD is a tangent to a circle centre O.
DC is a diameter of the circle.
Point B, on the circumference of the circle, is also the mid-point of AC.

Prove that ∠BDC = 45°. **(6 marks)**

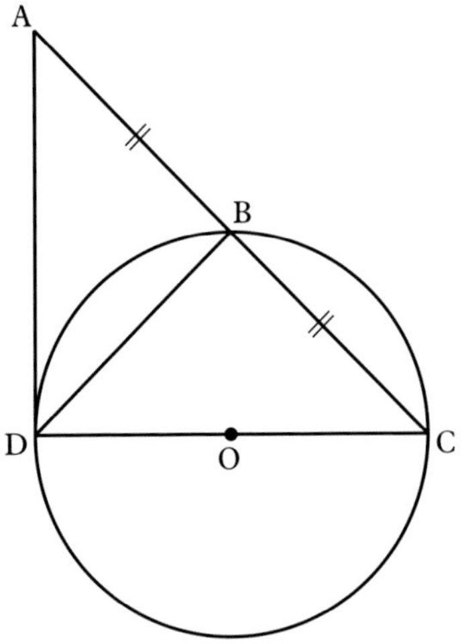

As DC is a diameter of the circle, ∠DBC = 90° (angle in a semi-circle). **1 mark with reason**

In triangles ABD and DBC.

side AB	=	side BC	(B is the mid-point of AC)
∠ABD	=	∠DBC	(both 90° as ∠DBC = 90°)
side DB	=	side DB	(common to both)

So triangles ABD and DBC are congruent
(two sides and included angle, SAS) **1 mark, and 1 mark for reasons**

Now ∠ADC = 90° (angle between tangent and a radius) **1 mark**
and ∠ADB = ∠BDC (congruent triangles) **1 mark**
so ∠BDC = 90° ÷ 2 = 45°.
1 final mark, dependent on previous work

1 ABCDEFGH is a regular octagon.
O is the centre of the octagon.

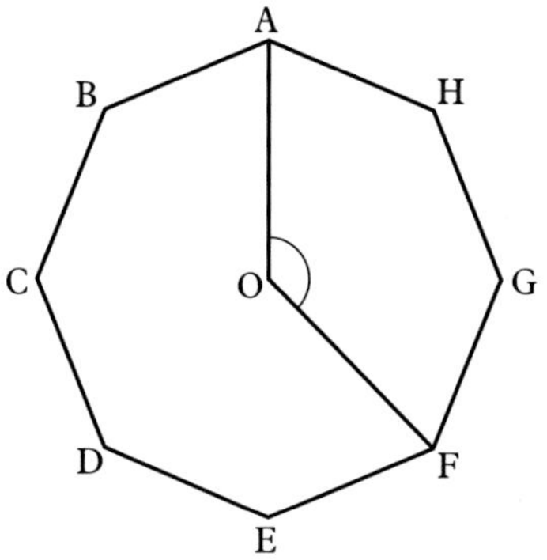

Diagram **NOT** accurately drawn.

Calculate the size of angle AOF. **(3 marks)**
[N2000 P6 Q2]

2 DE is parallel to BC
ADB and AEC are straight lines.

AD = 12 cm. BC = 12 cm.
AE = 8 cm. EC = 2 cm.

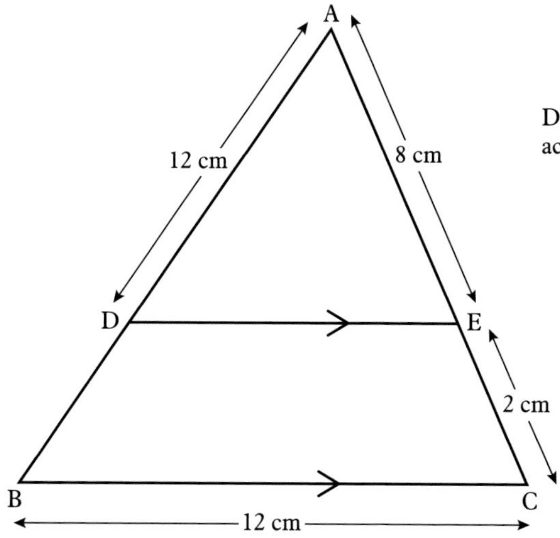

Diagram **NOT** accurately drawn.

Calculate the length of
(1) DE **(2)** DB **(4 marks)**
[N1998 P6 Q4]

3 AB is parallel to CD
The lines AD and BC intersect at point O.
AB = 11 cm AO = 8 cm OD = 6 cm

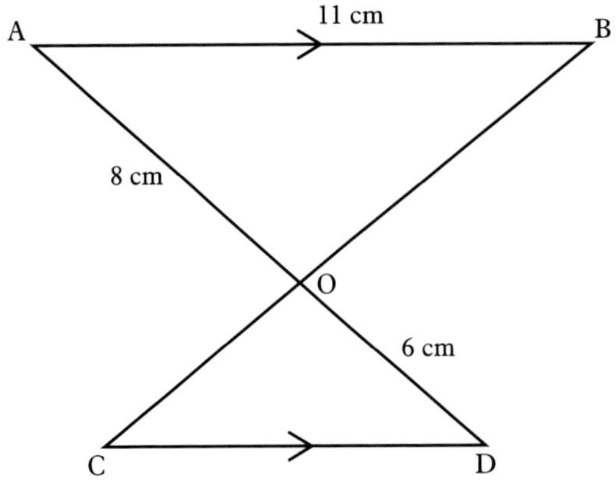

Diagram **NOT** accurately drawn.

Calculate the length of CD

(3 marks)
[N2001 P6 Q12]

4 Draw the locus of all points which are 3 cm away from the line AB.

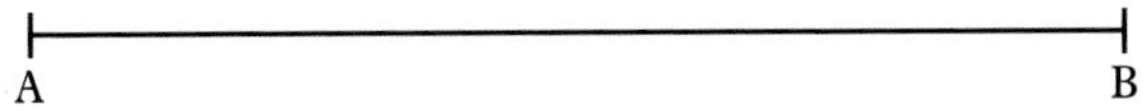

(3 marks)
[S1999 P5 Q4]

5 On the diagram, draw the locus of the points, **outside the rectangle,** that are 3 centimetres from the edges of this rectangle.

(3 marks)
[S2000 P5 Q11]

6 In the diagram, triangle ABQ and triangle BCP are equilateral.

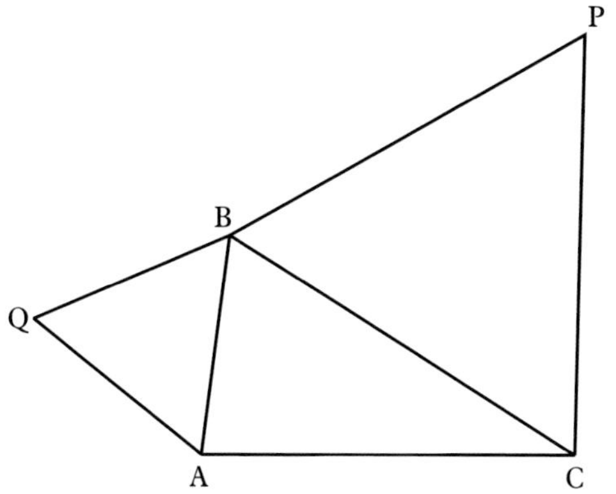

Diagram **NOT** accurately drawn.

Explain why triangle ABP and triangle QBC are congruent. **(3 marks)**

[S1998 P5 Q21]

7 A, B, C, D and E are points on the circumference of a circle.
SAT is the tangent to the circle at A.
AB = BC.
AE is parallel to CD.
Angle ECD = 48°.

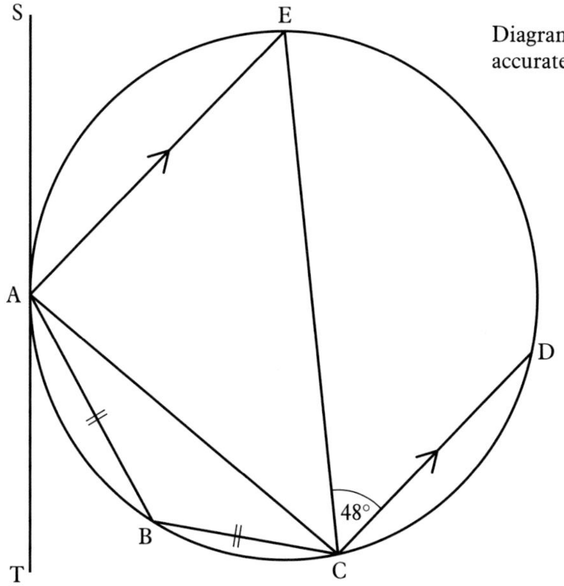

Diagram **NOT** accurately drawn.

a Calculate the size of angle ABC. Give reasons for your answer. **(2 marks)**
b Calculate the size of angle TAB. Give reasons for your answer. **(3 marks)**

[N1999 P5 Q14]

8 A, B, C and D are points on the circumference of a circle centre O.
A tangent is drawn from E to touch the circle at C.
Angle AEC = 36°. EAO is a straight line.

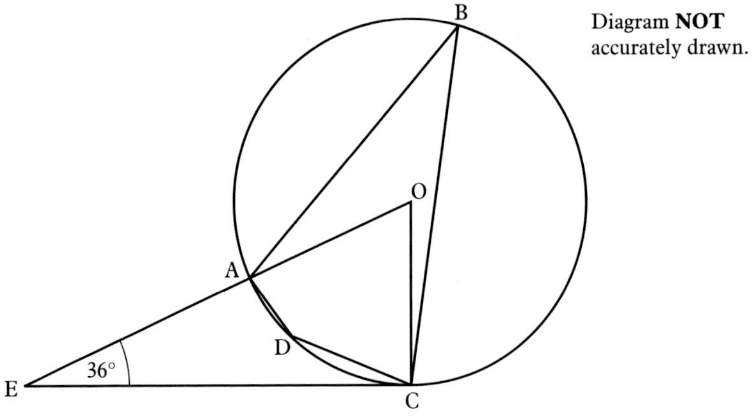

a Calculate the size of angle ABC.
Give reasons for your answer. **(4 marks)**

b Calculate the size of angle ADC.
Give reasons for your answer. **(2 marks)**

[N2001 P5 Q17]

9 Points A, B and C lie on the circumference of a circle with centre O.
DA is the tangent to the circle at A.
BCD is a straight line.
OC and AB intersect at E.
Angle BOC = 80°. Angle CAD = 38°.

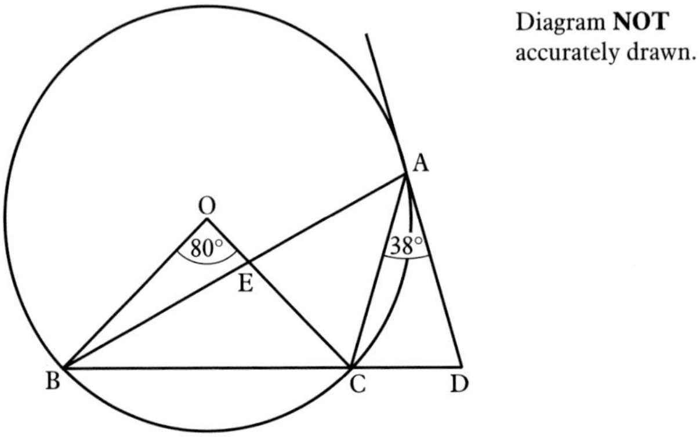

a Calculate the size of angle BAC. **(1 mark)**

b Calculate the size of angle OBA. **(3 marks)**

c Give a reason why it is not possible to draw a circle with diameter ED through the point A. **(1 mark)**

[N2000 P5 Q12]

11 Using trigonometry

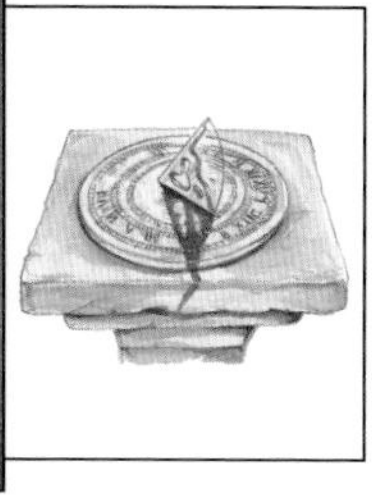

You need to know about:

- bearings
- using trigonometry in bearings problems
- angles of elevation and depression
- 3-D trigonometry
- the angle between a line and a plane

Bearing

A **bearing** is an angle.
Bearings are *always* measured clockwise starting from north.
A bearing must always have 3 figures.
If the angle is less than 100° put a zero as the first digit.

Bearing of B from A

The **bearing of B from A** means that you are at A. You need to face north and turn clockwise until you are facing B.
The **bearing of B from A** is the red angle.
You start at A.
You face north.
You turn clockwise until you are facing B.

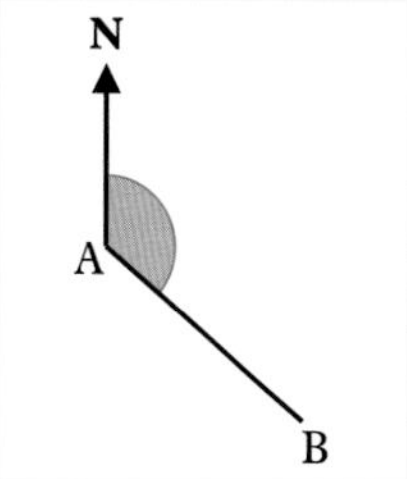

Bearings and trigonometry

When you are solving a problem:

- Start by drawing a diagram.
- Show all of the information that you are given.
- Draw a separate triangle that shows the relevant information.
- Then answer the question to the required accuracy.

Angles of elevation and depression

Start by looking horizontally.
When you look **up** at something, the angle is called an **angle of elevation.**
When you look **down** at something, the angle is called an **angle of depression.**
The angle is always measured from the horizontal.

3-D trigonometry

Solving problems in 3-D relies on you finding 2-D right-angled triangles to work in. A vertical line is perpendicular to a horizontal plane. So a vertical line is perpendicular to any line drawn in a horizontal plane.

When you are doing trigonometry in 3-D:

- think about lines that are perpendicular
- find a right-angled triangle to work in
- draw the triangle that you are using in 2-D so that you can see it clearly
- label the right angle and the sides or angles that you know in the triangle
- use the triangle to work out the side or the angle that you need.

Example

Find the angle a between AE and EC in this cuboid.

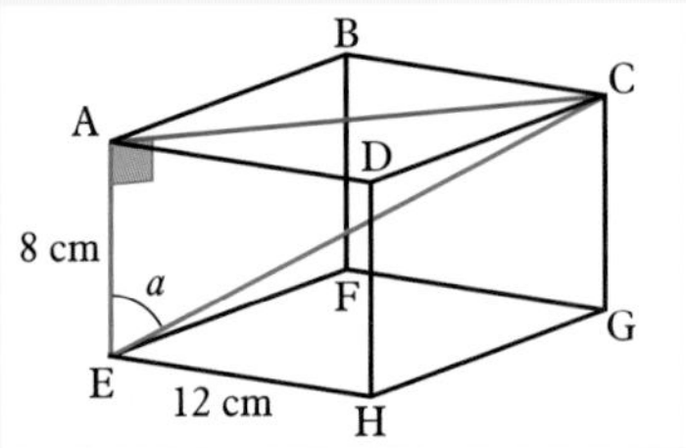

AE is perpendicular to the diagonal AC on the top face. You have to work out AC so that you can work out angle a.

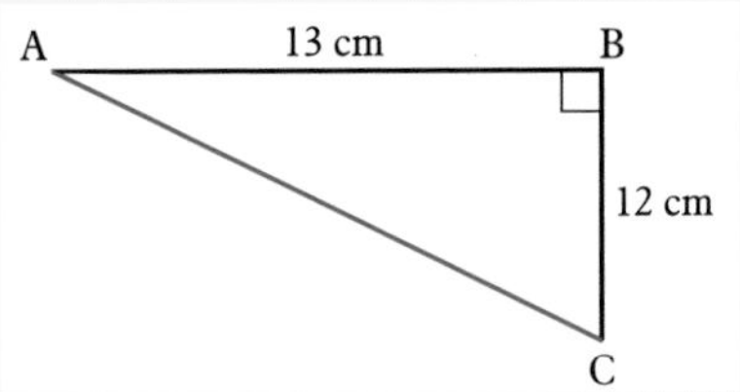

In triangle ABC, by Pythagoras' theorem:

$$AC^2 = AB^2 + BC^2$$
$$= 13^2 + 12^2$$
$$= 313$$
$$AC = \sqrt{313}\text{ cm}$$

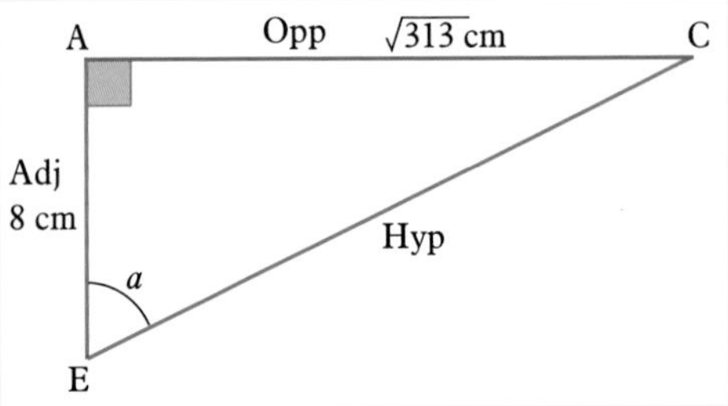

In triangle ACE, using trigonometry:

S Ø H	C A̸ H	T Ø A̸

$$\tan a = \frac{\sqrt{313}}{8}$$

$$a = 65.7° \text{ (1 dp)}$$

The angle between AE and EC is 65.7°.

The angle between a line and a plane

The angle between a line and a plane is the angle marked.

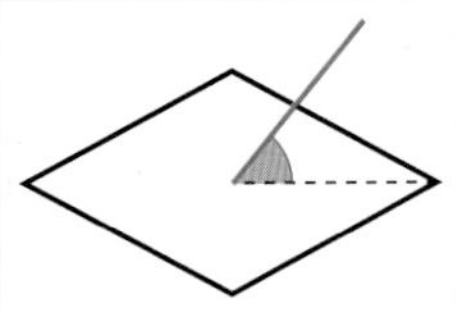

It is the angle between the line and the dashed line in the plane.

You can think of this dashed line as the shadow that you would get if you shine a light perpendicular to the plane.

When you do problems, you must find the shadow of the line on the plane that you need and then use a right-angled triangle as always.

1 In each of these find:

(1) the bearing of A from B
(2) the bearing of B from A.

a

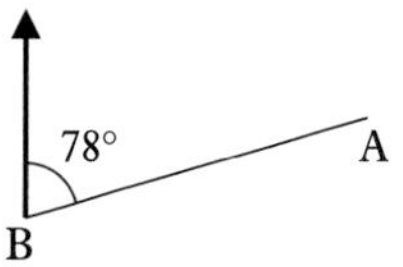

b

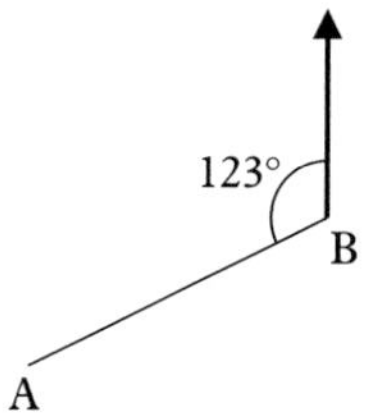

c

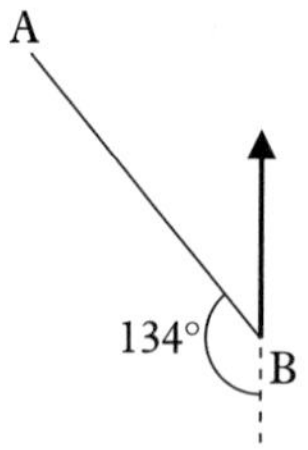

2 Ashok, Bill, Eileen and Ruksana are standing in a field at the corners of a square.
Bill is standing due north of Ashok.
Eileen is due east of Ruksana.

Write down the bearing of

a Eileen from Ashok

b Ruksana from Eileen

c Bill from Ruksana

d Ruksana from Ashok

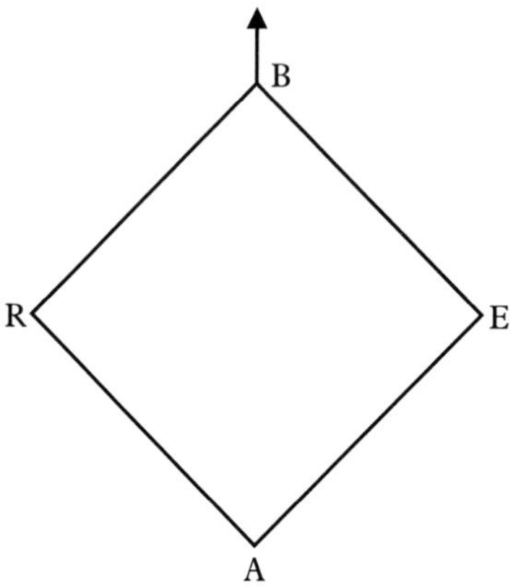

3 The diagram shows two ships A and B.
The bearing of B from A is 245°.

B is 27 km west of A.

a Copy the diagram and show the given information.

b How far south of A is B?

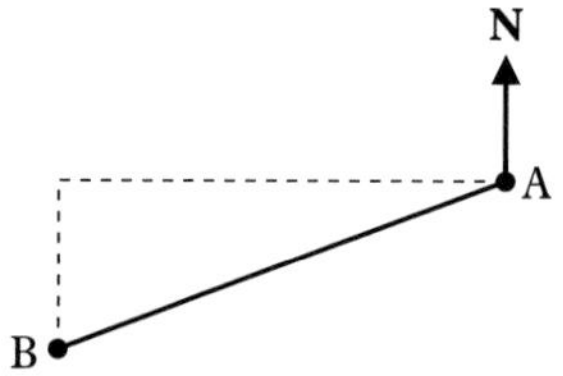

4 Denver International Airport has one of the world's tallest airport control towers.
The angle of elevation of the tower from a point 50 ft from the base is 81.3°.
Find the height of the tower in feet, correct to the nearest foot.

5 Suna is standing on a cliff looking down at the sea. The cliff is 123 m high. His friend Fran is on a boat below.
The angle of depression of Fran from Suna is 46°.
How far from the base of the cliff is Fran?

6 In this cuboid, find the angle between the diagonal EC and the plane EFGH.

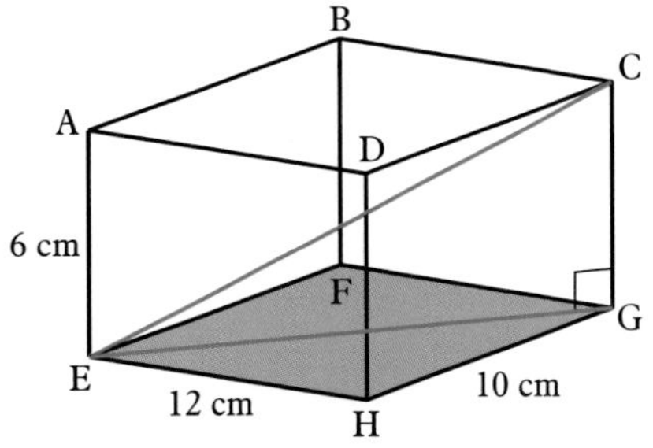

7 In this cuboid, find the angle between the diagonal EC and the plane ABCD.

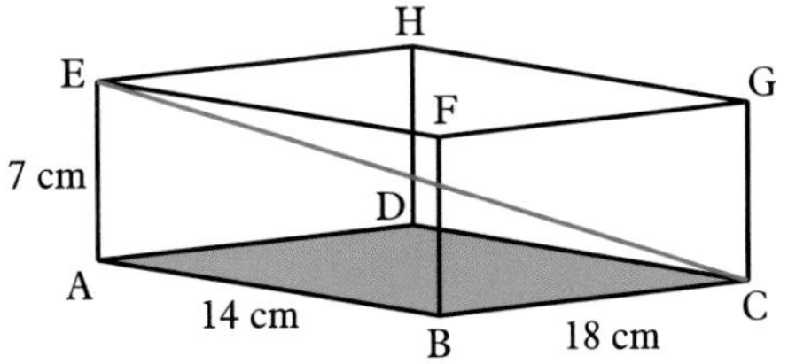

8 The diagram shows a tetrahedron PQRS. All of the sides of the tetrahedron are 10 cm long. M is the mid-point of RS.

a Find the length of QM.

b Write down the length of PM.

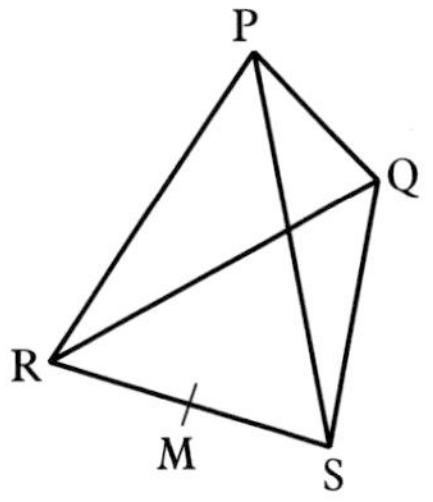

P is vertically above the point X, which is $\frac{2}{3}$ of the way along QM.

c Find the angle between PM and the plane QRS.

d Find the angle between PQ and the plane QRS.

9 The diagram shows a triangular prism.

a Find the angle between AC and AF.

b Find the angle between AF and the plane CDEF.

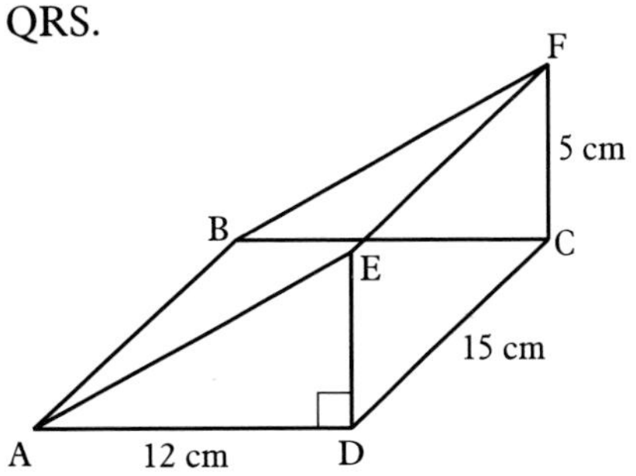

1

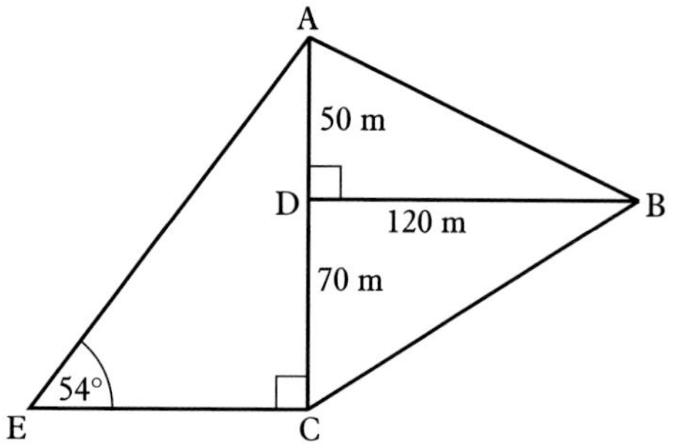

A surveyor makes the measurements shown in the diagram of a flat field ABCE, with right angles as marked.

a Calculate the bearing of B from A, giving your answer correct to one decimal place. **(3 marks)**

b Calculate the bearing of E from A, giving your answer correct to one decimal place. **(2 marks)**

c Calculate the distance EC, giving your answer correct to 3 significant figures. **(3 marks)**

a

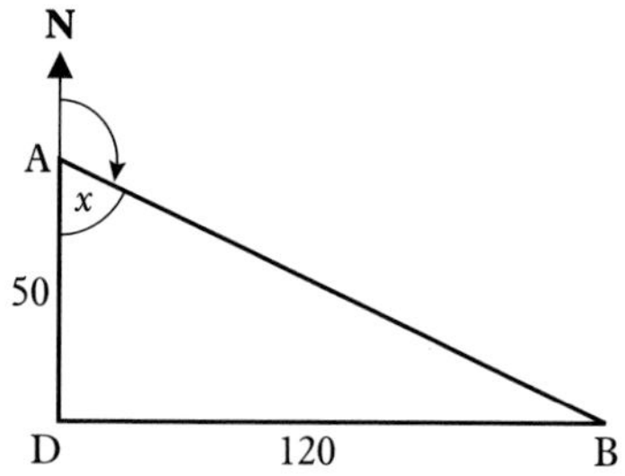

First find angle x.

$\tan x = \frac{120}{50} = 2.4$ **1 mark**

$x = 67.4°$ **1 mark**

The bearing is $180° - 67.4° = 112.6°$ (1 dp). **1 mark**

Don't forget to give the actual answer to the question.

b First find angle y.

$y = 180° - 90° - 54° = 36°$ **1 mark**

The bearing is $180° + 36° = 216°$. **1 mark**

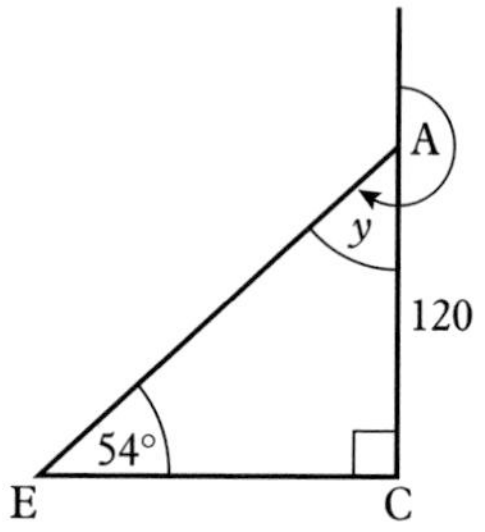

c $\tan 54° = \frac{120}{EC}$ **1 mark** $EC = \frac{120}{\tan 54°}$ **1 mark** $= 87.19 \text{ m} = 87.2 \text{ m}$ **1 mark**

2 ABCDE is a square-based pyramid.
Its base has sides of length 10 m.
Its slant edges have a length of 16 m.

a Calculate the height of the pyramid.
Give your answer correct to 3 significant figures. **(3 marks)**

b Find an angle between a face and the base.
Give your answer correct to 1 decimal place. **(3 marks)**

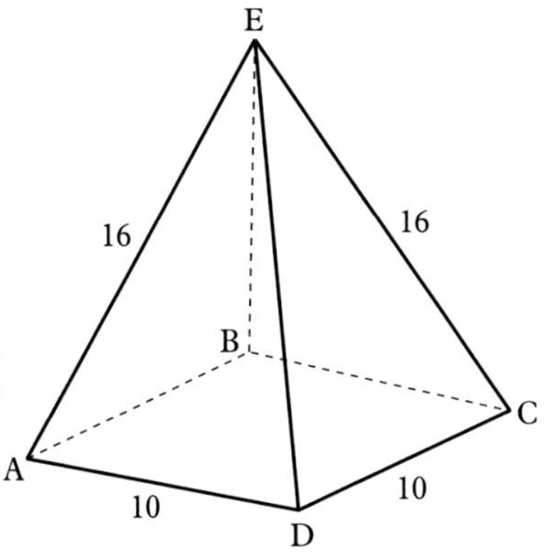

a

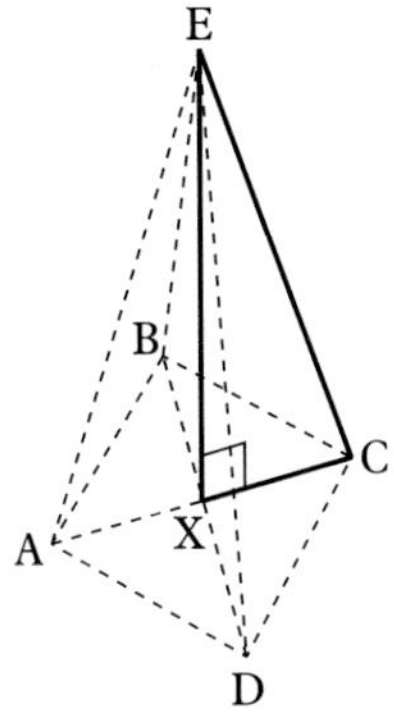

In order to calculate the height, use the right-angled triangle EXC.
The hypotenuse is 16 m. You need to find XC.
Use triangle ADC in the base to find it:

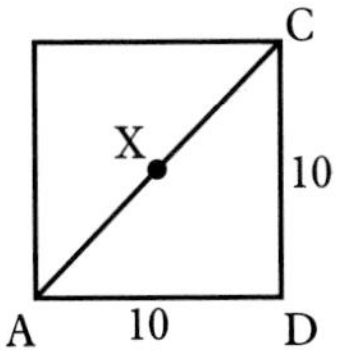

$AC = \sqrt{10^2 + 10^2}$
$= \sqrt{200} = 14.142$
$XC = \frac{1}{2}AC = 7.071\ldots$ **1 mark**

Keep the exact value in the calculator memory.

In triangle EXC:

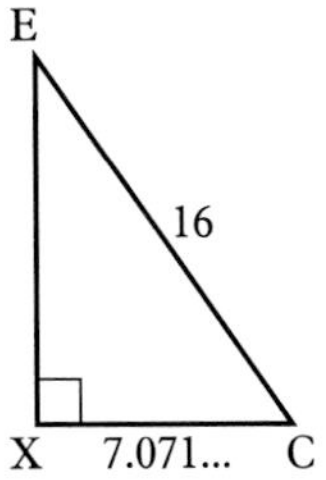

$EX = \sqrt{16^2 - 7.071^2} = \sqrt{256 - 50} = \sqrt{206}$ **1 mark**
$= 14.3527 = 14.4$ cm **1 mark**

Now keep this in the calculator memory, or write it out in full – you may need it again!

b

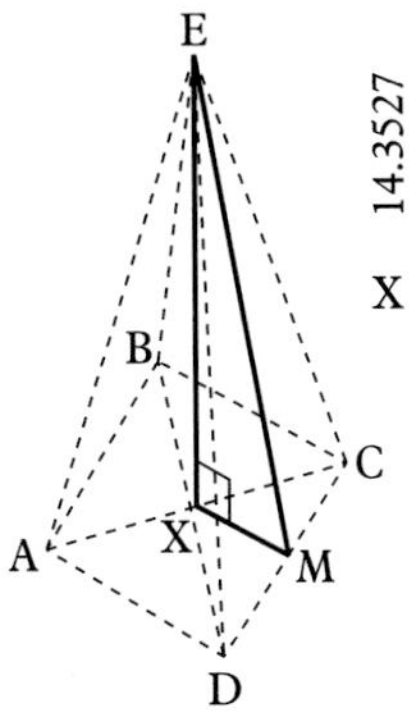

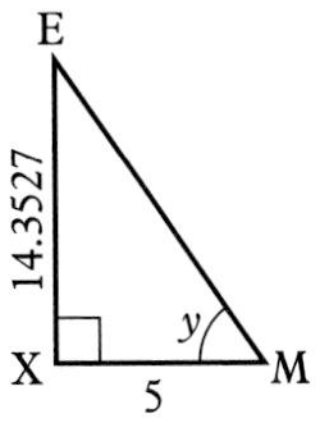

EXM is the right-angled triangle you will need to calculate angle y.
1 mark: selection of triangle

*This is the value you kept in the calculator from **a**.*

$\tan y = \frac{14.3527}{5} = 2.87\ldots$ **1 mark**

$y = 70.793\ldots = 70.8°$ **1 mark**

Give the accurate value first.

1 The diagram represents the positions of Wigan and Manchester.

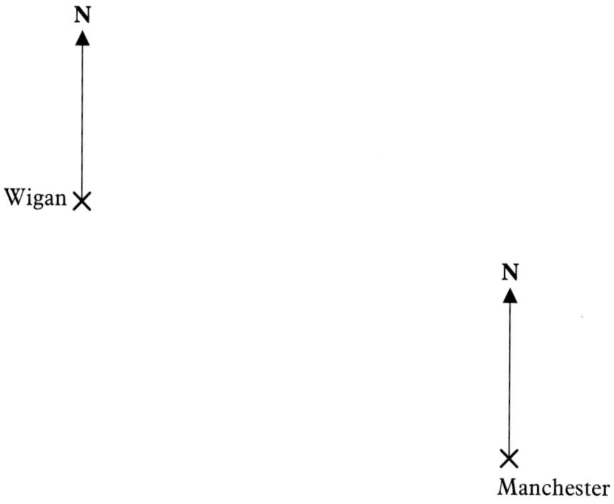

a Measure and write down the bearing of Manchester from Wigan. **(1 mark)**

b Find the bearing of Wigan from Manchester. **(2 marks)**

[S1998 P5 Q6]

2

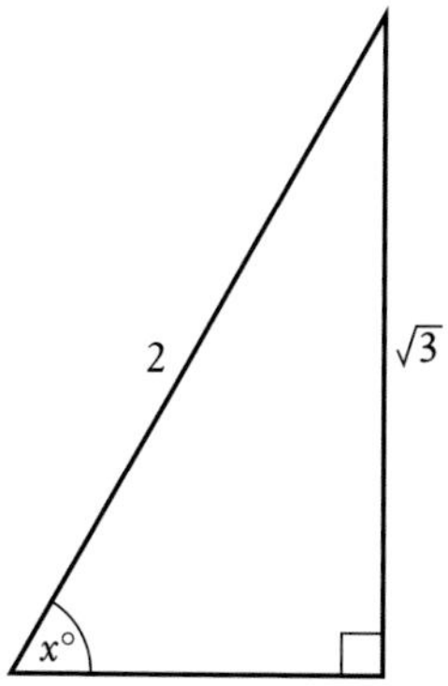

Diagram **NOT** accurately drawn.

$\sin x^\circ = \frac{\sqrt{3}}{2}$ and $0 < x < 90$

a Use the Theorem of Pythagoras for the above triangle and find the exact value of $\cos x^\circ$. **(1 mark)**

One value of x for which $\sin x^\circ = \frac{\sqrt{3}}{2}$ is 60.

b Find two values of y between 0 and 180 for which

$$\sin(2y)^\circ = \frac{\sqrt{3}}{2}$$

(3 marks)

[N2001 P5 Q19]

3

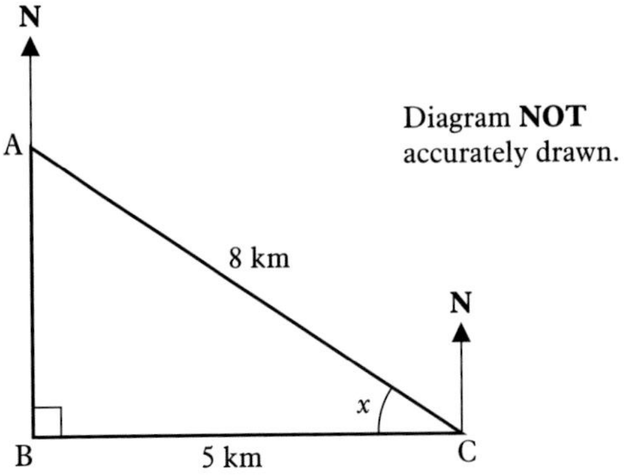

The diagram shows the postions of three telephone masts A, B and C.

Mast C is 5 kilometres due east of mast B.
Mast A is due north of mast B, and 8 kilometres from mast C.

a Calculate the distance of A from B.
Give your answer in kilometres, correct to three significant figures.
(3 marks)

b **(1)** Calculate the size of the angle marked $x°$.
Give your angle correct to one decimal place.
(2) Calculate the bearing of A from C.
Give your bearing correct to one decimal place.
(3) Calculate the bearing of C from A.
Give your bearing correct to one decimal place. **(5 marks)**
[S1999 P6 Q5]

4

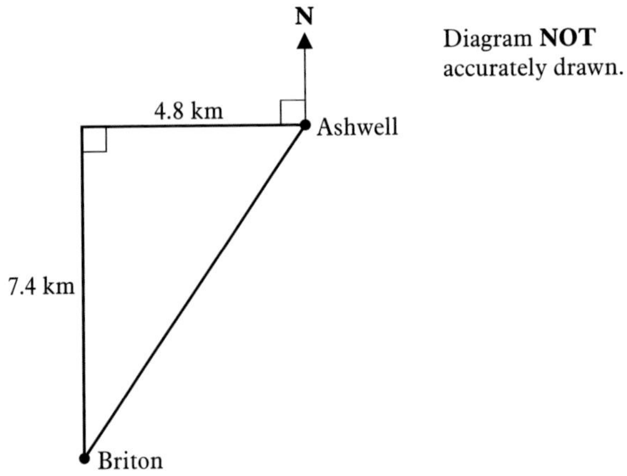

Paul flies his helicopter from Ashwell.
He flies due west for 4.8 km.
He then flies due south for 7.4 km to Birton.
Calculate the three figure bearing of Birton from Ashwell. **(4 marks)**
[N1999 P5 Q8]

5 The diagram shows the positions of three schools P, Q and R.

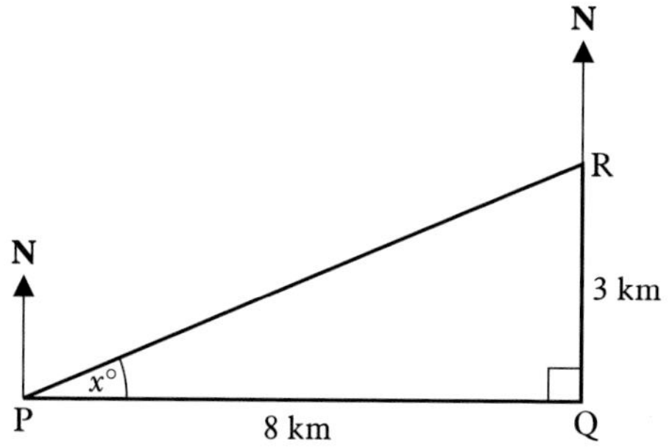

Diagram **NOT** accurately drawn.

School P is 8 kilometres due west of school Q.
School R is 3 kilometres due north of school Q.

a Calculate the size of the angle marked $x°$.
Give your answer correct to one decimal place. **(3 marks)**

Simon's house is 8 kilometres due east of school Q.

b Calculate the bearing of Simon's house from school R. **(2 marks)**

[S2001 P6 Q10]

6

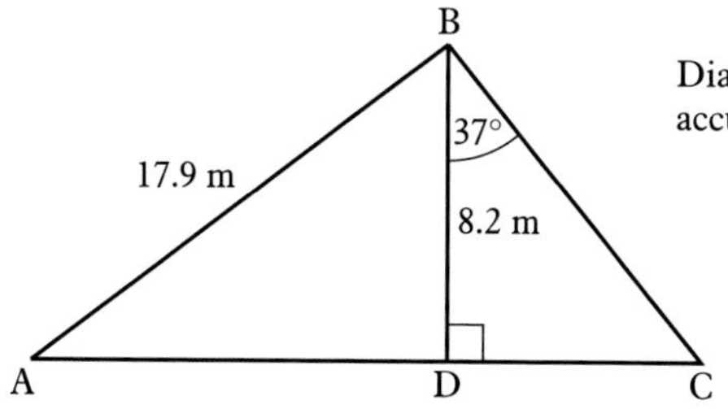

Diagram **NOT** accurately drawn.

In the diagram

AB = 17.9 m, BD = 8.2 m, angle CBD = 37° and angle BDC = 90°.
ADC is a straight line.

a Calculate the length of DC.
Give your answer, in metres, correct to 3 significant figures.
(3 marks)

b Calculate the size of angle DAB.
Give your answer correct to 1 decimal place. **(3 marks)**

[N1997 P5 Q9]

7

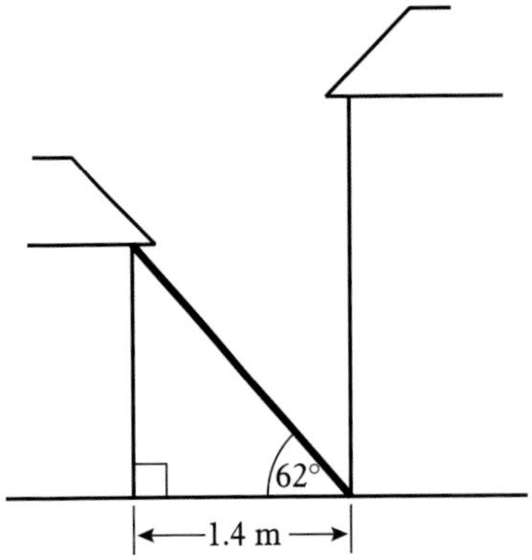

Diagram **NOT** accurately drawn.

The diagram shows a house and a garage on level ground.
A ladder is placed with one end at the bottom of the house wall.
The top of the ladder touches the top of the garage wall.
The distance between the garage wall and the house is 1.4 m.
The angle the ladder makes with the ground is 62°.

a Calculate the height of the garage wall.
Give your answer correct to 3 significant figures. **(3 marks)**

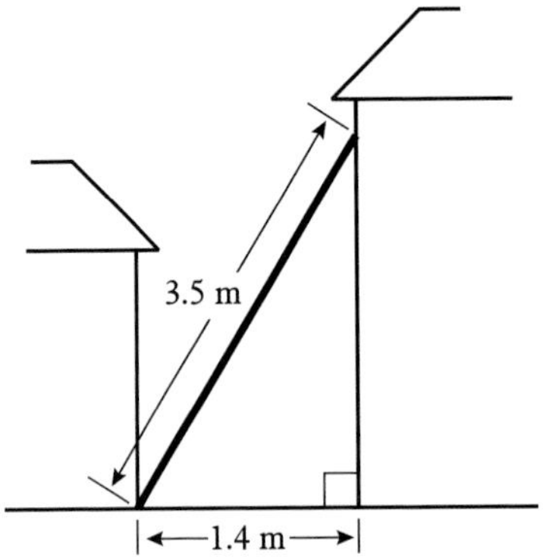

Diagram **NOT** accurately drawn.

A ladder of length 3.5 m is then placed against the house wall.
The bottom of this ladder rests against the bottom of the garage wall.

b Calculate the angle that this ladder makes with the ground.
Give your answer correct to 1 decimal place. **(3 marks)**
[S1997 P5 Q12]

12 Indices and standard form

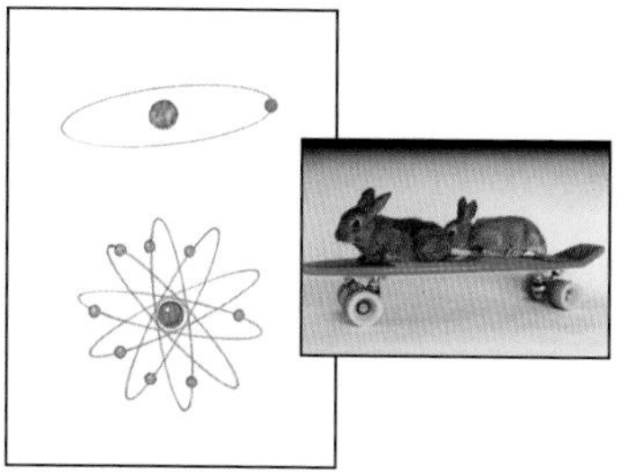

You need to know about:

- indices
- zero index
- reciprocals
- fractional indices
- exponential equations
- standard form

Base and index

In the expression x^n the number x is called the **base** and the number n is called the index. **Index** is another word for power. The plural of index is indices.

Index of 0

Any number with an **index of 0** is equal to 1.
$x^0 = 1$ for any value of x.

Reciprocal

The **reciprocal** of a number is 1 divided by that number.

So $\frac{1}{x}$ is the **reciprocal** of x.

An index of -1 is the reciprocal, i.e. $x^{-1} = \frac{1}{x}$.

The reciprocal of a fraction turns the fraction over.

$$\left(\frac{a}{b}\right)^{-1} = \frac{1}{\frac{a}{b}}$$

$$= 1 \times \frac{b}{a}$$

$$= \frac{b}{a}$$

So, for example, $\left(2\frac{1}{3}\right)^{-1} = \left(\frac{7}{3}\right)^{-1} = \frac{3}{7}$.

Negative index

A **negative index** of $-n$ is the reciprocal of a power of n.

The rule is $x^{-n} = \frac{1}{x^n}$.

So, for example, $8^{-2} = \frac{1}{8^2} = \frac{1}{64}$.

Multiplying numbers with powers

$$x^m \times x^n = x^{m+n}$$

You add the powers when you are multiplying terms with powers.

$3d^2 \times 5d^4 = 15d^6$ You multiply the numbers: $3 \times 5 = 15$

You add the indices: $2 + 4 = 6$

$$(x^m)^n = x^{mn}$$

You multiply the powers when you raise a power to another power.

$(2x^3)^4 = 16x^{12}$ Work out the number term and the algebra term separately: $(2x^3)^4 = 2^4 \times (x^3)^4$

$2^4 = 16$ and $(x^3)^4 = x^{3\times 4} = x^{12}$

Dividing numbers with powers

$$x^m \div x^n = x^{m-n}$$

You subtract the powers when you are dividing terms with powers.

$24e^{15} \div 8e^6 = 3e^9$ You divide the numbers: $24 \div 8 = 3$

You subtract the indices: $15 - 6 = 9$

So, $8a^2b^6c \div 16ab^7c = \dfrac{8a^{2-1}b^{6-7}c^{1-1}}{16} = \dfrac{ab^{-1}c^0}{2} = \dfrac{ab^{-1}}{2} = \dfrac{a}{2b}$

Indices that are fractions

$x^{\frac{1}{2}} = \sqrt{x}$ $x^{\frac{1}{n}} = \sqrt[n]{x}$ $x^{\frac{m}{n}} = \sqrt[n]{x^m}$ or $(\sqrt[n]{x})^m$

You should be able to work out all of the following examples without using your calculator. Questions like these can appear on the non-calculator paper.

$25^{\frac{1}{2}} = \sqrt{25} = 5$

$81^{\frac{1}{4}} = \sqrt[4]{81} = 3$

$16^{\frac{3}{4}} = (\sqrt[4]{16})^3 = 2^3 = 8$

$36^{-\frac{1}{2}} = \dfrac{1}{\sqrt{36}} = \dfrac{1}{6}$

$16^{-\frac{1}{4}} = \dfrac{1}{\sqrt[4]{16}} = \dfrac{1}{2}$

$27^{-\frac{2}{3}} = \dfrac{1}{(\sqrt[3]{27})^2} = \dfrac{1}{3^2} = \dfrac{1}{9}$

Exponent

Exponent is another word for power or index.

Exponential equations

Equations where the variable is in the index are called **exponential equations**.

$7^x = 49$ is an exponential equation.

Since $7^2 = 49$ the solution is $x = 2$.

Standard form

A large number can be written as **a number between 1 and 10** multiplied by **a positive power of 10.**

$34\,000 = 3.4 \times 10\,000 = 3.4 \times 10^4$

3.4 0 0 0.
1 2 3 4

The first part **must** be a number between 1 and 10.
A small number can be written as **a number between 1 and 10** multiplied by **a negative power of 10.**

$0.000076 = 7.6 \times 0.00001 = 7.6 \times 10^{-5}$

0.0 0 0 0 7. 6
5 4 3 2 1

Most calculators have an **EXP** key.
This key can help you deal with numbers in standard form.
You use this key to enter numbers in standard form into your calculator.

To enter 5×10^6 press these keys. **5** **EXP** **6**

Your calculator display will look like this: 5.X10^06 or 5.^06

You must use the **+/−** key to enter negative powers.

To enter 7×10^{-4} press these keys **7** **EXP** **+/−** **4**

Your calculator display will look like this: 7.X10^-04 or 7.^-04

You must also be able to do standard form calculations without your calculator.
To do this you work with the number parts and the power parts separately.

Examples

Work out $(2 \times 10^4) \times (6 \times 10^{-9})$.

$2 \times 6 = 12$ and $10^4 \times 10^{-9} = 10^{-5}$, so $(2 \times 10^4) \times (6 \times 10^{-9}) = 12 \times 10^{-5}$ but this is not in standard form. Notice that $12 = 1.2 \times 10^1$ so $12 \times 10^{-5} = 1.2 \times 10^1 \times 10^{-5} = 1.2 \times 10^{-4}$ which is the answer in standard form.

Work out $(6 \times 10^{-4}) \div (4 \times 10^{-9})$.

$6 \div 4 = 1.5$ and $10^{-4} \div 10^{-9} = 10^{-4-(-9)} = 10^5$, so $(6 \times 10^{-4}) \div (4 \times 10^{-9}) = 1.5 \times 10^5$ which is the answer in standard form.

1 Write down the value of each of these.

a 4^3 **d** 18^{-1} **g** 10^{-3} **j** $125^{-\frac{2}{3}}$

b 10^0 **e** 5^{-2} **h** $36^{\frac{1}{2}}$ **k** $256^{-\frac{3}{8}}$

c 2^{-1} **f** 7^0 **i** $16^{\frac{1}{4}}$ **l** $243^{-\frac{3}{5}}$

2 Write down the reciprocal of each of these.

a 6 **b** 20 **c** $\frac{4}{5}$ **d** $3\frac{3}{4}$

3 Simplify these.

a $a^3b^3 \times a^4b^{-4}$ **e** $(x^3y^6)^2$ **i** $a^6b^4 \div a^3b^2$

b $3a^2b^4 \times 2a^6b^{-6}$ **f** $(p^5q^{-3})^2$ **j** $12a^{-3}b^5 \div 6a^5b^3$

c $4x^3y^3 \times 2x^{-4}y^{-2}$ **g** $(3x^2y^{-3})^3$ **k** $8x^6y^{-2} \div 4x^{-1}y^{-4}$

d $x^{-3}y^{-2} \times 6x^6y^{-4}$ **h** $(4x^{-2}y^{-3})^4$ **l** $12x^{-4}y^{-6} \div 24x^{-7}y^{-3}$

4 Solve these equations.

a $2^x = 8$ **b** $3^x = 81$ **c** $3^y = \frac{1}{9}$ **d** $5^y = \frac{1}{125}$

5 Write these numbers in standard form.

a 42 000 000 **b** 603 **c** 0.000 041 **d** 0.002 34

6 These numbers are written in standard form.
Write them as ordinary numbers.

a 3.12×10^4 **b** 6.98×10^8 **c** 1.7×10^{-3} **d** 1.582×10^{-4}

7 The table shows the distances in km of the planets from the Sun.

a Which planet is nearest the Sun?

b How much further from the Sun is Earth than Mercury?
Give your answer in standard form.

c How many times further from the Sun is Pluto than Earth?
Give your answer correct to the nearest whole number.

Planet	Distance from the Sun (km)
Earth	1.5×10^8
Jupiter	7.78×10^8
Neptune	4.5×10^9
Mars	2.28×10^8
Mercury	5.8×10^7
Pluto	5.92×10^9
Saturn	1.43×10^9
Uranus	2.87×10^9
Venus	1.08×10^8

8 Mercury orbits the Sun at a mean distance of $5.790\,92 \times 10^7$ km and takes 87.9686 days to complete an orbit. It is the fastest planet in the Solar System.
Assume that the orbit is a circle with the given mean distance as the radius.

a Work out the distance that Mercury covers in a single orbit.
Give your answer in standard form, correct to 3 sf.

b Work out Mercury's average speed in km/h, giving your answer to the nearest thousand km/h.

9 A light year is defined as the distance that light travels in 1 year.
Light travels at 3×10^8 m/s. Work out a light year in km.
Give your answer in standard form, correct to 2 sf.

10 Work these out. Give your answers in standard form.
You must not use your calculator.

a $(3 \times 10^3) \times (2 \times 10^7)$
b $(5 \times 10^5) \times (3 \times 10^8)$
c $(4 \times 10^{-4}) \times (7 \times 10^9)$
d $(2 \times 10^{-3}) \times (8 \times 10^{-3})$
e $(9 \times 10^{12}) \div (3 \times 10^5)$
f $(6.2 \times 10^7) \div (3.1 \times 10^3)$
g $(7.5 \times 10^{-3}) \div (2.5 \times 10^{-6})$
h $(5 \times 10^{-3}) \div (2 \times 10^8)$

11 The number 10^{100} is called a googol.
Write these numbers in standard form.
You will not be able to use a calculator!

a 2 googols
b 155 googols
c 1000 googols
d 0.5 googols

12 The bacteria chlamydia has a length of 1 micron.
1 micron = 0.000 001 metre
The bacteria spirochaete has a length of 0.5 mm.

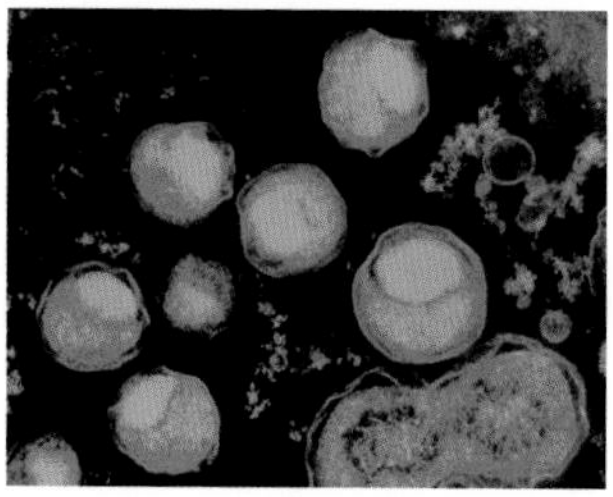

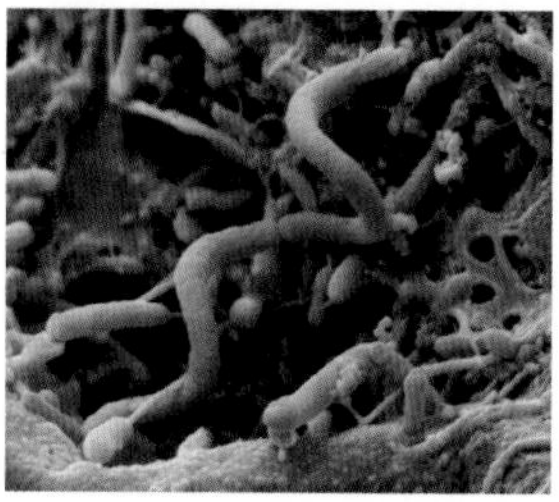

a Write the length of each bacteria in metres, in standard form.
b How many times larger is spirochaete than chlamydia?

13 The volume of a sphere of radius r is given by $V = \frac{4}{3}\pi r^3$.

Assume that the Earth is a sphere of diameter 12 756 km and that the Moon is a sphere of diameter 3476 km.
Find the value of n if

volume of the Earth : volume of the Moon = n : 1.

Give your answer correct to 2 sf.

1 Simplify:

a $(\frac{1}{2})^{-2}$ **b** $49^{-\frac{1}{2}}$ **c** $8^{\frac{4}{3}}$ **d** $25^{\frac{3}{2}}$ **(4 marks)**

a $(\frac{1}{2})^2 = \frac{1}{2} \times \frac{1}{2} = \frac{1}{4}$ So $(\frac{1}{2})^{-2} = \frac{1}{\frac{1}{4}} = 1 \div \frac{1}{4} = 4$ **1 mark**

The negative power means the reciprocal of.

b $49^{-\frac{1}{2}} = (\sqrt{49})^{-1} = \frac{1}{\sqrt{49}} = \frac{1}{7}$ **1 mark**

c $8^{\frac{4}{3}}$

The 4 is a power, the 3 is a root.
Find the root first, since this will help keep the numbers small.

$8^{\frac{4}{3}} = (\sqrt[3]{8})^4 = (2)^4 = 16$ **1 mark**

d $25^{\frac{3}{2}} = (\sqrt{25})^3 = (5)^3 = 125$ **1 mark**

2 The Earth is 9.296×10^7 miles from the Sun.
Light travels at 1.86×10^5 miles per second.
Calculate, in minutes and seconds, correct to the nearest second, the time taken for light to travel from the Sun to the Earth. **(3 marks)**

The calculation is $(9.296 \times 10^7) \div (1.86 \times 10^5)$. **1 mark**

You could do this as two separate calculations:

$9.296 \div 1.86 = 4.9978$ $10^7 \div 10^5 = 10^2$

or use the **EXP** *key:*

9 **.** **2** **9** **6** **EXP** **7** **÷** **1** **.** **8** **6** **EXP** **5** **=**.

So the answer is 4.9978×10^2 seconds, or 499.78 seconds **1 mark**

But remember the question asks for the answer in minutes and seconds.

$499.78 \div 60 = 8.3297$ minutes

The whole number part is 8 minutes. This is $8 \times 60 = 480$ seconds.

So $499.78 - 480 = 19.78$ seconds are left.

So the time is 8 minutes 20 seconds (to the nearest second). **1 mark**

3 A googol is 10^{100}.
Calculate the value of 25 googols, giving your answer in standard form. **(3 marks)**
This question could appear on a non-calculator paper.

$10^{100} = 1 \times 10^{100}$

25 googols is $25 \times 1 \times 10^{100} = 25 \times 100^{100}$ **1 mark**

25 can be written as $2.5 \times 10 = 2.5 \times 10^1$ in standard form. **1 mark**

So 25 googols is $2.5 \times 10^1 \times 10^{100} = 2.5 \times 100^{101}$ **1 mark**

4 The star Sirius is 81 900 000 000 000 km from Earth.

a Write 81 900 000 000 000 in standard form. **(1 mark)**

Light travels 3×10^5 km in 1 second.

b Calculate the number of seconds that light takes to travel from Sirius to the Earth.
Give your answer in standard form, correct to 2 significant figures. **(2 marks)**

a 81 900 000 000 000. $\rightarrow$ 8.1 900 000 000 000 $\times 10^{13}$

In standard form this number is 8.19×10^{13} **1 mark**

b The calculation is $(8.19 \times 10^{13}) \div (3 \times 10^5)$. **1 mark**

Enter this using the **EXP** *key on your calculator:*

8 **.** **1** **9** **EXP** **1** **3** **÷** **3** **EXP** **5** **=**

The answer is 273 000 000.

The question asked for the answer to be standard form, correct to 2 significant figures:

$= 2.7 \times 10^8$ **1 mark**

1 **a** Simplify
$x^3 \times x^5$ **(1 marks)**
b Simplify
$y^6 \div y^2$ **(1 marks)**
c Simplify
$\frac{8w^7}{2w^2 \times w^3}$ **(2 marks)**
[N2000 P5 Q8]

2 **a** Simplify
(1) $\frac{p^6}{p^2}$ **(2)** $q^3 \times q$ **(3)** $(4x^3)^2$ **(3 marks)**
b Simplify
(1) $3a^2 \times 2a^2b^3$ **(2)** $(2a^2)^{-3}$ **(2 marks)**
c Factorise completely
$9x^2y - 6xy^3$ **(2 marks)**
[S1999 P6 Q6]

3 **a** Expand and simplify
$2(x - 1) + 3(2x + 1)$ **(2 marks)**
b Expand and simplify
$(x + 3)(2x - 1)$ **(2 marks)**
c Factorise completely
$6a^3 - 9a^2$ **(2 marks)**
d Evaluate
(1) 5^{-2} **(2)** $8^{\frac{2}{3}}$ **(3)** $49^{-\frac{1}{2}}$ **(3 marks)**
[S2001 P5 Q3]

4 **a** Simplify $(3xy^3)^4$ **(1 mark)**
b $\sqrt{\left(\frac{x - 4}{5}\right)} = 2y$
Rearrange this formula to give x in terms of y. **(2 marks)**
[S2001 P5 Q14]

5 $p = 3^8$
a Express $p^{\frac{1}{2}}$ in the form 3^k, where k is an integer. **(1 mark)**
$q = 2^9 \times 5^{-6}$
b Express in the form $2^m \times 5^n$, where m and n are integers,
(1) $q^{\frac{1}{3}}$ **(2)** q^{-1} **(4 marks)**
[N1998 P5 Q17]

6 **a** Simplify
$14a^6 \div 2a^2$ **(2 marks)**

b Simplify
$c^9d^{-2} \times c^{-3}d$ **(2 marks)**

c Simplify
$(p^{-2})^{-4}$ **(1 mark)**

d Simplify
$(25m^6)^{\frac{1}{2}}$ **(2 marks)**

e Solve the equation
$x^2 - 6x - 27 = 0$ **(3 marks)**

f Factorise
$2y^2 - 11y + 12$ **(2 marks)**

[N1999 P5 Q13]

7 $p = 8 \times 10^3$,
$q = 2 \times 10^4$

a Find the value of $p \times q$.
Give your answer in **standard form.** **(2 marks)**

b Find the value of $p + q$.
Give your answer as an **ordinary number.** **(2 marks)**

[S2000 P5 Q2]

8 **a (1)** Write the number 5.01×10^4 as an ordinary number.
(2) Write the number 0.0009 in standard form. **(2 marks)**

b Multiply 4×10^3 by 6×10^5.
Give your answer in standard form. **(2 marks)**

[S2001 P5 Q6]

9 **a** Write 84 000 000 in standard form. **(2 marks)**

b Work out

$$\frac{84\,000\,000}{4 \times 10^{12}}$$

Give your answer in standard form. **(3 marks)**

[N2000 P5 Q7]

10 The mass of one electron is 0.000 000 000 000 000 000 000 000 91 grams.

a Write 0.000 000 000 000 000 000 000 000 91 in standard form. **(2 marks)**

b Calculate the mass of five million electrons.
Give your answer, in grams, in standard form. **(3 marks)**

[N1998 P6 Q6]

11 $v^2 = \frac{GM}{R}$

$G = 6.6 \times 10^{-11}$
$M = 6 \times 10^{24}$
$R = 6\,800\,000$

a Calculate the value of v. Give your answer in standard form, correct to 2 significant figures. **(4 marks)**

b Rearrange the formula $v^2 = \frac{GM}{R}$ to make M the subject. **(2 marks)**

[N1999 P6 Q7]

12 The time taken for light to reach Earth from the edge of the known universe is 14 000 000 000 years.

a Write 14 000 000 000 in standard form. **(2 marks)**

Light travels at the speed of 9.46×10^{12} km/year.

b Calculate the distance, in kilometres, from the edge of the known universe to Earth.
Give your answer in standard form. **(3 marks)**

[N1999 P6]

13 The volume of Elspeth's house is 379.6 m^3.

a Write 379.6 in standard form. **(2 marks)**

The heat, H, needed to keep a house warm is given by the formula

$$H = \frac{5544 \times 1.4 \times 3 \times 22.5 \times \text{Volume of house}}{2000}$$

b Calculate the value of H for Elspeth's house, giving your answer in standard form. **(3 marks)**

[N1999 P5]

14 The diameter of an atom is 0.000 000 03 m.

a Write 0.000 000 3 in standard form. **(2 marks)**

Using the most powerful microscope, the smallest objects which can be seen have diameters which are **one hundredth** of the diameter of an atom.

b Calculate the diameter, in metres, of the smallest objects which can be seen using this microscope.
Give your answer in standard form. **(2 marks)**

[S1996 P5]

15 Factorise completely

$2p^3q^2 - 4p^2q^3$ **(2 marks)**

[S1996 P5]

16 A US Centillion is the number 10^{303}
A UK Centillion is the number 10^{600}

a How many US Centillions are there in a UK Centillion?
Give your answer in standard form. **(2 marks)**

b Write the number 40 US Centillions in standard form. **(2 marks)**

[S1999 P6]

17 My computer can carry out 7.5×10^4 calculations in one hour.

Work out how many of these calculations my computer can carry out in one second.

Give your answer in standard form.

[N1997 P5]

18 The area of the Earth covered by sea is 362 000 000 km^2.

a Write 362 000 000 in standard form. **(2 marks)**

The surface area, A km^2, of the Earth may be found using the formula

$A = 4\pi r^2$

radius of earth $= 6.38 \times 10^3$ km

b Calculate the surface area of the Earth.
Give your answer in standard form, correct to 3 significant figures. **(2 marks)**

c Calculate the percentage of the Earth's surface which is covered by sea.
Give your answers correct to 2 significant figures. **(2 marks)**

[S1997 P6]

13 Graphs: moving on

You need to know about:

- transforming graphs
- combining transformations
- function notation
- using graphs to solve equations
- using straight line graphs for more complicated relationships

Transforming graphs

If you start with the graph of $y = f(x)$:

$y = f(x) + a$ is a translation of $y = f(x)$ through $\begin{pmatrix} 0 \\ a \end{pmatrix}$.

Adding a constant to the whole equation of the graph moves the graph **up** by this amount.

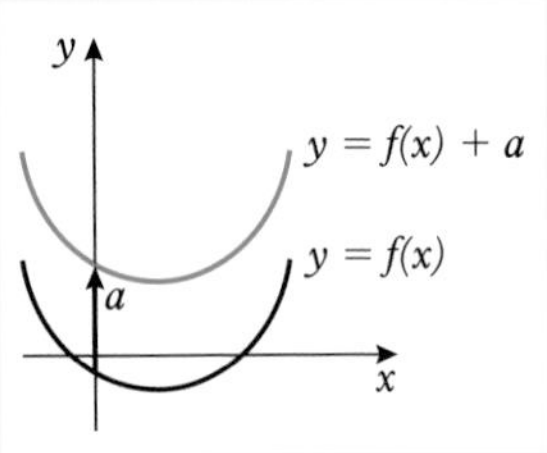

$y = f(x) - a$ is a translation of $y = f(x)$ through $\begin{pmatrix} 0 \\ -a \end{pmatrix}$.

Subtracting a constant from the whole equation of the graph moves the graph **down** by this amount.

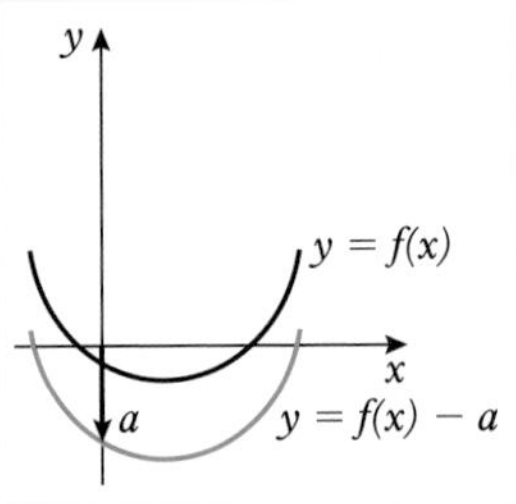

$y = f(x + a)$ is a translation of $y = f(x)$ through $\begin{pmatrix} -a \\ 0 \end{pmatrix}$.

Adding a constant to the x part of the equation moves the graph **left** by this amount.

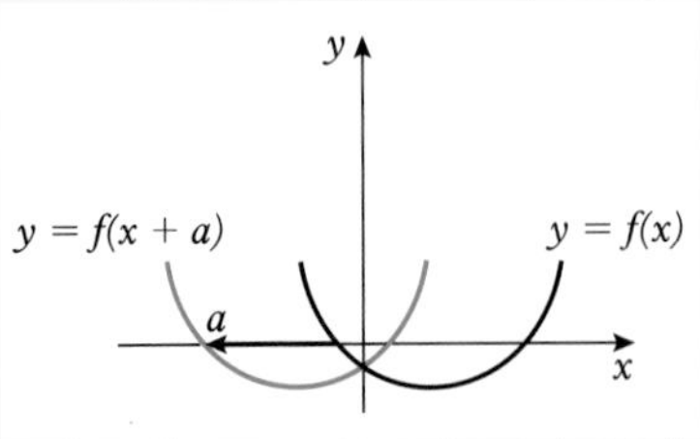

$y = f(x - a)$ is a translation of $y = f(x)$ through $\begin{pmatrix} a \\ 0 \end{pmatrix}$.

Subtracting a constant from the x part of the equation moves the graph **right** by this amount.

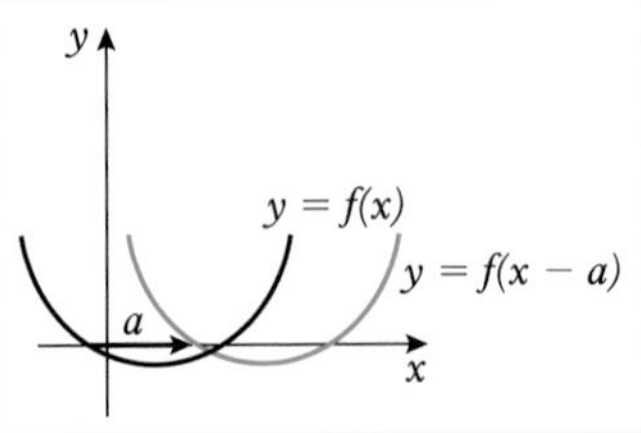

$y = -f(x)$ is a reflection in the x axis.

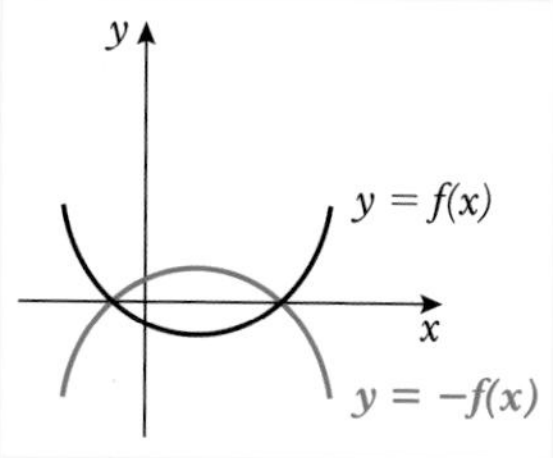

$y = f(-x)$ is a reflection in the y axis.

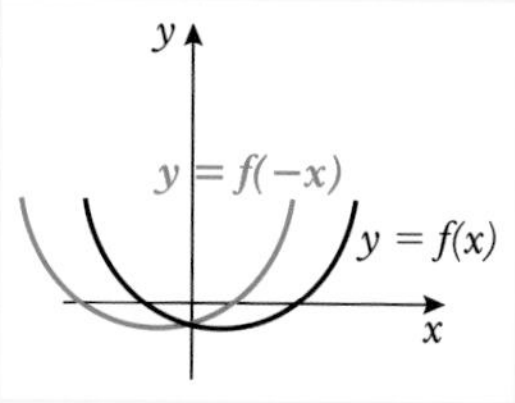

$y = kf(x)$ is a stretch in the y direction with factor k.

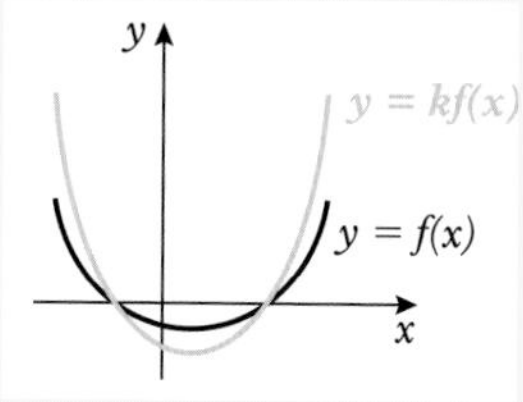

Combining transformations

The transformations given on the last page can be used together to build up more complicated graphs.

Example

Sketch the graph of $y = 2(x + 3)^2 + 4$

There are three transformations to do:

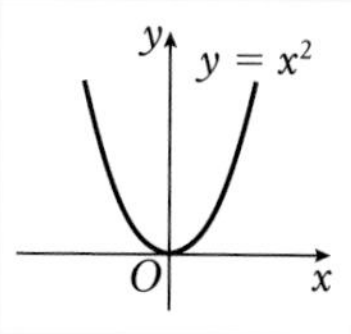

Start with the graph of $y=x^2$

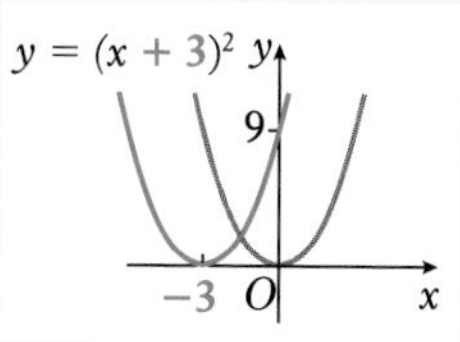

(1) The $+3$ moves the graph 3 to the left.

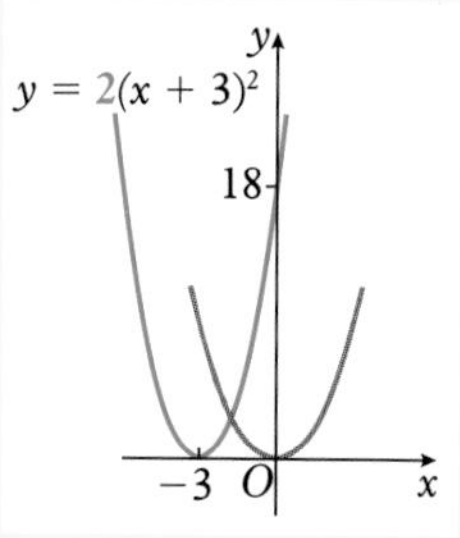

(2) The 2 stretches the graph by factor 2 in the y direction.

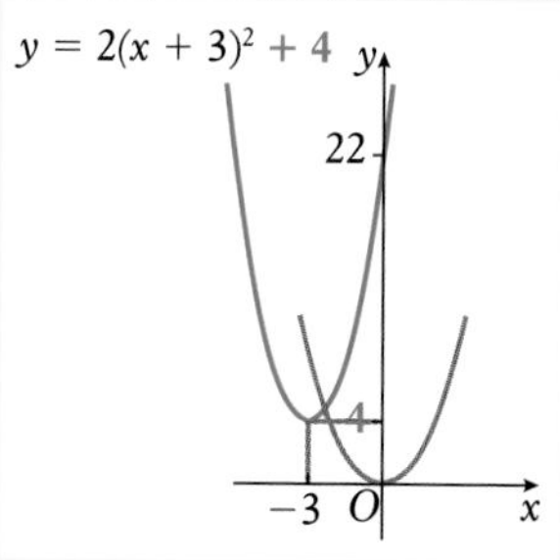

(3) The $+4$ moves the graph up 4.

Function notation

$y = x^2$ can be written in **function notation** as $f(x) = x^2$.
The value of y when $x = 3$ is then written $f(3)$.

Solving equations using graphs

If you have drawn $y = f(x)$ you can use this immediately to solve:

$f(x) = 0$ Find where the graph crosses the x axis.

$f(x) = a$ Find the x values where the graph crosses the horizontal line $y = a$.

$f(x) = g(x)$ Find the x values where the graph crosses the graph of $y = g(x)$.

If the equation you are trying to solve does not have exactly $f(x)$, you need to rearrange the equation to get $f(x)$ involved. To solve $x^2 - 4x + 1 = 0$ using the graph of $y = x^2 - 2x + 2$ you need to get the x parts of the equation of the graph you have drawn, $x^2 - 2x + 2$, involved in the equation you are trying to solve. So $x^2 - 4x + 1 = 0$ is the same as $x^2 - 2x + 2 - 2x - 1 = 0$, which is the same as solving $x^2 - 2x + 2 = 2x + 1$.

Now you draw $y = 2x + 1$ on the graph of $y = x^2 - 2x + 2$ that you have already drawn and find the x values where the graphs cross.

Using straight line graphs for non-linear relationships

You can use straight line graphs when dealing with relationships that actually have more complicated equations.

If you have an equation like $y = 3x^2 + 5$ and you draw a graph of y against x you will get a quadratic curve like this:

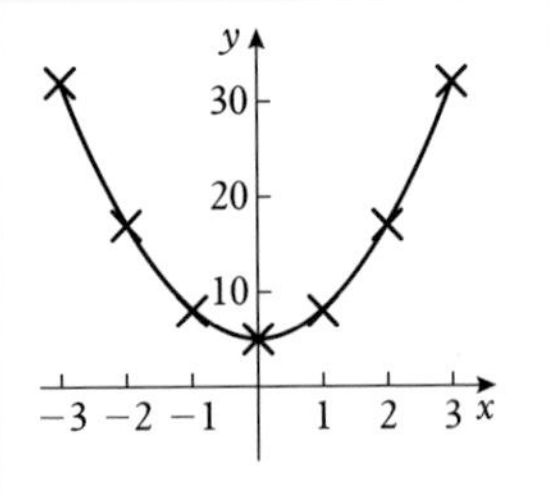

If you plot a graph of y against x^2, however, you will get a straight line like this:

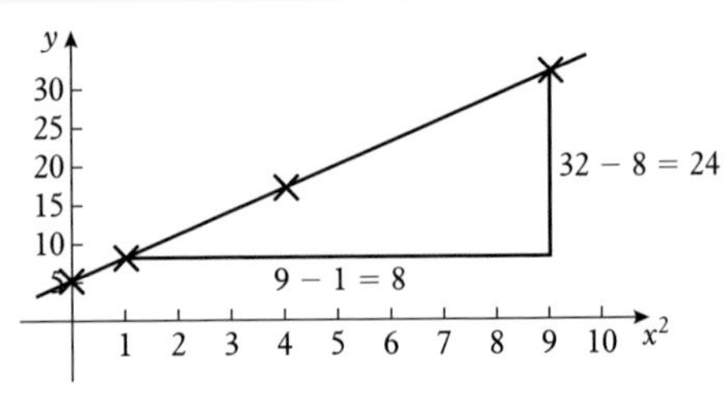

The gradient of the line is 3 and the y intercept of the line is +5.

If you suspect that data follows a relationship like $y = ax^2 + b$ draw a graph of y against x^2. If the graph is straight then this proves that your suspicion is correct and allows you to work out the values of a (the gradient) and b (the y intercept).

1 Sketch each of these graphs.

a $y = x^2$

b $y = x^2 + 3$

c $y = x^2 - 5$

d $y = (x + 1)^2$

e $y = (x - 3)^2$

f $y = 2x^2$

g $y = -x^2$

h $y = -3x^2$

2 Starting with the graph of $y = x^2$, list the transformations you would need to perform to produce the following graphs:

a $y = 3(x - 1)^2$

b $y = 2(x + 4)^2 - 3$

c $y = 2(x - 1)^2 + 4$

d $y = 3(x - 5)^2 - 2$

3 The diagram shows the graph of $y = f(x)$.

On the same axes draw the graph of:

a $y = f(x) + 1$

b $y = f(x + 1)$

c $y = -f(x)$

d $y = f(-x)$

e $y = 2f(x)$

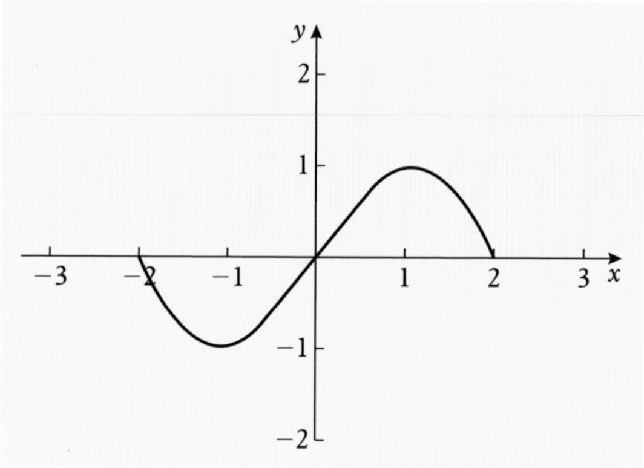

4 The function $f(x)$ is defined as $f(x) = x^2 - 4x + 3$.
Find:

a $f(2)$

b $f(-1)$

c $f(-3)$

d $f(a)$

e $f(-p)$

f $f(2a)$

5 The function $g(x)$ is defined as $g(x) = x^2 - x + 2$.
Find the values of x for which $g(x) = 4$.

6 $f(x) = x - 4 \quad g(x) = 2x^2 - 8x + 3$

a Find the values of x for which $f(x) = g(x)$.

b Find the values of $f(x)$ for your values of x from **a**.

c Write down the points of intersection of the graphs of $y = f(x)$ and $y = g(x)$.

7

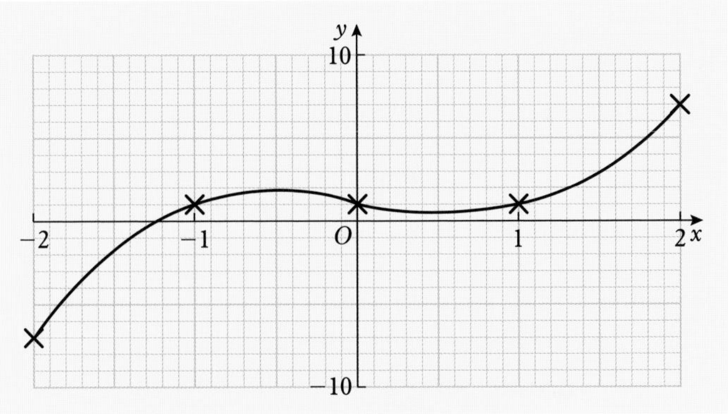

The diagram shows part of the graph of $y = x^3 - x + 1$.
Copy the graph and use your copy to solve the following equations.
You will need to draw additional lines on your graph.

a $x^3 - x + 1 = 0$

b $x^3 - x + 1 = 3$

c $x^3 - x + 2 = 0$

d $x^3 - 2x + 4 = 0$

e $-\frac{3}{x} = x^2 - 1$

8 Bobby measures the length of a pendulum and how long it takes to complete a full swing from A to B and back.
When the length of the pendulum is l metres, the time taken is T seconds.

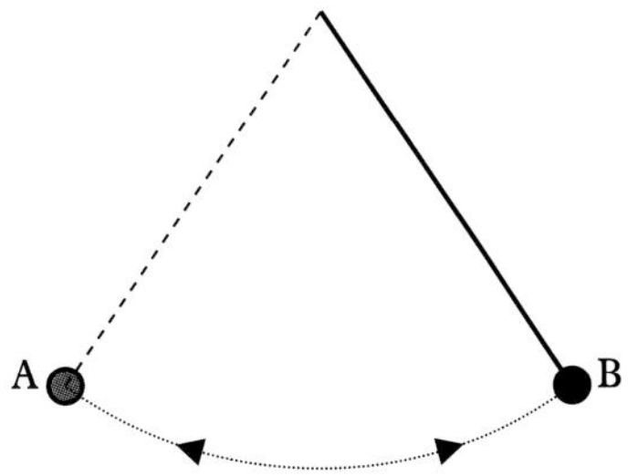

l	0.2	0.4	0.6	0.8	1	1.2	1.4
T	0.89	1.26	1.55	1.79	2.00	2.19	2.37

Bobby has been told to investigate whether there is a relationship of the form $T = a\sqrt{l} + b$.
By drawing a suitable graph on 2 cm graph paper, show that there is a relationship of this form and find the values of a and b.

1 This is a graph of the equation $y = 2x^2 - x - 10$.

Use the graph to solve these equations.

a $2x^2 - x - 10 = 0$

b $2x^2 - x - 3 = 0$

c $2x^2 - 2x - 6 = 0$

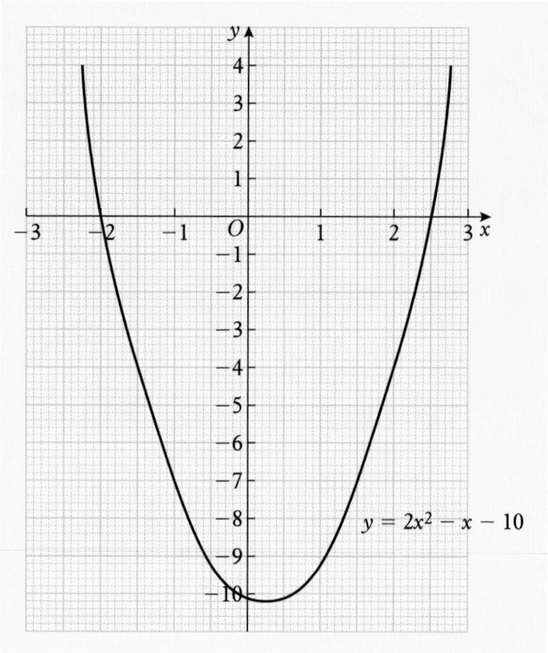

(5 marks)

a $2x^2 - x - 10 = 0$ ← *This is where the graph cuts the x axis. (y = 0 is the x axis.)*

Read off the values: $x = 2.5$ and $x = -2$. **(1 mark)**

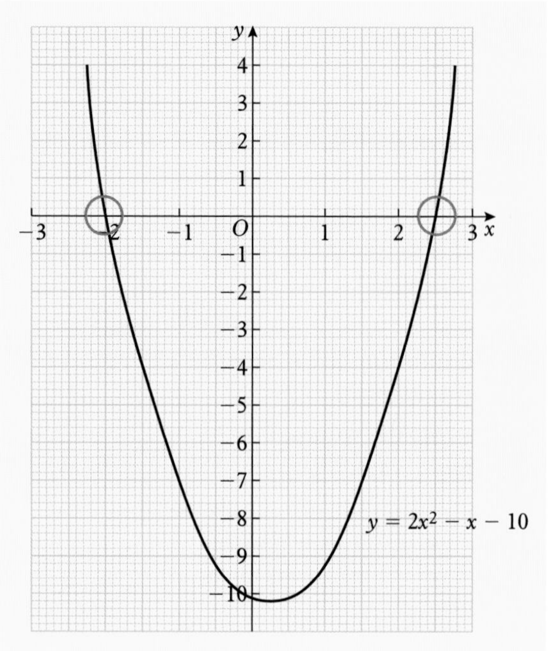

b $2x^2 - x - 3 = 0$ is not the equation of the graph, so the equation will have to be rearranged.

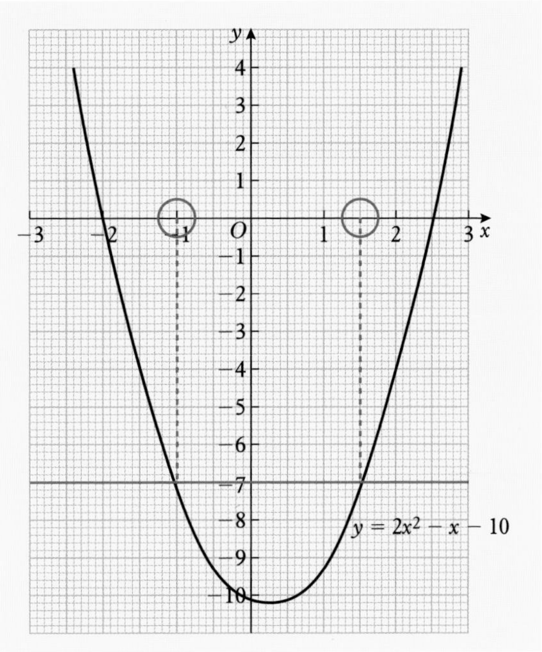

This is the equation of the graph.

$$\overbrace{2x^2 - x - 10}\underbrace{+\ 7} = 0$$

This is needed to get the equation in the question.

So $2x^2 - x - 10 = -7$. The solutions are where $y = -7$.

1 mark

Read off the values: $x = 1.5$ and $x = -1$.

1 mark

Make sure you show the lines on the graph where you do the reading off.

c $2x^2 - 2x - 6 = 0$. Again this is not the equation of the graph, so the equation will have to be rearranged.

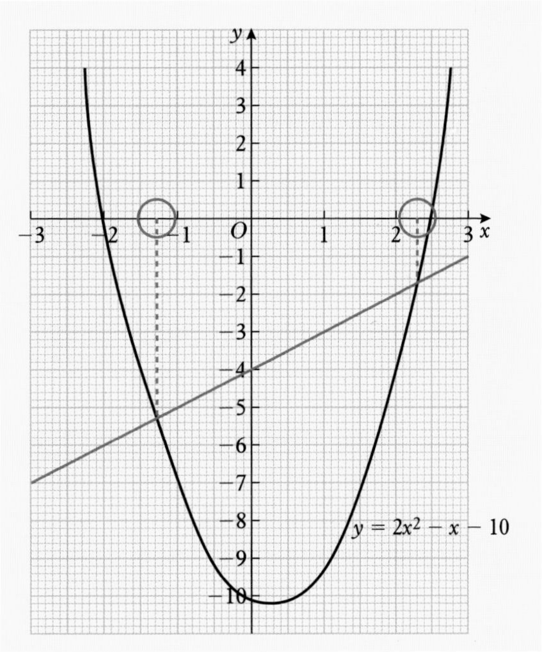

This is the equation of the graph.

$$\overbrace{2x^2 - x - 10}\underbrace{-\ x + 4} = 0$$

This is needed to get the equation in the question.

So $2x^2 - x - 10 = x - 4$. Draw the line $y = x - 4$ on the diagram.

1 mark

The solutions are where the graph intersect.

Read off the values: $x = 2.3$ and $x = -1.3$.

1 mark

Make sure you show the lines on the graph where you do the reading off.

2 This is the sketch of a function.

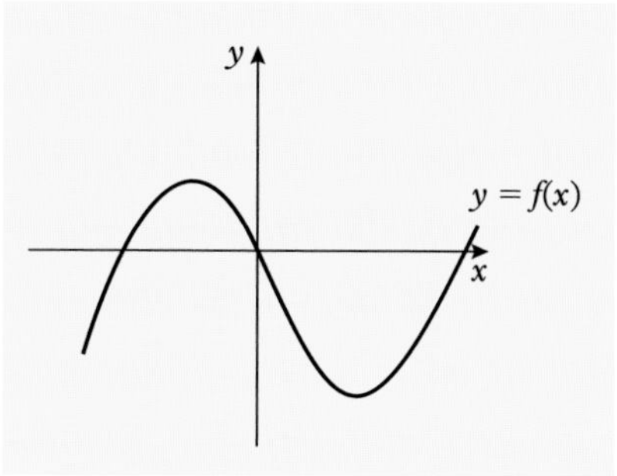

Sketch graphs of

a $y = f(x) + a$

b $y = f(x + a)$

c $y = -f(x)$

d $y = f(-x)$

where a is positive.

(4 marks)

a $f(x) + a$ is a translation of a upwards:

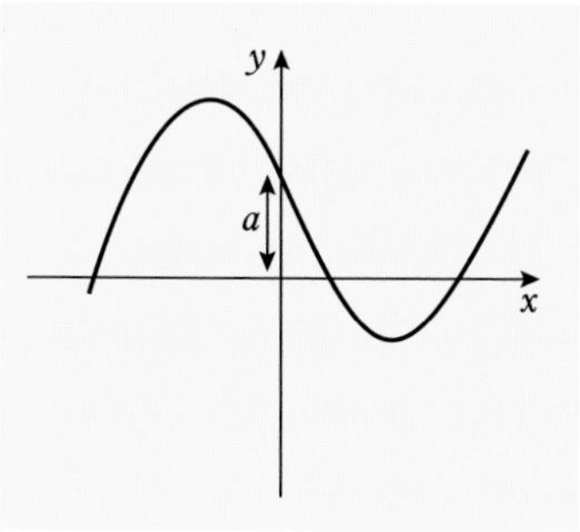

1 mark

b $f(x + a)$ is a translation of a to the left.

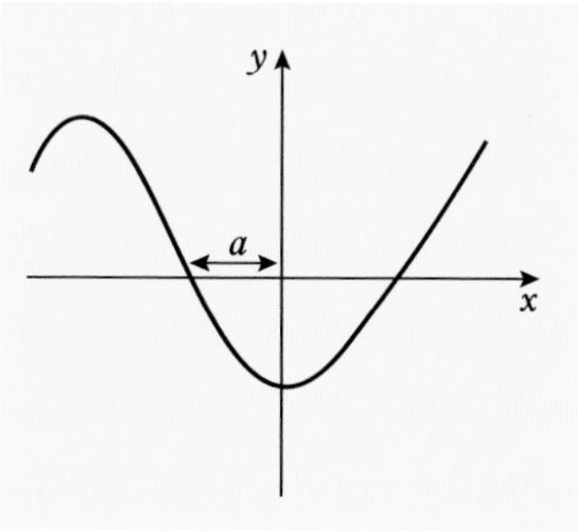

1 mark

*It is good to label **a** on the axis to make it clear how far the graph is being moved.*

c $y = -f(x)$ is a reflection in the x axis:

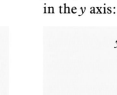

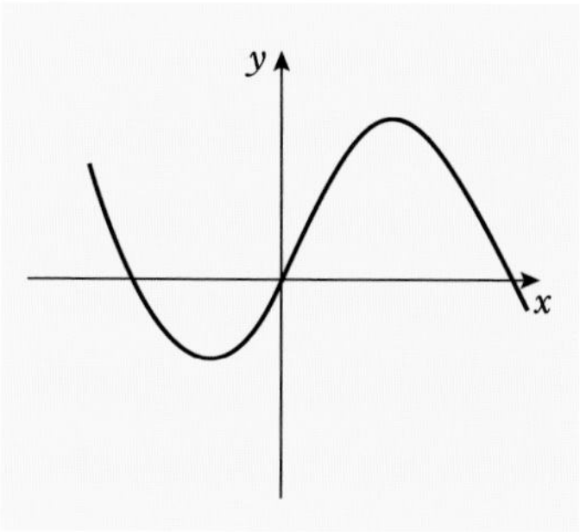

1 mark

d $y = f(-x)$ is a reflection in the y axis:

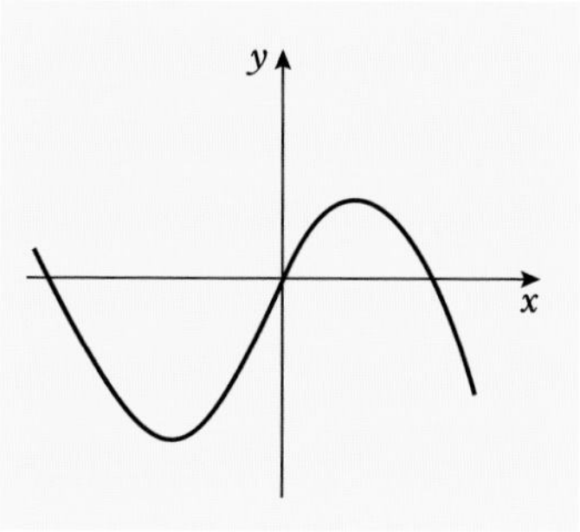

1 mark

Even though these are sketch graphs, try to get them as accurate as you can.

1 The diagram shows the curve with equation $y = f(x)$, where $f(x) = x^2 - 2x - 3$.

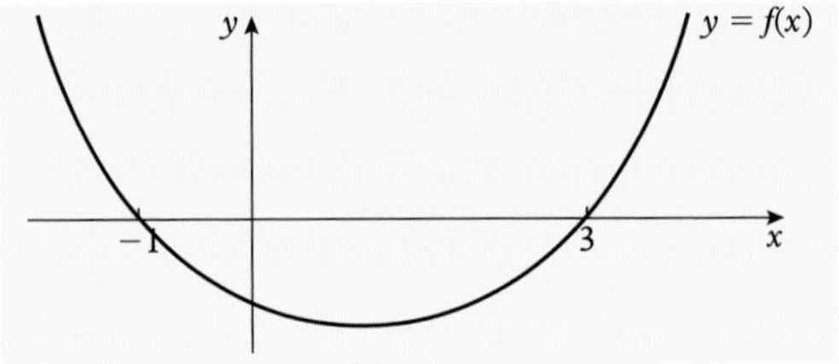

a On the same diagram sketch the curve with equation $y = f(x - 2)$. Label the points where this curve cuts the x axis. **(2 marks)**

The curve with equation $y = f(x)$ meets the curve with equation $y = f(x - a)$ at the point **P**.

b Calculate the x co-ordinate of the point **P**. Give your answer in terms of a. **(4 marks)**

The curve with equation $y = x^2 - 2x - 3$ is reflected in the y axis.

c Find the equation of this new curve. **(2 marks)**

[S1998 P6 Q16]

2 Here is the graph of $y = f(x)$.

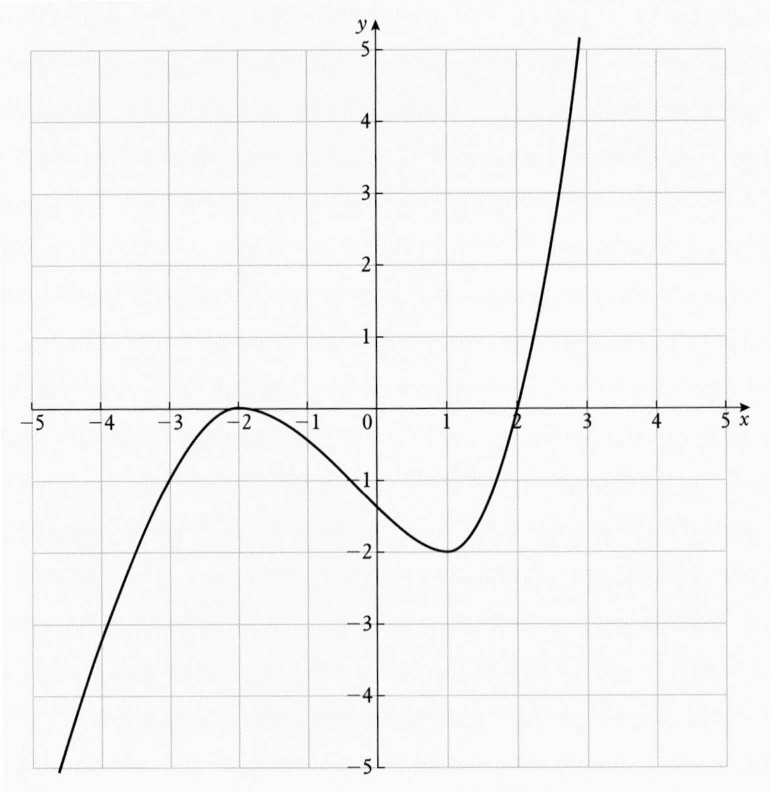

(1) On the grid, sketch the graph of $y = f(x + 2)$.

(2) On the grid, sketch the graph of $y = 2f(x)$. **(4 marks)**

3 A transformation has been applied to the graph of $y = x^2$ to give the graph of $y = -x^2$.

a Describe fully the transformation. **(1 mark)**

For all values of x,

$x^2 + 4x = (x + p)^2 + q$

b Find the the values of p and q. **(4 marks)**

A transformation has been applied to the graph of $y = x^2$ to give the graph of $y = x^2 + 4x$.

c Using your answer to part **b**, or otherwise, describe fully the transformation. **(2 marks)**

[S2001 P6 Q20]

4 This is a sketch of the curve with equation $y = f(x)$.

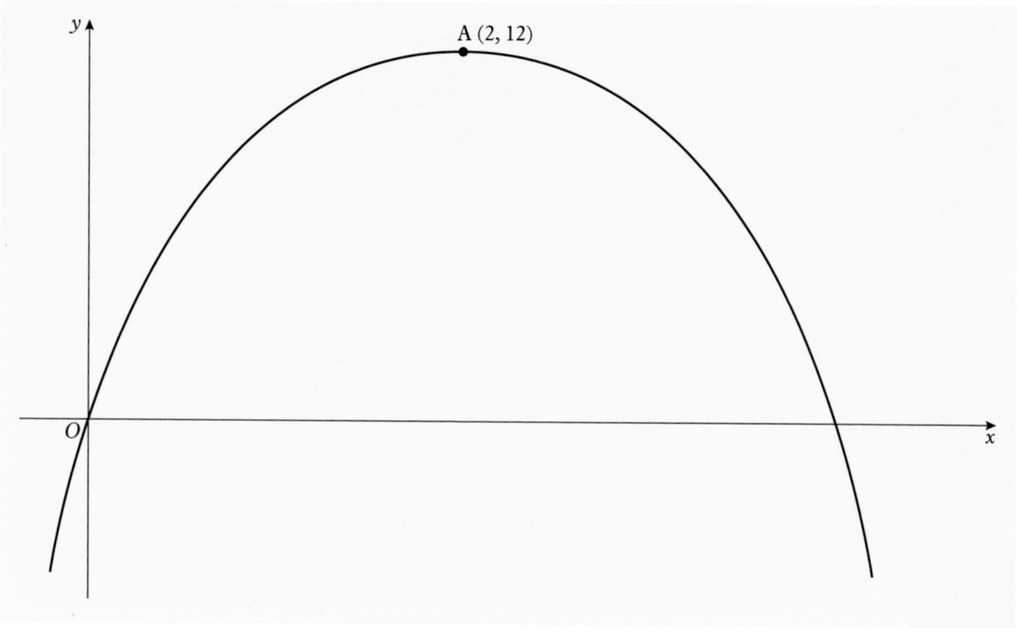

The vertex of the curve is A(2, 12).
Write down the co-ordinates of the vertex for each of the curves having the following equations.

a $y = f(x) + 6$ **(1 mark)**

b $y = f(x + 3)$ **(1 mark)**

c $y = f(-x)$ **(1 mark)**

d $y = f(4x)$ **(1 mark)**

[S1999 P5 Q12]

5

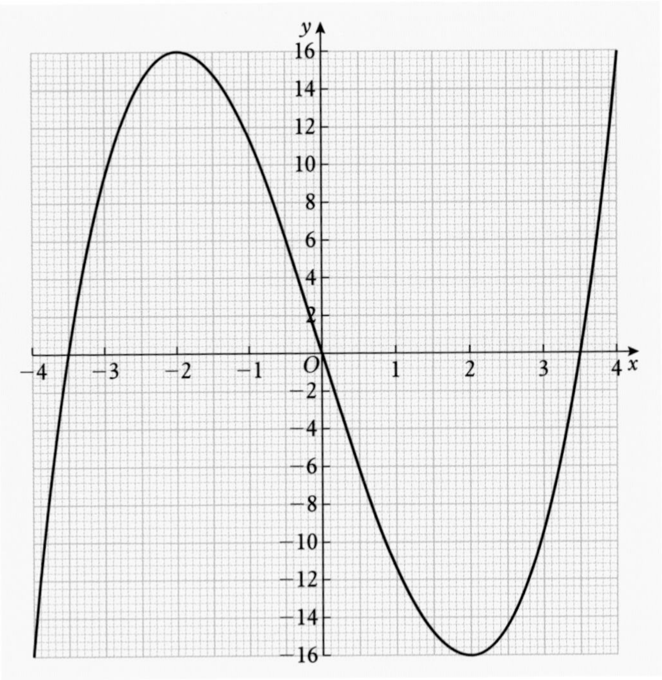

The diagram above shows the graph of

$y = x^3 - 12x$

for values of x from -4 to 4.

a Use the graph to find estimates of the 3 solutions of the equation

$x^3 - 12x = 0.$ **(1 mark)**

b By drawing a suitable straight line on the grid, find estimates of the solutions of the equation

$x^3 - 12x - 5 = 0.$

Label clearly the straight line that you have drawn. **(2 marks)**

c By drawing a suitable straight line on the grid, find estimates of the solutions of the equation

$x^3 - 14x + 5 = 0.$

Label clearly the straight line that you have drawn. **(3 marks)**

[N1999 P6 Q17]

6 This diagram shows the graph of $y = x^2 + 3 + \frac{2}{x}$

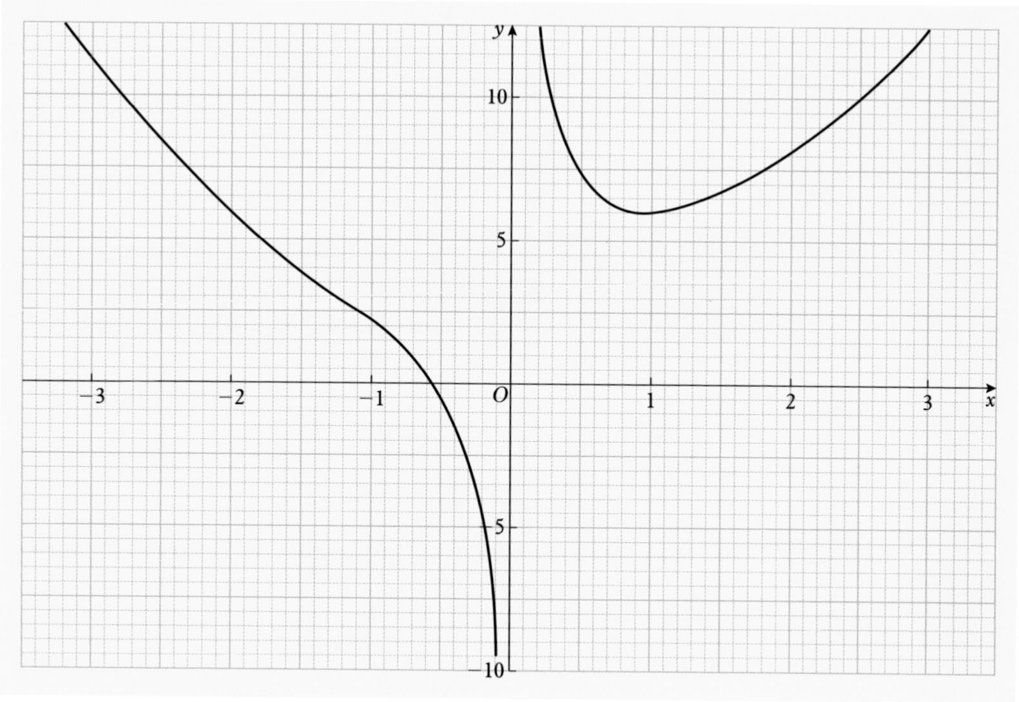

a By drawing a suitable straight line on the grid, find estimates of **all** the solutions of

$$x^2 - 5 + \frac{2}{x} = 0$$

Label your line with its equation. **(3 marks)**

b Write down the equation of the line you would need to draw to solve the equation

$$x^2 - 3x + \frac{2}{x} = 0$$

using the given graph. **(2 marks)**

[N2000 P5 Q17]

14 Probability: in theory

You need to know about:

- the probability of events happening more than once
- expected number
- relative frequency as an estimate of probability
- using two-way tables
- methods of finding probabilities
- more complicated probabilities

Probability of events happening more than once

The idea of independent events from chapter 7 can be extended to more than two events. If you roll a dice three times, the probability of getting three sixes is

$$P(6) \times P(6) \times P(6) = \frac{1}{6} \times \frac{1}{6} \times \frac{1}{6} = \frac{1}{216}.$$

Expected number

The **expected number** of successes when doing a probability experiment is the probability of a success multiplied by the number of times you do the experiment.

expected number = number of trials × probability of success

Example

A dice is rolled 40 times. Find the expected number of sixes.

$P(6) = \frac{1}{6}$, so expected number of sixes $= 40 \times \frac{1}{6} = 6\frac{2}{3}$.

So you would expect 7 sixes in 40 throws of a dice.

Frequency

The **frequency** of an event is the number of times that it happens.

Relative frequency

The **relative frequency** of an event is the number of times that it happens divided by the total number of trials.

Relative frequency $= \dfrac{\text{frequency}}{\text{total frequency}}$

The **relative frequency** of an event gives you an estimate of the probability. As you perform the experiment more times, the relative frequency gives a better estimate of the probability.

Two-way tables

A two-way table allows you to see information broken down into two categories. You will need to use different parts of the table to work out probabilities in questions.

Example

This table shows the members of a tennis club. One member of the club is chosen at random.

Membership of a tennis club

	male	female
child	23	34
adult	33	50

a Find the probability that the person chosen is

(1) a boy **(2)** an adult female **(3)** male

b Given that a male is chosen, find the probability that the person chosen is an adult.

a **(1)** There are $23 + 33 + 34 + 50 = 140$ members.

There are 23 boys. Probability $= \frac{23}{140}$

(2) There are 50 adult friends. Probability $= \frac{50}{140} = \frac{5}{14}$

(3) There are $23 + 33 = 56$ males. Probability $= \frac{56}{140} = \frac{2}{5}$

b Given that a male is chosen, there are now only 56 members to choose from.

Probability of male being an adult $= \frac{33}{56}$.

There are three methods for finding or estimating probability.

Equally likely outcomes

This is how to find the probability of getting a 6 on a dice.

Survey or experiment

This is how to estimate the probability that a piece of toast will land butter side up.

Research data

This is how to estimate the probability that it will snow at Christmas.

Only equally likely outcomes, which uses theory, gives the actual probability. The other methods use relative frequency to estimate the probability and can be improved by conducting an experiment with more trials or finding more data.

Conditional probability

When the outcome of an event affects the outcome of another event, the probability of the second event is conditional on what has already happened. This is an example of **conditional probability**.
You still multiply the probabilities for successive events but the actual values will change at each stage.

Example

A box of chocolates contains 12 cream centres and 8 hard centres.
Charlotte wants a cream centre and stops taking chocolates when she gets one, but she eats all the chocolates that she takes! She picks chocolates at random.

a Find the probability that Charlotte gets a cream chocolate as **(1)** her first chocolate **(2)** her second chocolate.

b What is the probability that Charlotte needs to take at least two chocolates before she gets a cream centre?

a **(1)** $P(\text{cream first}) = \frac{12}{20} = \frac{3}{5}$

(2) Because Charlotte needs a second chocolate it means that she gets a hard centre first. This leaves 19 chocolates for her second choice of which 12 are creams.

$P(\text{cream second})$
$= P(\text{hard centre first and cream second})$

$= \frac{8}{20} \times \frac{12}{19} = \frac{24}{95}$

b At least two chocolates are needed unless Charlotte gets a cream centre on her first choice.
So $P(\text{at least two chocolates needed})$
$= 1 - P(\text{cream first}) = 1 - \frac{3}{5} = \frac{2}{5}$

You can show conditional probabilities using a tree diagram. For this problem you don't need to carry the tree on past the second chocolate.

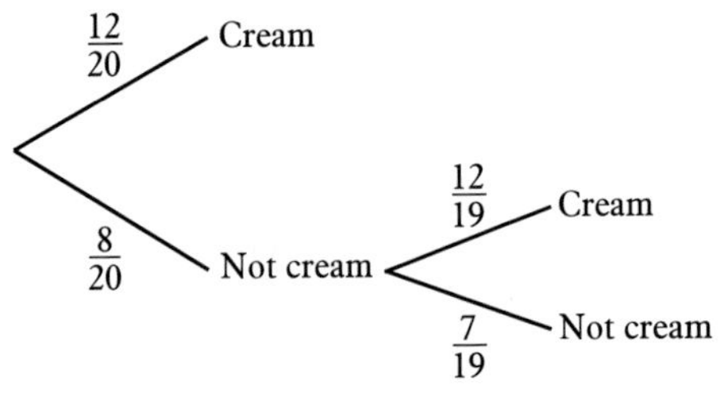

1 Parveen throws 7 dice at a fete.
If she gets 7 sixes she wins a car!
Work out the probability that Parveen wins the car.
Give your answer as a fraction.

2 Beth throws a pair of dice 100 times and records if she scores a double.

a Find the probability of scoring a double.

b Find the expected number of doubles that Beth will get.

3 An archery target has three zones, coloured black, red and yellow as shown.
Winston fires two arrows at the target.
The probabilities that he will hit the coloured regions with any arrow are:

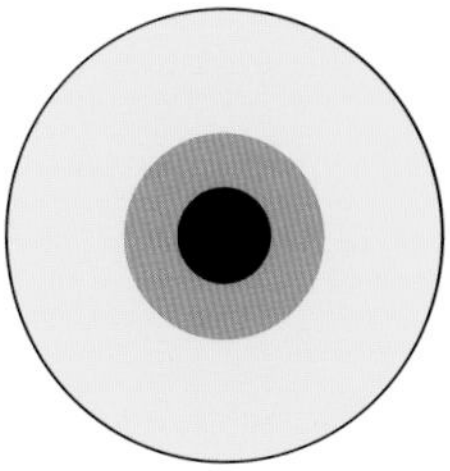

black $\frac{1}{20}$, red $\frac{7}{20}$, yellow $\frac{23}{40}$.

Find the probability that Winston

a misses the target with a single arrow

b will miss the target with both arrows

c scores two blacks

d scores one red and one black

e misses the target with one of his arrows

f hits the target with at least one arrow.

4 A fair spinner has eight equal sections labelled 1–8.
The spinner is spun 20 times and these are the results.

5 6 3 8 2 4 1 3 6 6 3 7 1 4 6 8 6 1 3 6

a Find the relative frequency of a 6 for these results.

b As the spinner is spun more times, what do you expect will happen to the relative frequency of a 6?

5 Richard has done a survey of hair colour and type for a sample of 100 pupils. These are his results.

	Blonde	Brown	Black	Red
Curly	5	5	4	1
Wavy	12	17	6	2
Straight	15	20	9	4

Richard chooses one person from his school.

a Use the sample data from the table to estimate the probability that the person chosen has hair that is

(1) straight blonde

(2) blonde

(3) black

(4) wavy.

b Given that a red-haired person is chosen, estimate the probability that the person does not have straight hair.

6 Describe how to estimate the probability that a biased spinner with 4 sections coloured red, blue, green and yellow will land on green.

7 Nadia has two stages to her journey to get home from school. For the first part, she either gets a lift or she takes a bus.
Then she completes her journey by walking or taking a different bus.

The probability that she gets a lift is 0.4.
If she gets a lift, the probability that she walks the rest of the journey is 0.7
If she has taken a bus for the first part, the probability that she walks the rest of the way home is 0.3.

a Copy and complete this tree diagram to show the information.

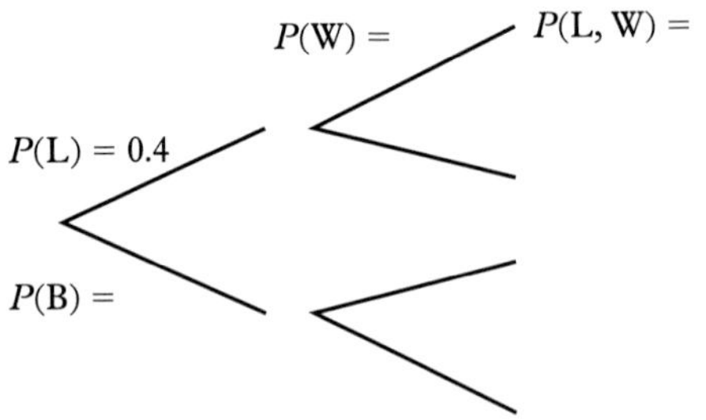

b Find the probability that Nadia

(1) takes the bus all the way home

(2) does not walk

(3) uses at least one bus.

1 Jane has six white socks and two blue socks in a drawer.
She takes one sock out, puts it back, then picks another sock.
What is the probability that she has picked

a two white socks

b one blue sock? **(5 marks)**

$P(\text{white sock}) = \frac{6}{8}$

$P(\text{blue sock}) = \frac{2}{8}$

Work out the probabilities first.
The sock is replaced, so the probability doesn't change for the second sock.

a $P(\text{two white socks}) = \frac{6}{8} \times \frac{6}{8} = \frac{3}{4} \times \frac{3}{4} = \frac{9}{16}$

1 mark **1 mark**

b P(one blue sock) means that Jane picked **either** one blue and one white sock or one white and one blue sock.

$\frac{2}{8} \times \frac{6}{8} + \frac{6}{8} \times \frac{2}{8}$

1 mark: 1st multiplication
1 mark: 2nd multiplication added

$= \frac{1}{4} \times \frac{3}{4} + \frac{3}{4} \times \frac{1}{4}$

This could be a non-calculator question.

$= \frac{3}{16} + \frac{3}{16}$

$= \frac{6}{10}$

$= \frac{3}{8}$

1 mark: answer

Always cancel the final answer after showing all the working.
Probabilities are always written as percentages, fractions or decimals.

2 A coin is spun three times.

a Calculate the probability of getting

(1) three heads

(2) only one head. **(5 marks)**

The coin is spun 200 times.

b Estimate the number of times you would expect to get

(1) two heads

(2) three heads. **(4 marks)**

a **(1)** $P(\text{3 heads}) = \frac{1}{2} \times \frac{1}{2} \times \frac{1}{2} = \frac{1}{8}$ **1 mark: $\frac{1}{2}$s**
1 mark: answer

(2) P(1 head) means **either** HTT or THT or TTH

Identify all possible permutations (orders).

$$= \frac{1}{2} \times \frac{1}{2} \times \frac{1}{2} + \frac{1}{2} \times \frac{1}{2} \times \frac{1}{2} + \frac{1}{2} \times \frac{1}{2} \times \frac{1}{2}$$

$$= 3 \times \left(\frac{1}{2} \times \frac{1}{2} \times \frac{1}{2}\right)$$ **1 mark: $\frac{1}{2}$s**

$$= \frac{3}{8}$$ **1 mark: ×3**
1 mark: answer

b **(1)** $P(\text{two heads}) = \frac{1}{2} \times \frac{1}{2} = \frac{1}{4}$

So the **number** of times

$= \frac{1}{4} \times 200 = 50$ ← *This answer is a number, not a probability.*

1 mark **1 mark**

(2) $P(\text{three heads}) = \frac{1}{2} \times \frac{1}{2} \times \frac{1}{2} = \frac{1}{8}$

So the **number** of times

$= \frac{1}{8} \times 200 = 25$

1 mark **1 mark**

3 There are 40 apples in a box.
An apple is picked from the box and then replaced.
A second apple is then picked from the box.
The probability that they are both red apples is $\frac{9}{64}$.
How many red apples must there be in the box? **(3 marks)**

The probability of picking two red apples is $\frac{9}{64}$.
If the probability of picking one red apple is p

Then $p \times p = \frac{9}{64}$

So $p^2 = \frac{9}{64}$

$p = \sqrt{\frac{9}{64}}$

So the probability of picking one red apple is $\sqrt{\frac{9}{64}} = \frac{3}{8}$. **1 mark**

So there are $\frac{3}{8} \times 40$ apples $= 15$ red apples in the box.

1 mark **1 mark**

1 The diagram shows the board for a game played in a maths lesson.

FINISH	−3	−2	−1	0	+1	+2	+3	FINISH

Rules of the Game

Start with a counter on **0**.
Throw a coin.
If the coin shows heads, then move the counter one space to the right.
If the coin shows tails, then move the counter one space to the left.
The game ends when the counter reaches **'Finish'**.

Jim places a counter at **0**.
He throws a fair coin 3 times.

a Calculate the probability that the counter will be at
(1) +3, **(2)** 0. **(3 marks)**

Ann places a counter at **0**.
She has a biased coin. The probability that the coin shows heads is 0.7.
Ann throws the biased coin twice.

b Calculate the probability that her counter will be back at 0. **(3 marks)**

Billy places a counter at **0**.
He has a biased coin. The probability that the biased coin shows heads on any throw is 0.7.
Billy throws the biased coin 3 times.

c Calculate the probability that after 3 throws
(1) the counter will be at **+1**,
(2) the counter will have been at **+1** at least once. **(5 marks)**

[N1999 P6 Q10]

2 In a bag there are 10 counters.
4 of the counters are red and 6 of the counters are blue.
Ann and Betty are going to play a game.
Ann is going to remove 2 counters at random from the bag. She will not put them back.

If both counters are the same colour, Ann will win the game.

a Calculate the probability that Ann will win the game. **(3 marks)**

If the counters are different colours, it will be Betty's turn.
Betty will remove **one** counter at random from the 8 counters still in the bag. If the counter is red, Betty will win the game.
If the counter is blue, the result will be a draw.

b Calculate the probability that the result will be a draw. **(5 marks)**

[N2000 P5 Q18]

3 Robin has 20 socks in a drawer.
Twelve of the socks are red.
Six of the socks are blue.
Two of the socks are white.
He picks two socks at random from the drawer.
Calculate the probability that he chooses two socks of the same colour.
(4 marks)
[S2000 P5 Q13]

4 There are 25 beads in a bag.
Some of the beads are red.
All the other beads are blue.
Kate picks two beads at random without replacement.
The probability that she will pick 2 red beads is 0.07.
Calculate the probability that the two beads she picks will be of different colours.
(6 marks)
[N1998 P5 Q22]

5

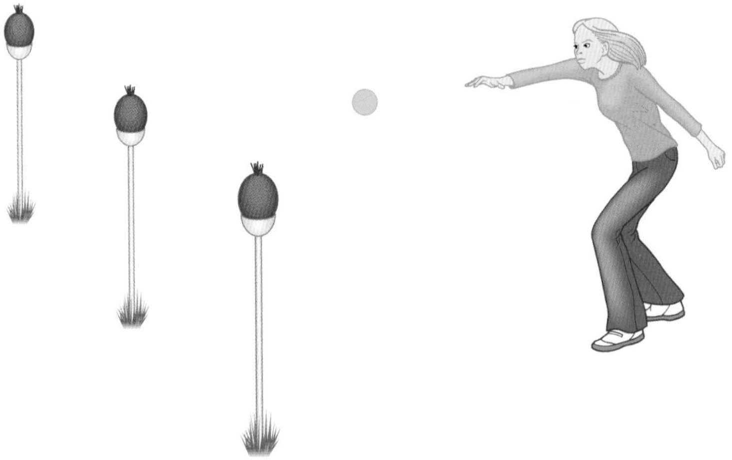

Helen tries to win a coconut at the fair.
She throws a ball at a coconut.
If she knocks a coconut off its stand, she wins the coconut.
Helen has two throws.
The probability that she will win a coconut with her first throw is 0.2.
The probability that she will win a coconut with her second throw is 0.3.

Work out the probability that, with her two throws, Helen will win
(1) 2 coconuts, **(2)** exactly 1 coconut.
(5 marks)
[N1998 P5 Q12]

6 The probability that a team will win a game is always 0.8.
The team plays n games.
The probability that the team will win every game is less than $\frac{1}{4}$.
Calculate the smallest possible value of n. **(3 marks)**
[S1998 P5 Q21]

7

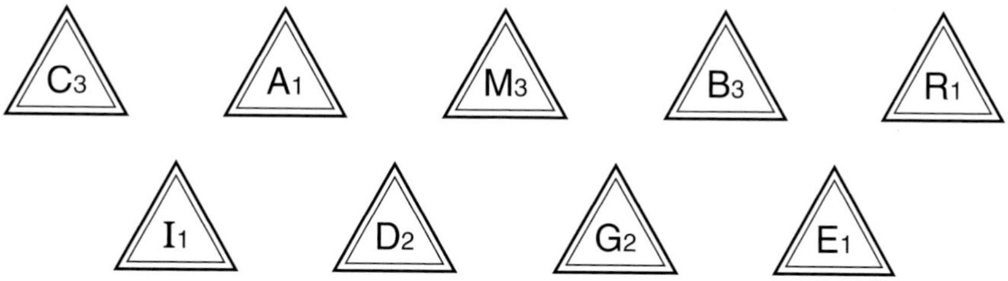

The diagram shows nine tiles from a word game.
On each tile is a letter and a score.

For example, this tile has a letter C and a score of 3.
Caroline puts all nine tiles in a bag.
She picks at random two tiles without replacement.
Calculate the probability that

(1) the two tiles will have equal scores,
(2) the scores on the two tiles will have a sum of 5 or more. **(6 marks)**

[S2001 P6 Q21]

8 The probability of a person having brown eyes is $\frac{1}{4}$.
The probability of a person having blue eyes is $\frac{1}{3}$.
Two people are chosen at random.
Work out the probability that

(1) both people will have brown eyes
(2) one person will have blue eyes and the other person will have brown eyes. **(5 marks)**

[N1998 P6 Q10]

9 Darren estimates the probability of his passing a test at the first attempt as 0.6.
If he fails he can take the test again.
He estimates the probability of passing the test on the second attempt as 0.7.

a Using a tree diagram, or otherwise, calculate the probability that he will pass

(1) at the second attempt,

(2) at the first or second attempt. **(4 marks)**

Darren estimates the probability of his girlfriend passing the test at the first attempt as 0.8.

b Calculate the probability of them **both** passing the test, at the first attempt. **(2 marks)**

[N1996 P6]

10 A fair coin is to be tossed three times.
Find the probability of throwing two tails and a head, in any order.
(2 marks)

[N1996 P6]

11 Peter and Asif are both taking their driving test for a motor cycle for the first time.
The table below gives the probabilities that they will pass the test at the first attempt or, if they fail the first time, the probability that they will pass at the next attempt.

	Probability of passing at first attempt	Probability of passing at next attempt if they fail the first attempt
Peter	0.6	0.8
Asif	0.7	0.7

On a particular day 1000 people will take the test for the first time.
For each person the probability that they will pass the text at the first attempt is the same as the probability that Asif will pass the test at the first attempt.

a Work out an estimate for how many of these 1000 people are likely to pass the test at the first attempt. **(2 marks)**

b Calculate the probability that both Peter and Asif will pass the test at the first attempt. **(2 marks)**

c Calculate the probability that Peter will pass the test at the first attempt and Asif will fail the test at the first attempt. **(2 marks)**

d Calculate the probability that Asif will pass the test within the first two attempts. **(3 marks)**

[S1996 P5]

15 Statistics: about average

You need to know about:

- the mean, the median and the mode
- finding averages from frequency tables
- finding averages from grouped frequency tables
- sampling
- moving averages and trends
- cumulative frequency graphs
- box and whisker plots

Mode

The **mode** is the most common or most popular data value. The mode is useful when one value occurs much more often than any other.
In a frequency table, the modal value is the value with the highest frequency.
In a grouped frequency table you can only give the modal group, which is the group with the highest frequency.

Median

The **median** is the middle data value when the data is in order.
For n values, the median is the $\frac{n+1}{2}$th data value.
When there are extreme values in the data, the median is unaffected by these values. In a **frequency table** you need to decide where the median value is by counting up the frequency column until you come to the median position and give the data value for this position. In a **grouped frequency table**, you can decide which group contains the median. Count the frequencies until you find the group in which the median value lies.

Mean

The **mean** is the total of the data values divided by the number of data values. The mean is the only average which uses all of the data values. Extreme values can have too great an effect upon the mean.
For a **frequency table** with data values, x, and corresponding frequencies, f, the mean of a total of n data values is $\bar{x} = \frac{\Sigma xf}{n}$.
Multiply each value by its frequency, add these values together and divide by the total number of values.
For a **grouped frequency** table use the **mid-point of each group** as the x values and use the same formula. In this case, the method only gives you an **estimate** of the mean as you cannot use the exact data values.

Sampling

A **random sample** is one chosen so that every member of the population has an equal chance of being selected.
A **stratified random sample** is one chosen so that the sample contains the same proportion of people of each type as the population.

Moving averages

A **moving average** uses a few data values at a time in order to establish a trend in the data. For a 4-point moving average use the first 4 values and work out their mean. Then remove the first value and use the 5th value to give the next mean and so on until you reach the last 4 values. By plotting the moving average on a graph, a line of best fit can be drawn through these values to show the trend of the data. This line is called a trend line.

Example

This is a diagram showing company profits in each quarter of the year for 3 years. Notice that the red cross, which is the first 4-point moving average, is plotted in the middle of the 4 values that were used, so it is plotted between the second and third quarters. The blue cross is the mean of quarters 2, 3, 4 and 5, and is plotted between quarters 3 and 4. The trend line shows that profits are slowly increasing.

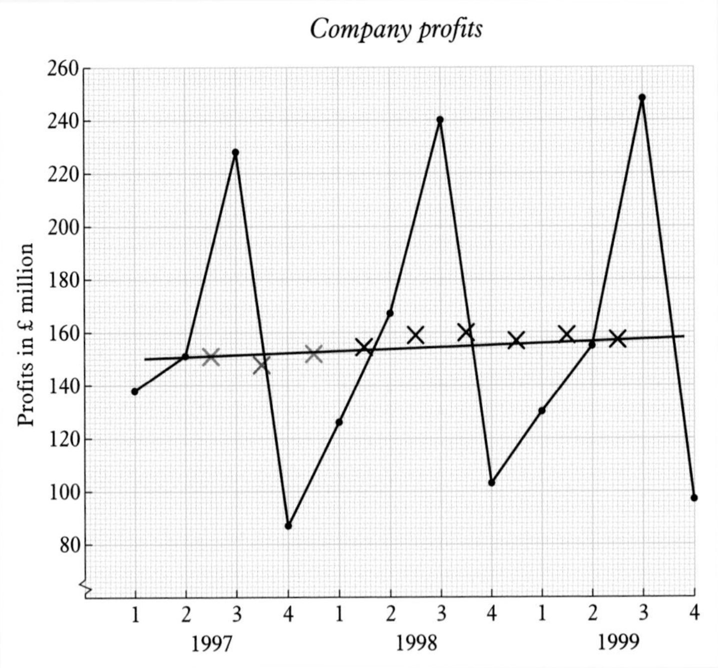

Range

The **range** is the largest data value minus the smallest data value.

Cumulative frequency

Cumulative frequency is a running total. A cumulative frequency diagram is a plot of the cumulative frequencies against the upper end of each group.

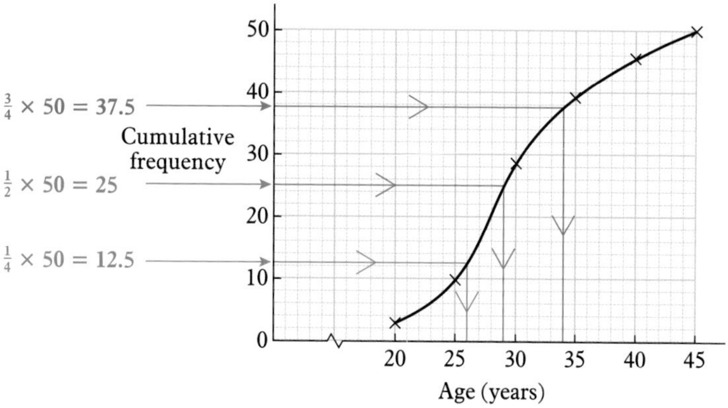

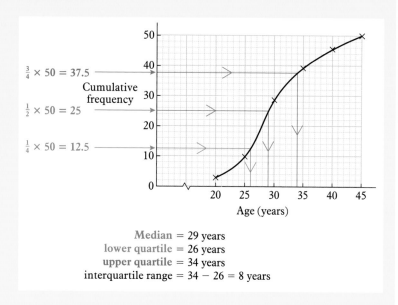

Median = 29 years
lower quartile = 26 years
upper quartile = 34 years
interquartile range = 34 − 26 = 8 years

Box and whisker plots

A **box and whisker plot** shows the distribution of a set of data. They can be used to compare two sets of data by drawing one diagram above the other using the same scale.

Box and whisker plot showing train journey times in minutes

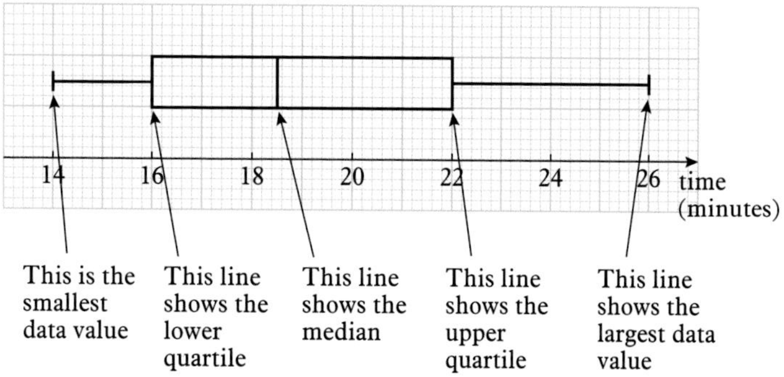

1 These are the attendances at a theme park for 10 days during the holidays.

5445	5506	4992	5523	5434
5656	5267	5422	956	4855

a Work out the mean attendance.
b Work out the median attendance.
c Which gives the better average?
Explain your answer.

2 The mean of the numbers x, $(x + 5)$ and $(x + 4)$ is 12.
Find the value of x.

3 Miss Anderton's class has 33 pupils.
When the class takes a test Bushra is absent.
The mean score of the remaining pupils is 68.5.
When Bushra takes the test, Miss Anderton decides not to tell Bushra her result but tells her that the mean for the whole class is now exactly 69.
How many marks did Bushra get?

4 **a** Write down the 3 consecutive whole numbers that have a median of n, where n is a whole number.
b Show that the mean of the 3 numbers in **a** is also n.
c Write down and simplify the squares of the 3 numbers in **a**.
d Write down the median of the 3 numbers in **c**.
e Find the mean of the 3 numbers in **c**.

5 The head teacher of a school claims that the GCSE results in her school are getting better. She has the data for the last 12 years of the percentage of pupils who achieved 5 or more GCSE A*–C grades.

Year	1990	1991	1992	1993	1994	1995	1996	1997	1998	1999	2000	2001
%	33.5	34.5	34.1	24.6	35.6	42.6	46.5	35.7	46.8	47.4	32.1	35.1

a Plot this data on a graph.
b Work out a 5-point moving average for the data.
c Plot these values on your graph.
d Draw a trend line on your graph and explain if the head can justify her claim.

6 The times taken for two taxi firms to get to 20 customers is shown below. The times are in minutes, correct to the nearest minute.

Tel's Taxis

12	13	11	9	12	10	4	2	14	13
10	22	12	11	14	10	8	9	20	11

Speedy Steve's

1	4	24	26	11	14	6	20	33	1
13	5	2	1	18	15	22	23	2	2

a For each company work out
(1) the median
(2) the range
(3) the lower and upper quartiles of the times taken.
b Draw a box and whisker plot for the two companies on the same diagram.
c Write about the differences between the two companies.
d Which company would you call if you wanted a taxi? Explain your answer.

7 The table shows the amount that 100 families spend each week on shopping.

Amount, A(£)	Number of families
$0 < A \leqslant 20$	4
$20 < A \leqslant 40$	14
$40 < A \leqslant 60$	41
$60 < A \leqslant 100$	35
$100 < A \leqslant 200$	6

a Write down the modal group.
b Work out an estimate for the mean amount spent per family.
c Explain why your answer in **b** is only an estimate.
d What do you notice about the size of the mean compared with the modal group? Using the data in the table, explain your answer.
e Draw a cumulative frequency graph to show the spending.
f Find the median amount spent.
g Describe how the median amount compares with the mean.
h Find the interquartile range.

1 The captain of a passenger ferry noted down the number of passengers carried on each journey.

103 124 118 140 125 124 114 119 101 139
112 136 119 124 118 131 104 148 103 122

Draw a stem and leaf diagram to show these figures. **(3 marks)**

A stem and leaf diagram has two parts: a stem and a leaf.

The stem can be the first two digits: 10, 11, 12, 13 and 14.

The leaf part can be the 3rd digit of each number.

Put the numbers into a diagram:

Stem	Leaf
10	3 1 4 3
11	8 4 9 2 9 8
12	4 5 4 4 2
13	9 6 1
14	0 8

Make sure the columns are labelled.

1 mark

Now you should put all the numbers in numerical order:

Stem	Leaf
10	1 3 3 4
11	2 4 8 8 9 9
12	2 4 4 4 5
13	1 6 9
14	0 8

Spread the numbers out in clear columns.

Make sure you include a key, even if it is not asked for!

1 mark

Key 10 | 1 means 101 passengers.

1 mark: key

2 The table below shows the heights of tomato plants in a garden.

Height in cm, h	$0 < h \leqslant 5$	$5 < h \leqslant 10$	$10 < h \leqslant 15$	$15 < h \leqslant 20$	$20 < h \leqslant 25$	$25 < h \leqslant 30$
Frequency	1	4	10	19	12	4

a Draw a cumulative frequency table. **(2 marks)**

b Draw a cumulative frequency diagram. **(2 marks)**

c Calculate the interquartile range. **(2 marks)**

a Draw a cumulative frequency table: *Cumulative class intervals.*

Height in cm, h	$h \leqslant 5$	$h \leqslant 10$	$h \leqslant 15$	$h \leqslant 20$	$h \leqslant 25$	$h \leqslant 30$
Frequency	1	5	15	34	46	50

1 mark for class intervals, 1 mark for cumulative frequencies

b *Draw cumulative frequency diagram:*

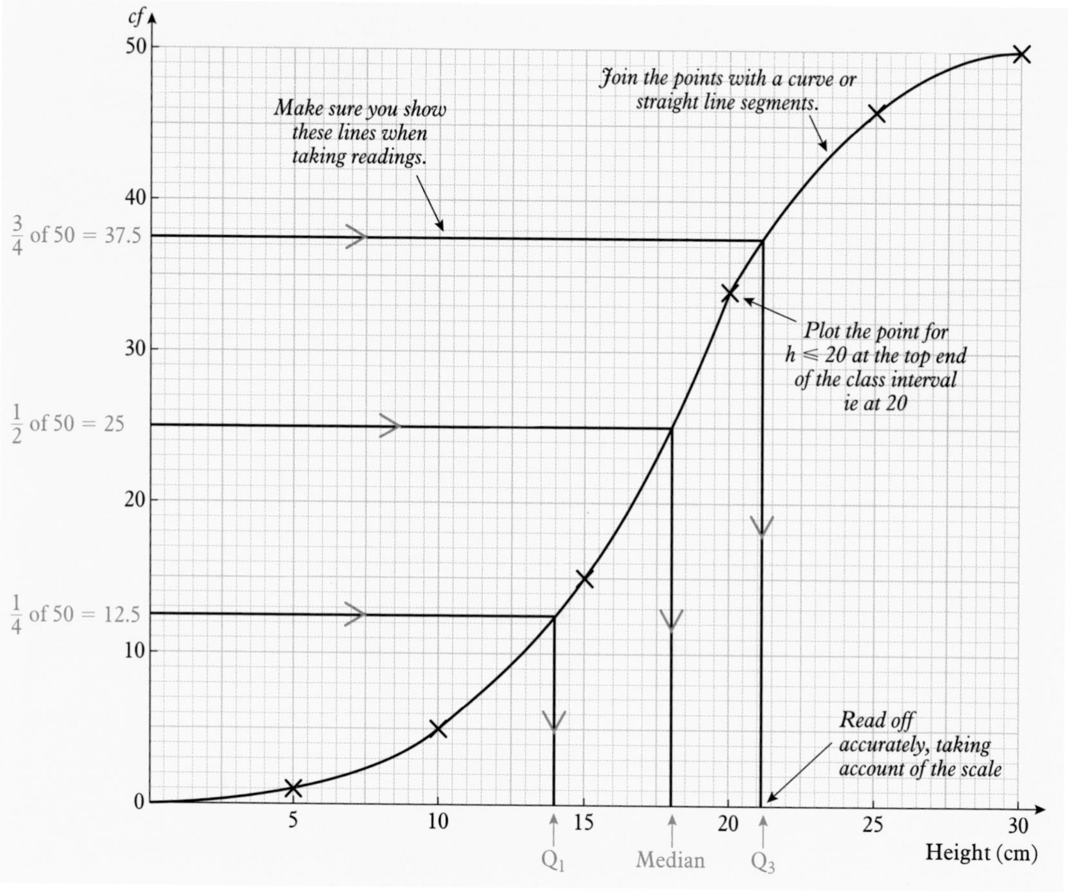

1 mark for points, 1 mark for curve or polygon

c Interquartile range = $Q_3 - Q_1$
where Q_3 is the upper quartile, and Q_1 is the lower quartile.
From your graph allow Q_3 from 20.5 to 21.5
Q_1 from 13.5 to 14.5
Interquartile range = 21.2 − 14 = 7.2

Show these figures as working out.

1 mark for Q_1, Q_3 values or lines on graph
1 mark for answer, follow through from your Q_1 and Q_3

1 75 boys took part in a darts competition. Each boy threw darts until he hit the centre of the dartboard.
The numbers of darts thrown by the boys are grouped in this frequency table.

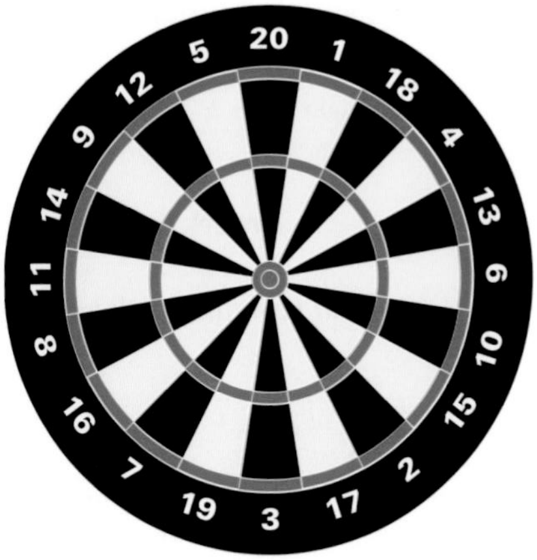

Number of darts thrown	Frequency
1 to 5	10
6 to 10	17
11 to 15	12
16 to 20	4
21 to 25	12
26 to 30	20

a Work out the class interval which contains the median. **(2 marks)**

b Work out an estimate for the mean number of darts thrown by each boy. **(4 marks)**

[N2000 P6 Q6]

2 A class took a test. The mean mark for the 20 boys in the class was 17.4.
The mean mark for the 10 girls in the class was 13.8.

a Calculate the mean mark for the whole class. **(2 marks)**

5 pupils in another class took the test.
Their marks, written in order, were 1, 2, 3, 4 and x.
The mean of these 5 marks is equal to twice the median of these 5 marks.

b Calculate the value of x. **(3 marks)**

[S1998 P6 Q5]

3 A shop employs 8 men and 2 women.
The mean weekly wage of the 10 employees is £396.
The mean weekly wage of the 8 men is £400.
Calculate the mean weekly wage of the 2 women. **(4 marks)**

[N2000 P5 Q4]

4 Alan is doing a survey of the heights of boys and girls in Year 7. He first takes a random sample of 70 boys from Year 7.

a Suggest a suitable method that Alan could use to take a random sample. **(2 marks)**

The table and the incomplete histogram show information about the boys' heights in this sample of 70 boys.

Height of boys, h centimetres	Frequency
$140 \leq h < 145$	10
$145 \leq h < 148$	15
$148 \leq h < 150$	20
$150 \leq h < 154$	16
$154 \leq h < 157$	9

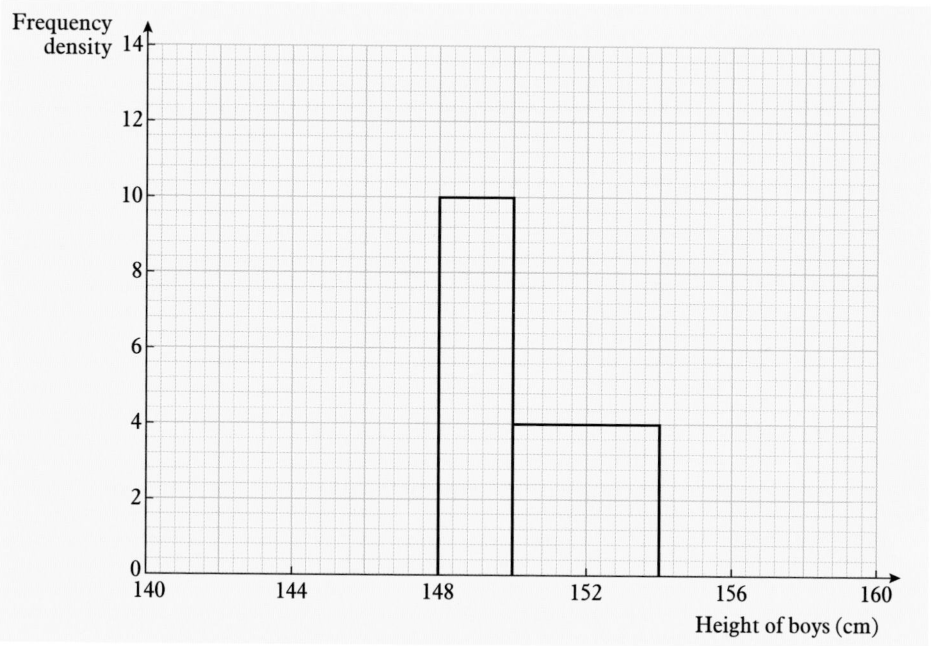

b Use the information in the table to complete the histogram. **(3 marks)**

5 The table shows the number of students in each year group at Mathstown High School.

Ben is carrying out a survey about the students' favourite television programmes.

He uses a stratified sample of 50 students according to year group.

Year group	Number of students
9	300
10	290
11	340
12	210
13	180

Calculate the number of year 12 students that should be in his sample. **(3 marks)**

[N1998 P5 Q15]

6 The table gives information about the ages, in years, of 100 aeroplanes.

Age (t years)	Frequency
$0 < t \leqslant 5$	41
$5 < t \leqslant 10$	26
$10 < t \leqslant 15$	20
$15 < t \leqslant 20$	10
$20 < t \leqslant 25$	3

a Work out an estimate of the mean age of the aeroplanes. **(4 marks)**

b Complete the cumulative frequency table.

Age (t years)	Cumulative frequency
$t \leqslant 5$	
$t \leqslant 10$	
$t \leqslant 15$	
$t \leqslant 20$	
$t \leqslant 25$	

(1 mark)

c On the grid, draw a cumulative frequency graph for your table.

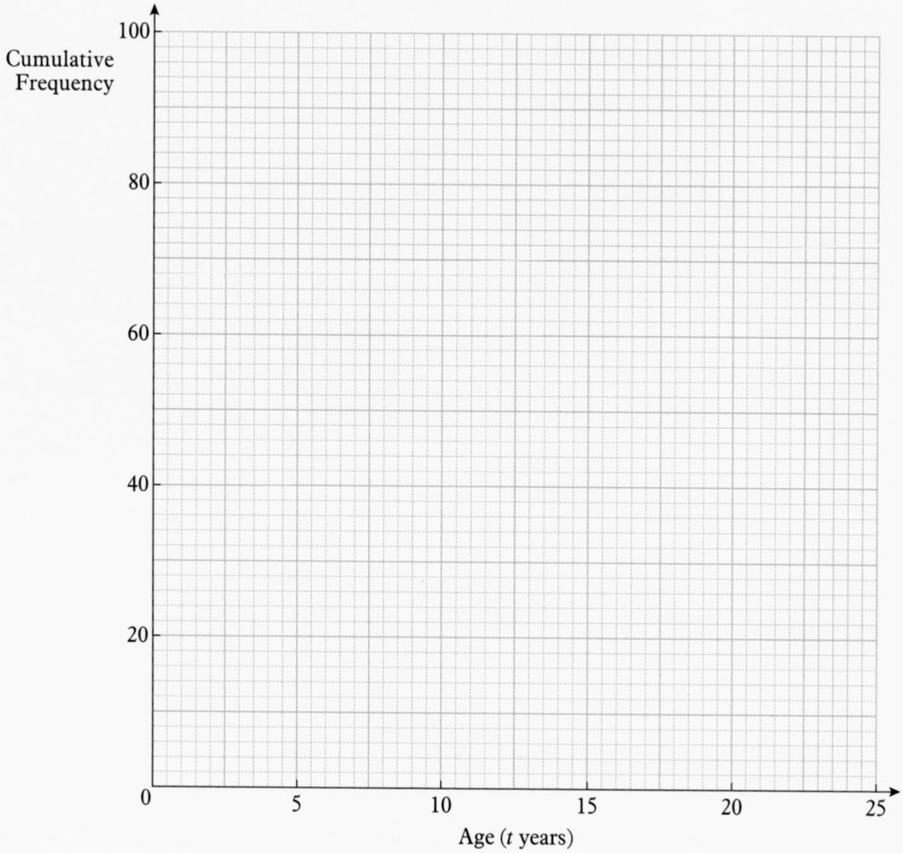

d Use your graph to find an estimate of the upper quartile of the ages.
Show your method clearly. **(2 marks)**

[N1998 P5 Q9]

7 The mean of the five consecutive integers 1, 2, 3, 4, 5 is 3.
So the square of the mean of consecutive integers 1, 2, 3, 4, 5 is 9.
The mean of the squares of those five consecutive integers 1^2, 2^2, 3^2, 4^2, 5^2 is 11.

Show algebraically that the square of the mean of **any** five consecutive integers is **always** 2 less than the mean of the squares of those five consecutive integers. **(6 marks)**

[N2000 P6 Q21]

16 Shape: the final frontier

You need to know about:

- circle formulas
- sectors and segments
- the area of a triangle
- prism formulas
- pyramid formulas
- cone formulas
- sphere formulas
- area and volume unit conversion
- areas and volumes of similar figures
- dimensional analysis

Circles

You need to know the formulas for circles.

Circumference

Circumference = $\pi \times$ diameter

$$C = \pi d$$

Arc length

A length around part of the circumference of a circle is called an **arc length**.
If the angle in a sector of a circle is θ then the **arc length,** s, is given by

$$s = \frac{\theta}{360} \times \pi d.$$

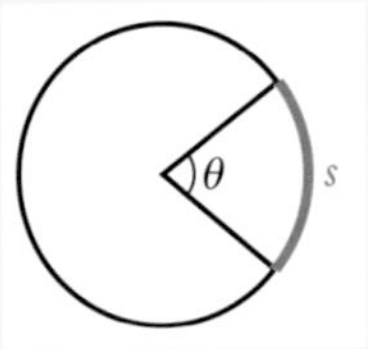

Area

Area of circle = $\pi \times$ radius^2

$$A = \pi r^2$$

Area of a sector

A sector is a region of a circle bounded by two radii and an arc.
If the angle in a sector of a circle is θ then the **area of the sector,**

A, is given by $A = \frac{\theta}{360} \times \pi r^2$.

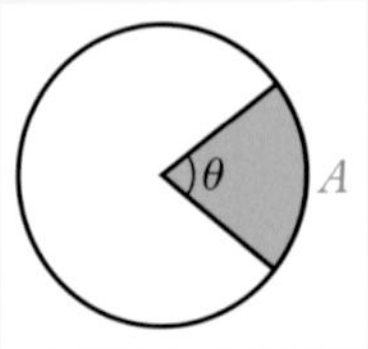

Area of a triangle

The area of this triangle is $\frac{1}{2}ab \sin C$

The angle used in this formula must be between the two sides used.

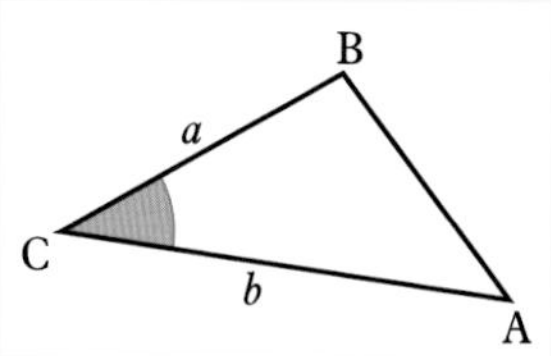

Area of a segment

The **area of a segment** contained in a sector of a circle of radius r with angle θ is

$$\frac{\theta}{360} \times \pi r^2 - \frac{1}{2}r^2 \sin \theta$$

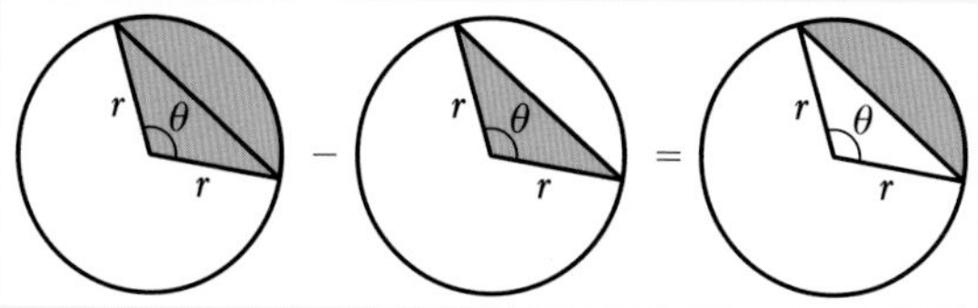

Volume of a prism

A solid shape with a constant cross section is a prism.
Volume of a prism = area of cross section × length

Surface area of a prism

To find the **surface area of a prism** work out the area of each face and add these areas together. There is no single formula to learn.

Cylinder

A **cylinder** is like a prism but its cross section is a circle.

Volume, $V = \pi r^2 h$

Curved surface area $= 2\pi rh$ or πdh

Total surface area $= 2\pi rh + \pi r^2 + \pi r^2$
$= 2\pi rh + 2\pi r^2$

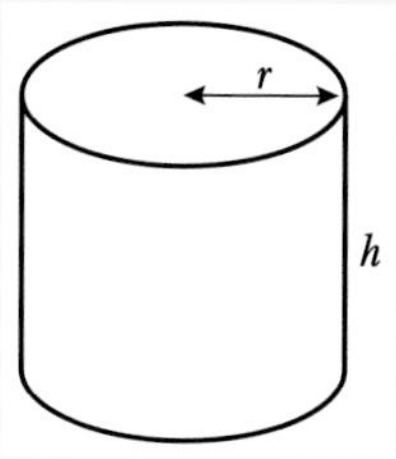

Pyramid

Volume, $V = \frac{1}{3} \times$ base area $\times$ height

Cone

Volume, $V = \frac{1}{3} \times$ base area $\times$ height
$= \frac{1}{3}\pi r^2 h$

Curved surface area $= \pi rl$

Total surface area $= \pi rl + \pi r^2$

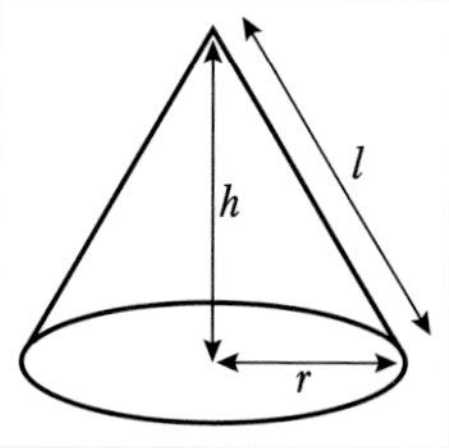

Sphere

Volume, $V = \frac{4}{3}\pi r^3$

Surface area $= 4\pi r^2$

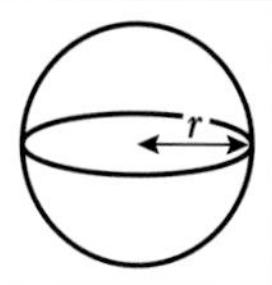

Area and volume units

1 cm = 10 mm $1\ \text{cm}^2 = 10^2 = 100\ \text{mm}^2$
$1\ \text{cm}^3 = 10^3 = 1000\ \text{mm}^3$
1 m = 100 cm $1\ \text{m}^2 = 100^2 = 10\,000\ \text{cm}^2$
$1\ \text{m}^3 = 100^3 = 1\,000\,000\ \text{cm}^3$

For the **area** conversion factor, **square** the linear conversion factor.
For the **volume** conversion factor, **cube** the linear conversion factor.

Area and volume of similar figures

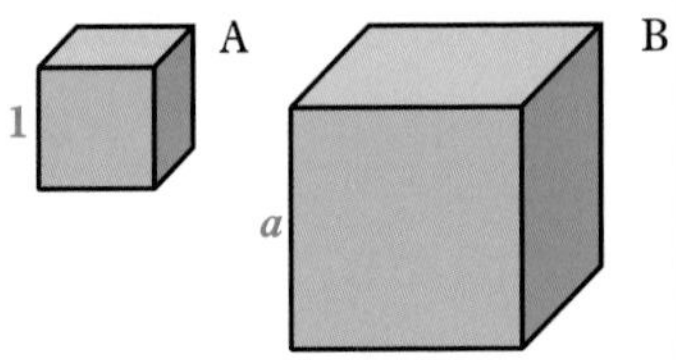

A and B are similar shapes.
If the **linear** scale factor from shape A to shape B is a,
the **area** scale factor from shape A to shape B is $a \times a = a^2$
and the **volume** scale factor from shape A to shape B is $a \times a \times a = a^3$.

Remember that mass depends on volume, so if you are told the mass of one similar shape and asked for the mass of the other you need to multiply by the volume conversion factor.

The amount of paint needed to cover a solid depends on its surface area, so if you are told the amount of paint needed to paint one similar shape and asked for the amount needed for the other you need to multiply by the area conversion factor.

This diagram summarises how to get from one scale factor to another. You always find the linear scale factor first if you're not given it.

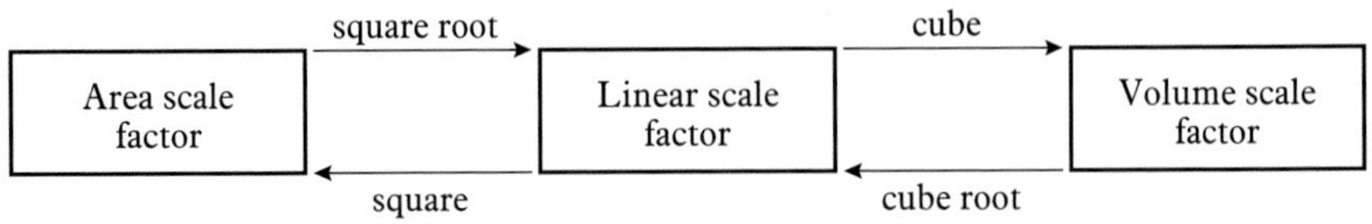

Density

$$\textbf{Density} = \frac{\text{mass}}{\text{volume}}$$

Dimension

The **dimension** of a formula is the number of lengths that are multiplied together.

Constant

A **constant** has no dimension. It is just a number.
Some formulas have more than one part.
When this happens, all of the parts must have the same dimension if the formula is for length (dimension 1) or area (dimension 2) or volume (dimension 3).

1 In each of these, find:

(1) the length of the red arc

(2) the area of the blue sector.

a

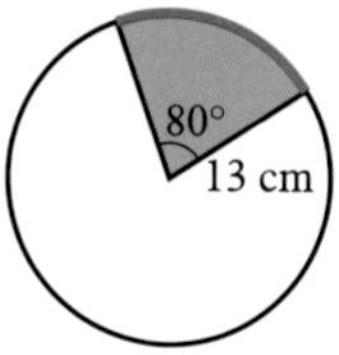

b

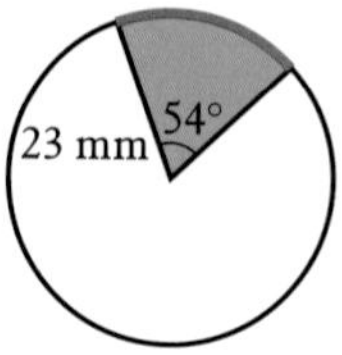

c

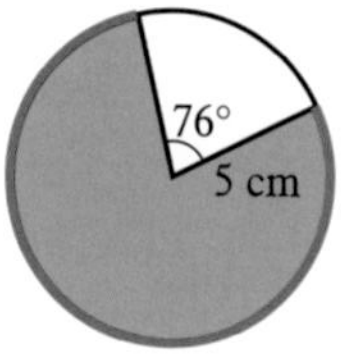

2 Find the area of the blue segment.
Give your answer to 3 sf.

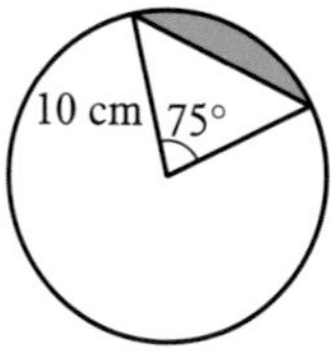

3 Find the area of the green segment.
Give your answer in terms of π.

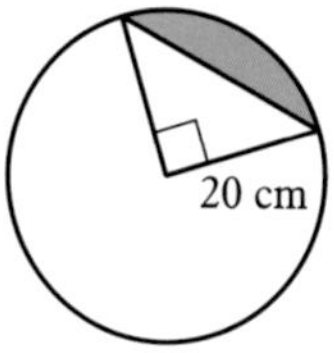

4 Find the area of this triangle.

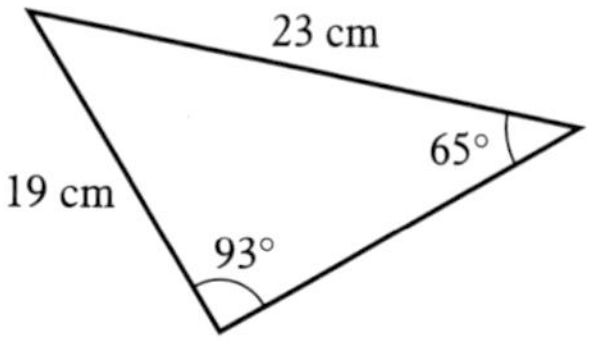

5 For each of these shapes, find in terms of π:

(1) the surface area

(2) the volume.

a

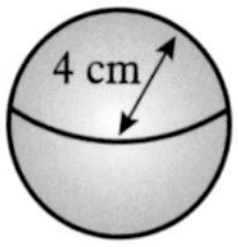

sphere

b

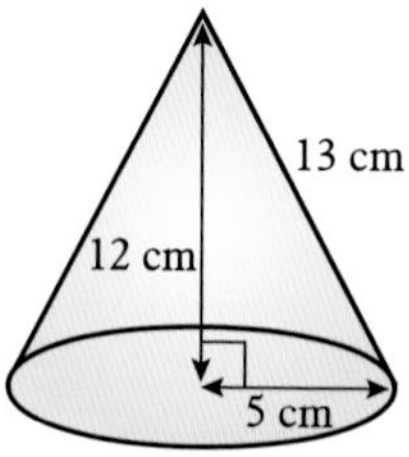

cone

c

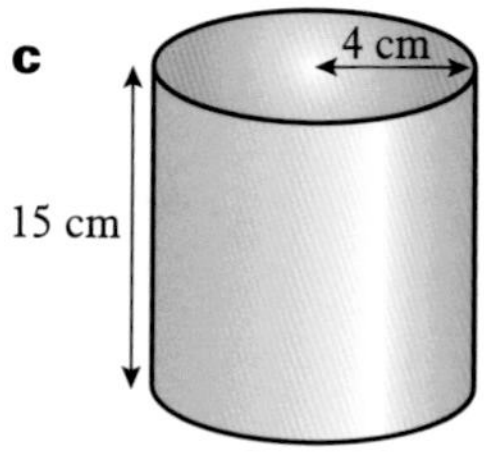

cylinder

6 The diagram shows the entrance to a road tunnel
The entrance is rectangular with a semi-circular arch on top.

a Find the cross-sectional area of the tunnel.

The tunnel is 140 m long.

b Find the volume, in m^3, of earth that had to be removed to form the tunnel.

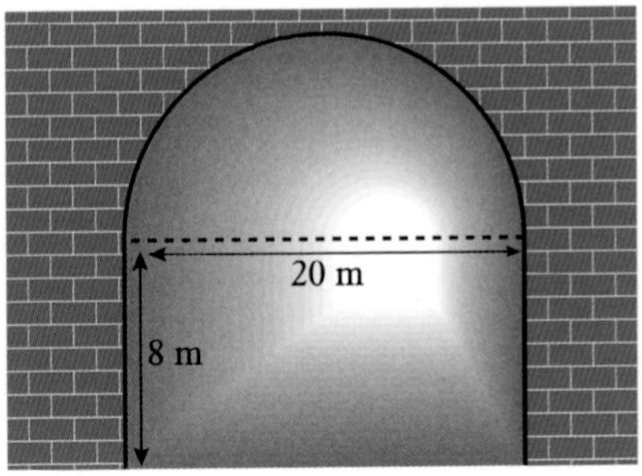

7 These two jugs are similar.
The capacity of the small jug is 0.5*l* and of the large jug is 4*l*.
The base area of the smaller jug is 46 cm^2.

a Find the base area of the larger jug in cm^2.

b Find the base area of the larger jug in mm^2.

8 This lampshade is a frustrum of a cone.
Its height is 40 cm.
The shade is covered in material.
0.6 m^2 of material is needed to cover this shade.
A similar shade is made with height 10 cm.
What area of material will be needed to cover it?
Give your answer in cm^2.

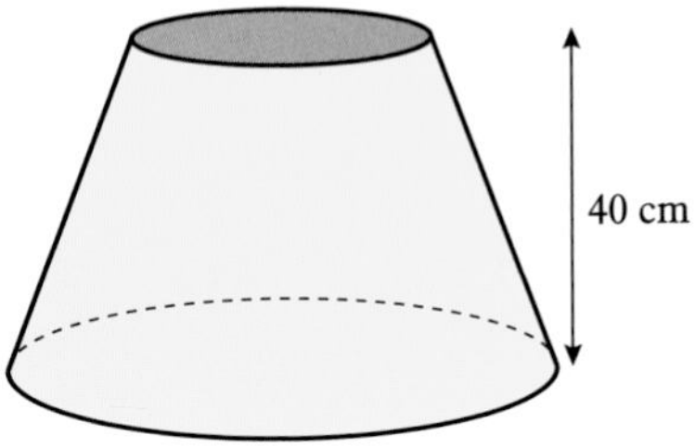

9 In this question, x, y and z are lengths, and a, c and k are constants.
Write down what each of these formulas could represent.
Choose from *length*, *area* or *volume*.

a kx^2z

b $\frac{ax^2y}{kz}$

c $kxz + axz$

d $\frac{kx^2y}{a} + \frac{az^3}{k}$

e $ckx^2y + \frac{12\pi x^3y}{z}$

f $\frac{23\pi z^3}{xy}(xy + z^2)$

1 Two statues are similar and are made from the same material.
The smaller statue has a mass of 8 kg. The larger one has a mass of 125 kg.
The smaller statue has a surface area of 12 cm^2.

a Calculate the surface area of the larger statue in cm^2. **(3 marks)**

b Write your answer in mm^2. **(2 marks)**

a The mass is related to the volume scale factor.
Length scale factor = $\sqrt[3]{\text{Volume scale factor}}$

$$\text{So } \frac{\text{Length of larger}}{\text{Length of smaller}} = \sqrt[3]{\frac{\text{Volume of larger}}{\text{Volume of smaller}}}$$

Write down the relationship.

$$= \sqrt[3]{\frac{\text{mass of larger}}{\text{mass of smaller}}}$$

$$= \sqrt[3]{\frac{125}{8}}$$

Show the figures ***before*** *you do the calculation.*

$$= \frac{\sqrt[3]{125}}{\sqrt[3]{8}} = \frac{5}{2}$$ **1 mark**

Now, Area scale factor = (length scale factor)^2

$$\text{So } \frac{\text{Area of larger}}{\text{Area of smaller}} = \frac{5^2}{2^2} = \frac{25}{4}$$ **1 mark**

So the surface area of the larger statue is $\frac{25}{4} \times 12 = 75$ cm^2. **1 mark**

b 10 mm = 1 cm **1 mark: clear explanation shown**
10^2 mm^2 = 1^2 cm^2
100 mm^2 = 1 cm^2
So 75 cm^2 becomes $75 \times 100 = 7500$ mm^2. **1 mark:** follow through from your answer to **a**

2

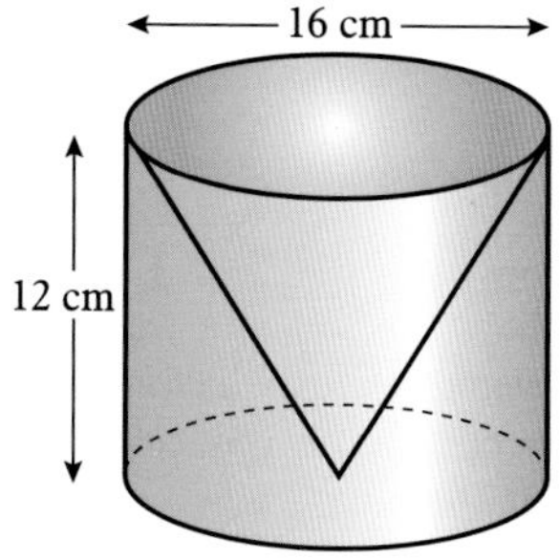

A symmetrical solid vase consists of a circular cylinder from which a cone of the same height and base radius has been removed.

Find the volume of material used in making the vase.
Give your answer as a multiple of π.
(3 marks)

This means that you must not use the decimal value of π.

This is a multiple of π.

Volume of cylinder = $\pi r^2 h = \pi \times 8^2 \times 12 = \pi \times 64 \times 12 = 768\pi$
1 mark

$$\text{Volume of cone} = \frac{1}{3}\pi r^2 h = \frac{1}{3} \times \pi \times 8^2 \times 12 = \frac{\pi \times 64 \times 12}{3} = 256\pi$$

1 mark

Volume of vase $= 768\pi - 256\pi = 512\pi\ \text{cm}^3$ **1 mark**

3 A pyramid is made in which all the edges are 8 cm in length. It is made with a material which has a density of 8.5g/cm^3.

Find the mass of the pyramid. Give your answer correct to 3 significant figures.

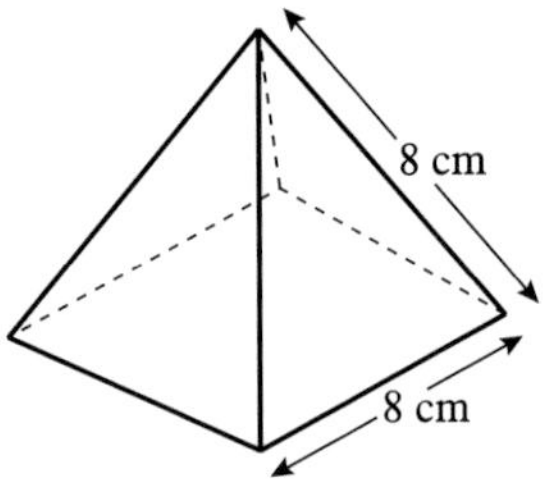

(5 marks)

First find the height of the pyramid.

Your working out must include diagrams to show what you are doing.

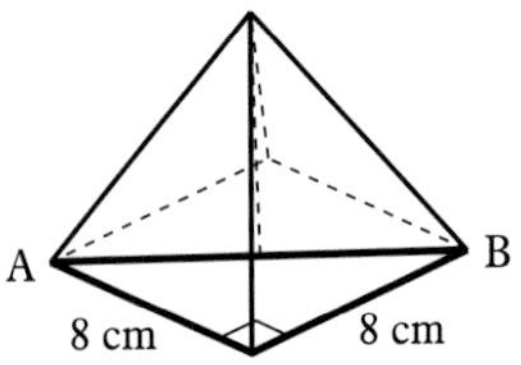

$$AB^2 = 8^2 + 8^2$$
$$\text{So } AB = \sqrt{64 + 64}$$
$$= \sqrt{128}\ \text{cm}$$

1 mark

Answers can be left like this rather than rounding off decimals, which could lead to inaccuracy. If you do use decimals, keep the exact value in your calculator memory and use that value in subsequent parts.

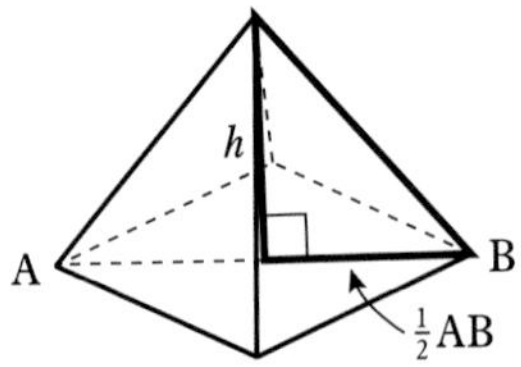

$$h^2 = 8^2 - (\sqrt{128} \div 2)^2$$
$$= 64 - 32$$
$$h = \sqrt{32}\ \text{cm}$$

1 mark

$$\text{Volume of pyramid} = \frac{1}{3} \times \text{area of base} \times \text{height}$$
$$= \frac{1}{3} \times 64 \times \sqrt{32} = 120.68\ldots\ \text{cm}^3$$

1 mark

Show the volume before multiplying by the density. Keep the exact answer in your calculator to use for the next part

Mass $= 8.5 \times 120.68$ **1 mark**

$= 1025.776$ g

$= 1030$ g (3 sf) **1 mark**

Show the accurate answer before rounding.

1 The diagram shows the shape PQRST.

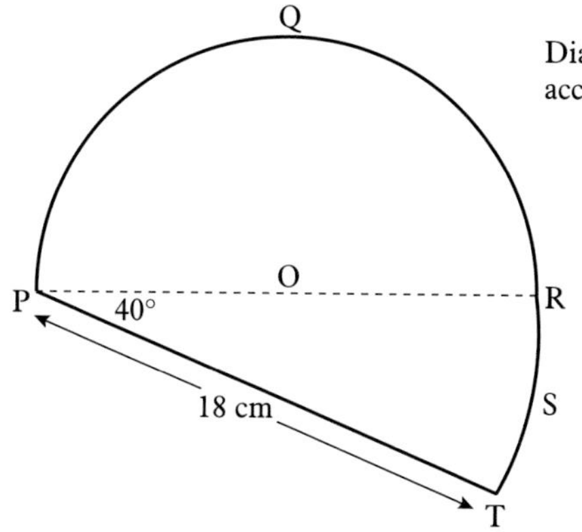

Diagram **NOT** accurately drawn.

RST is a circular arc with centre P and radius 18 cm.
Angle RPT = 40°.

a Calculate the length of the circular arc RST.
Give your answer correct to 3 significant figures. **(2 marks)**

PQR is a semicircle with centre O.

b Calculate the **total** area of the shape PQRST.
Give your answer correct to 3 significant figures. **(3 marks)**

[N2000 P6 Q18]

2 Diagram 1 and Diagram 2 show two circles with radii R and r, where $R > r$.

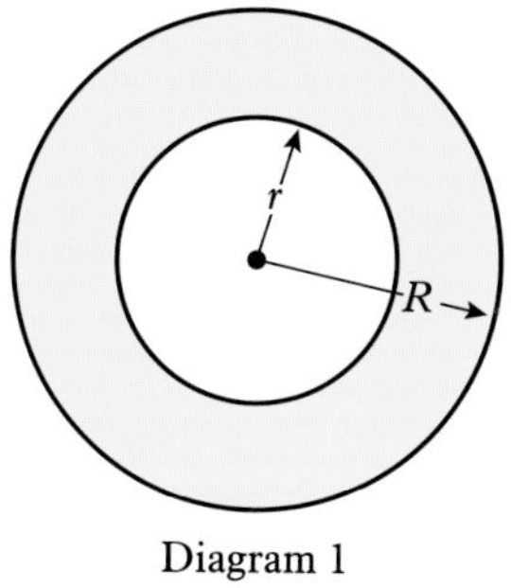

Diagram 1

Diagram **NOT** accurately drawn.

The area of the shaded region is equal to the area of the smaller circle.

a Express R in terms of r.

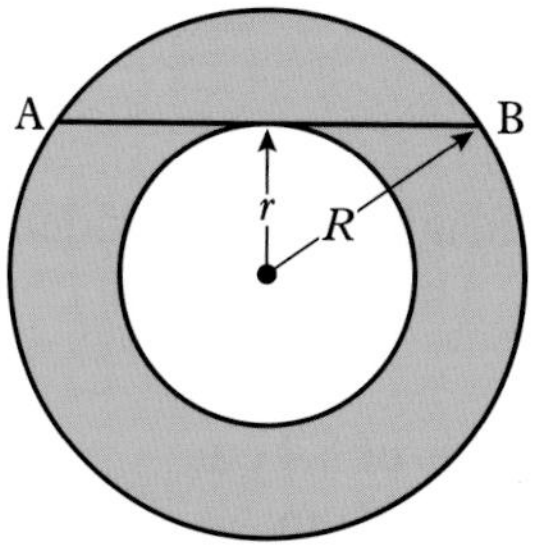

Diagram 2

Diagram **NOT** accurately drawn.

(3 marks)

AB is a chord of the larger circle and a tangent to the smaller circle.

b Show that AB = $2r$. **(2 marks)**

[N1998 P5 Q18]

3 **a** Factorise $2x^2 + 19x - 33$. **(2 marks)**

A cone fits exactly on top of a hemisphere to form a solid toy.
The radius, CA, of the base of the cone is 3 cm.
AB = 5 cm.

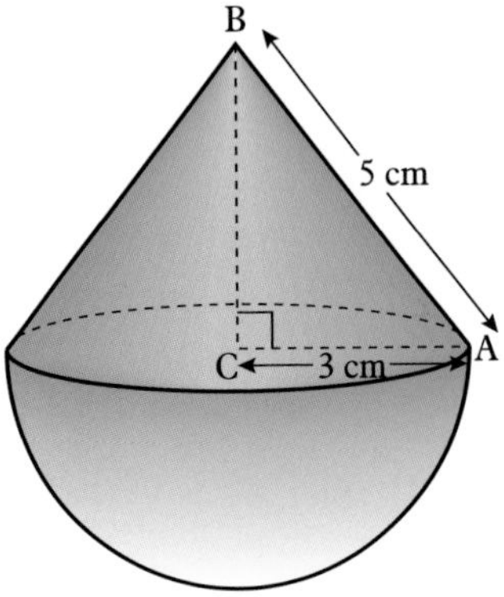

Diagram **NOT** accurately drawn.

b Show that the **total** surface area of the toy is $33\pi\,\text{cm}^2$. **(2 marks)**

The radius of the base of a cylinder is x cm.
The height of the cylinder is 9.5 cm longer than the radius of its base.
The area of the **curved** surface of the cylinder is equal to the **total** surface area, $33\pi\,\text{cm}^2$, of the toy.

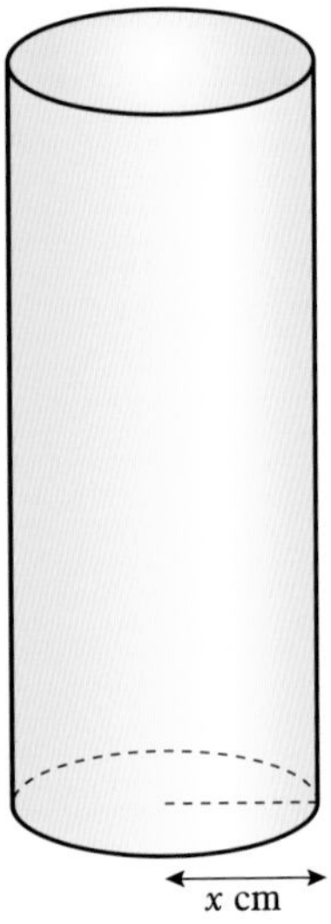

Diagram **NOT** accurately drawn.

c Calculate the height of the cylinder. **(6 marks)**

[N2000 P5 Q14]

4 A sphere has a radius of 5.4 cm.
A cone has a height of 8 cm.
The volume of the sphere is equal to the volume of the cone.
Calculate the radius of the base of the cone.
Give your answer, in centimetres, correct to 2 significant figures.

(3 marks)

[S2001 P6 Q16]

5 The diagram shows a triangular prism.
A cross section of the prism is the triangle ABC.
The width AB of the triangle is 3.6 m.
The height BC of the triangle is 5 m.
Angle ABC = 90°.
The length of the prism is L metres.
The volume of the prism is 22.5 m^3.

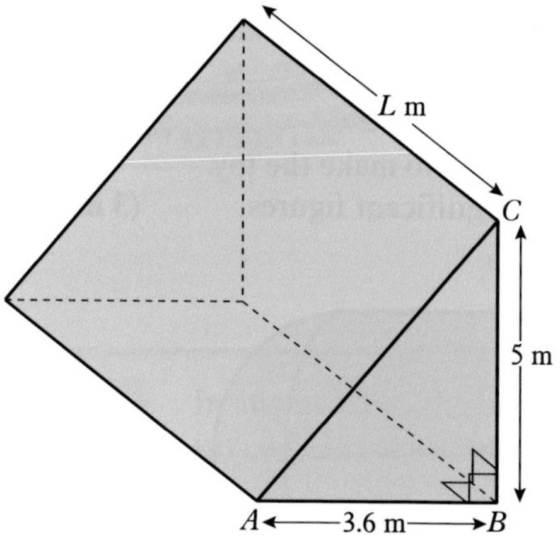

Diagram **NOT** accurately drawn.

a Calculate the value of L. **(3 marks)**

A cross section of a similar prism is the triangle DEF.
The width, DE, of the prism is 14.4 cm.

Diagram **NOT** accurately drawn.

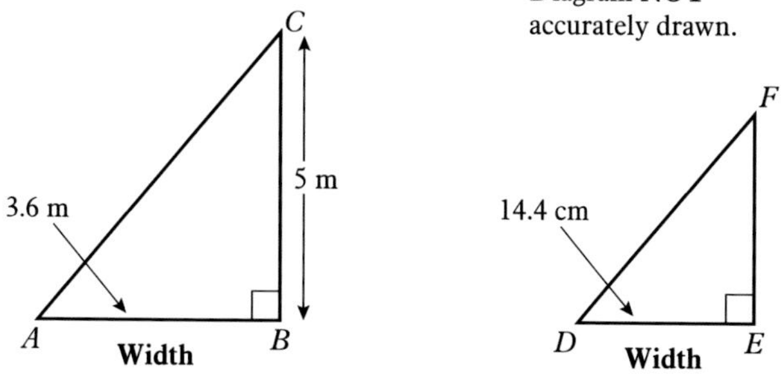

b Calculate the height, EF, of the prism. Give your answer in centimetres. **(3 marks)**
[N1999 P6 Q8]

Estimating square roots

Round the numbers to the nearest square number to get a simple estimate.
So $\sqrt{23} \approx \sqrt{25} = 5$ for example.
You may also choose to estimate $\sqrt{23}$ as a little less than 5 so $\sqrt{23} \approx 4.8$ say.
If you round $\sqrt{23}$ to 1 significant figure you get $\sqrt{20}$.
This is just as difficult to work out as the question!

Percentage error

You always work out a percentage error as a percentage of the exact value.

$$\textbf{Percentage error} = \frac{\textbf{error}}{\textbf{exact value}} \times 100\%$$

Example

Work out the percentage error when a length of 3.473 m is measured as 3.45 m.

$$\text{Percentage error} = \frac{3.473 - 3.45}{3.473} \times 100\%$$
$$= 0.66\% \text{ to 2 sf}$$

Sensible accuracy

You will sometimes need to choose a sensible number of significant figures to use.
For an exact length given in cm you would give no more than 1 dp in your answer.
This would then be correct to the nearest mm.
For a length in m you would give no more than 2 dp in your answer.
This would then be correct to the nearest cm.
The numbers that you are given in questions will also be a clue.
If numbers are only given to 3 sf then you should not give more than 3 sf in your answers.

Upper and lower bounds

A length of 30 mm to the nearest mm means that it is not necessarily exactly 30 mm.
The length could be anywhere from 29.5 mm right up to, but not including, 30.5 mm.
You can show this on a number line like this.

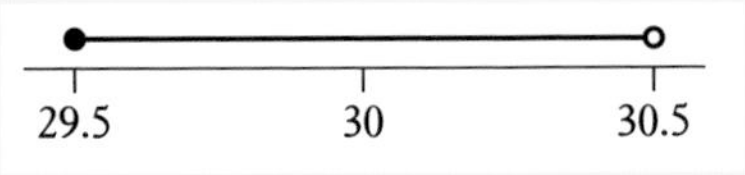

$$29.5 \leqslant \text{length} < 30.5$$

1 A farmer has a rectangular field ABCD
He paces out the dimensions, and finds them to be 60 metres and 50 metres.

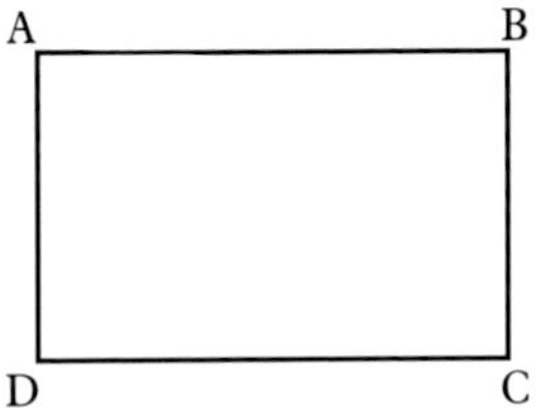

Find the distance from D to B. **(4 marks)**

Use Pythagoras' theorem:

$DB = \sqrt{60^2 + 50^2} = \sqrt{3600 + 2500} = \sqrt{6100} = 78.102\,496\,76$ m

1 mark **1 mark**

You have been given no guidance on the answer.
The question will give marks not only for the numerical answer, but also for

- *how suitable the answer is, in terms of accuracy*
- *correct units stated with the answer.*

As the farmer has paced the dimensions, the measurements will be accurate to ***only*** *the nearest metre,* ***or*** *the nearest 10 metres.*

So round 78.102 496 76 to 78 or 80. **1 mark**

The units used are metres, so the complete answer is 78 m or 80 m.
1 mark for stating the units

2
$$\frac{35.8 \times 24.7}{8.96 \times 4.83}$$

Explain how you would find the most appropriate estimate for this calculation. **(3 marks)**

Method 1 Rounding numbers to one significant figure would give:

$$\frac{40 \times 20}{9 \times 5} = \frac{800}{45} \approx \frac{800}{50} = 16$$

But the most appropriate method to use is:

Method 2 Round the denominator to one significant figure:

$$\frac{35.8 \times 24.7}{9 \times 5}$$ **1 mark**

Write the numerator in numbers that can be cancelled ($35.8 \approx 36$, $24.7 \approx 25$):

$$\frac{36 \times 25}{9 \times 5} = 4 \times 5 = 20$$ **1 mark: identify factors**
1 mark: answer

(*NB: The accurate answer is 20.43 which shows how poor the estimate in Method 1 is compared with Method 2.*)

7 $P = 2.57$ to 3 sf, $Q = 4.2$ to 1 dp and $R = 0.7$ to 1 sf.
Work out the lower and upper bounds for these.

a $P + Q$ **c** QR **e** $2P + 3Q$

b $P - Q$ **d** $\frac{P}{R}$ **f** $\frac{Q + R}{P}$

8 Russia has a total area of 1.71×10^7 km^2 to 3 sf.
It has a population of 1.48×10^8 to 3 sf.

a Write down the lower and upper bounds for the area.
Give your answers in standard form.

b Between what limits does the population lie?

c Population density is defined as the number of people per square km.
Find the lower and upper bounds for the population density of Russia.
Give your answers to 1 dp.

d Russia has 70 times the area of the United Kingdom.
Given that this number is correct to 2 sf find the lower and upper bounds for the area of the UK.

e Russia has 2.5 times the population of the United Kingdom.
Given that this number is correct to 2 sf find the lower and upper bounds for the population of the UK.

f Use your answers to **d** and **e** to find the range of possible values of the population density of the UK. Give your final answer in the form $a <$ population density $< b$, where a and b are integers.

9 The fastest moving glacier is the Columbia Glacier, between Anchorage and Valdez in Alaska.
It moves at 20 m per 24 hours.
Assuming this data is accurate to 2 sf, find the lower and upper bounds for the speed of the glacier in metres per hour, to 3 sf.

1 Round each of these numbers to the given accuracy.

a 345.67 to 2 sf
b 12.572 to 1 sf
c 23 345.67 to 3 sf
d 1 231 645 to 4 sf
e 1.863×10^5 to 2 sf
f 6.163×10^{-7} to 1 sf

2 Estimate the answer to each of these questions by rounding each number to 1 sf.
Give your answers in standard form.

a $(2.63 \times 10^5) \times (1.83 \times 10^4)$
b $(1.16 \times 10^9) \times (4.64 \times 10^{-3})$
c $(6.863 \times 10^9) \times (4.14 \times 10^{-4})$
d $(5.663 \times 10^8) \times (3.643 \times 10^{-3})$
e $(6.13 \times 10^7) \div (1.65 \times 10^{-3})$
f $(7.863 \times 10^5) \div (1.86 \times 10^{-5})$
g $\dfrac{4.67 \times 10^6}{1.78 \times 10^{-3}}$
h $\dfrac{6.7 \times 10^{-5}}{4.13 \times 10^7}$

3 For each part:
(1) Estimate the answer by rounding each number to 1 sf.
(2) Work out the exact answer.
(3) Find the percentage error of the estimate.

a 34.89×815.72
b $\dfrac{82.78 \times 3.89}{44.392}$
c $\dfrac{3.78 \times 5.89}{4.132}$

4 For each part:
(1) Estimate the answer, showing clearly how you obtain your estimate.
(2) Work out the exact answer.
(3) Find the percentage error of the estimate.

a $\dfrac{23.78 \times 7.89}{4.392}$
b $\dfrac{\sqrt{102.5} \times 8.733}{16.8}$
c $\dfrac{12.78 \times 74.89}{34.92 \times 5.91}$
d $\sqrt{\dfrac{61.78 \times 8.34}{6.392 \times 3.89}}$

5 **a** Work out the exact value of the area of this trapezium.
b Estimate the area by rounding each length to 1 sf.
c Work out the percentage error in using this estimate.
d Estimate the area by rounding each length to the nearest cm.
e Work out the percentage error in using this estimate.

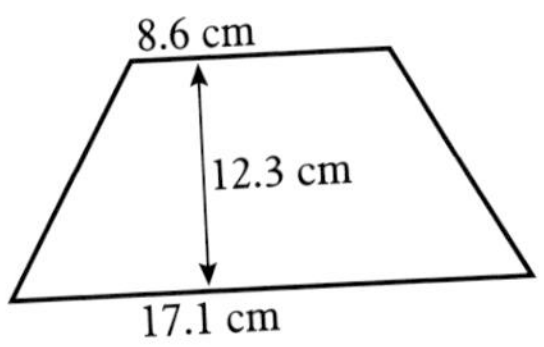

6 The diagram shows a rectangular field.
Usman takes a short cut across this field from A to C.
How much shorter is his journey than if he walks around the perimeter via B?
Give your answer to a sensible level of accuracy.

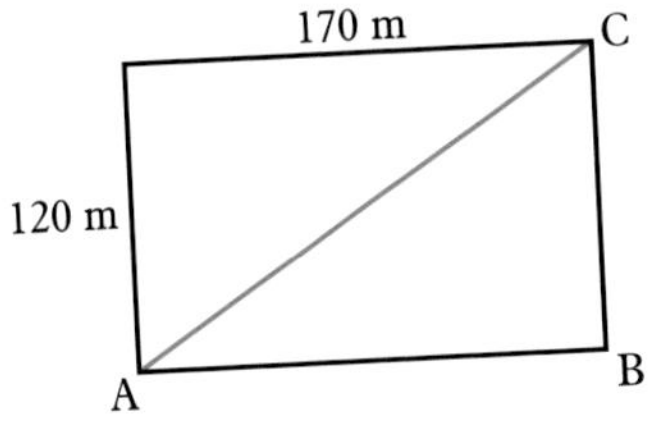

of 0.1 cm, measured to the nearest 0.1 cm.

are cross section of side 10 cm and a length of 20 cm,
cm.

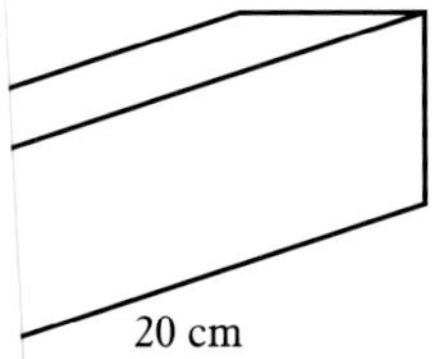

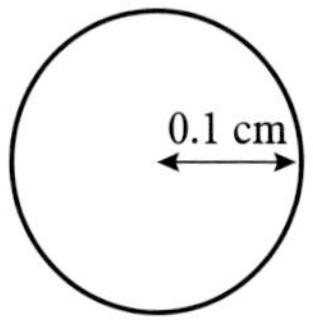

wn and used to make a metal bar.
e maximum number of balls needed. **(7 marks)**

that will be needed:

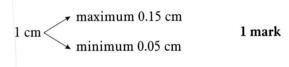

1 cm → maximum 0.15 cm, minimum 0.05 cm **1 mark**

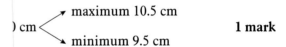

) cm → maximum 10.5 cm, minimum 9.5 cm **1 mark**

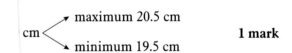

cm → maximum 20.5 cm, minimum 19.5 cm **1 mark**

to be done is:

$\dfrac{\text{ume of bar}}{\text{of ball } (\frac{4}{3}\pi r^3)} = \dfrac{10 \times 10 \times 20}{\frac{4}{3} \times \pi \times (0.1)^3}$

of balls: For the maximum number of balls:

$\dfrac{\text{maximum bounds}}{\text{minimum bounds}}$

mark $= \dfrac{10.5 \times 10.5 \times 20.5}{\frac{4}{3} \times \pi \times (0.05)^3}$ **1 mark**

$= \dfrac{2260.125}{0.000\,523\,6}$

mark $= 4\,316\,521$ balls **1 mark**

1 Use your calculator to work out the value of

$$\frac{\sqrt{12.3^2 + 7.9}}{1.8 \times 0.17}.$$

Give your answer correct to 1 decimal place. **(3 marks)**
[S2000 P6 Q1]

2 Use your calculator to find the value of

$$\sqrt{47.3^2 - 9.1^2}$$

a Write down all the figures on your calculator display. **(2 marks)**

b Write your answer to part **a** correct to 2 significant figures. **(1 mark)**
[N2000 P6 Q3]

3 Work out an estimate for the value of

$$\frac{29.91 - 2 \times 10.03}{29.91^2 - 10.03^2}$$

Give your answer as a fraction in its simplest form. **(3 marks)**
[S2001 P5 Q7]

4 Sally estimates the value of $\frac{42.8 \times 63.7}{285}$ to be 8.

a Write down three numbers Sally could use to get her estimate.

$$\frac{\ldots\ldots\ldots \times \ldots\ldots\ldots}{\ldots\ldots\ldots}$$ **(2 marks)**

b Without finding the exact value of $\frac{42.8 \times 63.7}{285}$, explain why it must be **more** than 8. **(2 marks)**
[N1998 P5 Q7]

lled the **lower bound.**
mallest number that will round up to 30 mm.

lled the **upper bound.**
st number that will round down to 30 mm is
99…
onvenient to use this value, so 30.5 is used as the upper
en though 30.5 would round up to 31 to the nearest
nber.

bering how to work out the upper and lower bounds:
the nearest 10 then
nore than this number
ss than this number. 5 is half of 10
the nearest unit then
more than this number
ess than this number. 0.5 is half of 1
the nearest 0.1
ect to 1 dp) then
more than this number
less than this number. 0.05 is half of 0.1
upper bound and take away to get the lower bound is
uracy.

ound of a sum $a + b$ or a product ab is found by
per bound of a and b.
ound of a sum $a + b$ or a product ab is found by
ver bound of a and b.

ound for a difference $a - b$ or a quotient $\frac{a}{b}$ is found
upper bound of a and the lower bound of b.
und for a difference $a - b$ or a quotient $\frac{a}{b}$ is found
ower bound of a and the upper bound of b.
, watch out for speed questions and density
uiring

nce / ne and Density = $\frac{\text{Mass}}{\text{Volume}}$

5 $F = \dfrac{ab}{a - b}$

Imran uses this formula to calculate the value of F.
Imran estimates the value of F without using a calculator.

$a = 49.8$ and $b = 30.6$

a **(1)** Write down approximate values for a and for b that Imran could use to estimate the value of F.
(2) Work out the estimate for the value of F that these approximations give.
(3) Use your calculator to work out the accurate value for F.
Use $a = 49.8$ and $b = 30.6$.
Write down all the figures on your calculator display. **(4 marks)**

Imran works out the value of F with two new values for a and b.

b Calculate the value of F when

$a = 9.6 \times 10^{12}$ and $b = 4.7 \times 10^{11}$.

Give your answer in standard form, correct to two significant figures. **(3 marks)**

[S1999 P6 Q4]

6 The formula

$$Q = \frac{f^2 + 2g^2}{f - 3g}$$

can be used to calculate the value of Q.

$f = 9.04$ and $g = 1.8$.

Jane uses the formula to estimate the value of Q without using a calculator.

(1) Write down approximate values for f and g that Jane could use to estimate the value of Q.
(2) Work out the estimate for the value of Q that these approximate values give. **(3 marks)**

[N1999 P5 Q6]

7 **a** Work out an estimate for

$$\frac{3.08 \times 693.89}{0.47}$$

(3 marks)

The length of a rod is 98 cm correct to the nearest centimetre.

b **(1)** Write down the maximum value that 98 cm could be.
(2) Write down the minimum value that 98 cm could be. **(2 marks)**

[S2000 P5 Q3]

8 The label on a jar of coffee says 'This 200 gram jar of coffee makes approximately 110 cups of coffee.'

200 grams is correct to 3 significant figures.
110 cups is correct to 2 significant figures.

a For the weight of the coffee in the jar write down
 (1) the least upper bound
 (2) the greatest lower bound. **(2 marks)**

b For the number of cups of coffee write down
 (1) the least upper bound
 (2) the greatest lower bound. **(2 marks)**

c Calculate the least upper bound for the mean weight of coffee in each cup. Write down all the figures on your calculator display. **(2 marks)**

[N1998 P6 Q17]

9 $x = 3$, correct to 1 significant figure.
$y = 0.06$, correct to 1 significant figure.

Calculate the greatest possible value of

$$y - \frac{x - 7}{x}$$

(2 marks)

[S2001 P6 Q17]

10 The table shows the number of hours of sunshine and the rainfall, in centimetres, in each of six places one day in December.

	Number of hours of sunshine	Rainfall in cm
Anglesey	5.7	0.06
Birmingham	4.4	0.35
Folkestone	3.6	0.42
Guernsey	7.2	0.08
Jersey	6.5	0.17
Torquay	6.9	0.04

The number of hours of sunshine is given correct to 1 decimal place.
The rainfall is given correct to the nearest 0.01cm.

a Write down
 (1) the lower bound of the number of hours of sunshine in Anglesey,
 (2) the upper bound of the rainfall in Torquay. **(2 marks)**

b Calculate the lower bound of the sum of the number of hours of sunshine in Birmingham and in Folkestone. **(2 marks)**

c Calculate the greatest possible difference between the rainfall in Guernsey and the rainfall in Jersey. **(2 marks)**

[N1999 P5 Q19]

11 Bill has a rectangular sheet of metal.
The length of the rectangle is **exactly** 12.5 cm.
The width of the rectangle is **exactly** 10 cm.

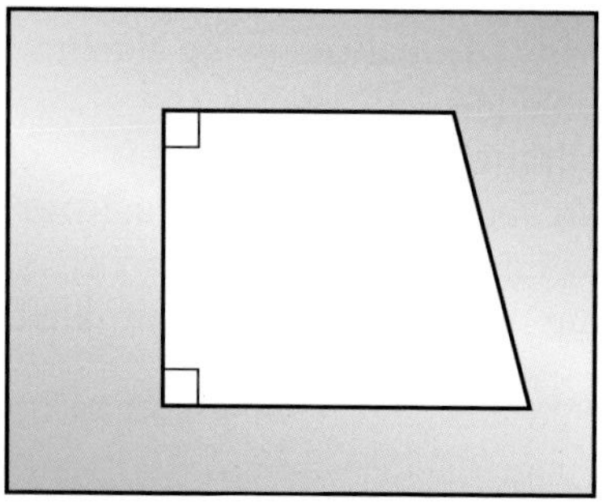

Diagram **NOT** accurately drawn.

Bill cuts out a trapezium.
Its dimensions, shown in the diagram, are correct to the nearest millimetre.
He throws away the rest of the metal sheet.

Calculate the **greatest** possible area of the rectangular sheet that he throws away. **(5 marks)**
[N2000 P6]

12 Kim is doing an experiment using a pendulum. She uses the formula

$$g = \frac{40L}{T^2}$$

where g is a constant acceleration, L the length of the pendulum, and T is the time for one full swing of the pendulum.
In Kim's experiment the length L is 1 metre, correct to the nearest centimetre.
She measured the value of T to be 2 seconds, correct to the nearest 0.2 of a second.
Calculate the upper bound and the lower bound of Kim's values for g.
Give your answer in metres per second per second correct to two decimal places. (6 marks)
[N2000 P6 Q19]

18 Inequalities

You need to know about:

- showing inequalities on number lines
- solving inequalities using algebra
- double inequalities
- quadratic inequalities
- using regions to show inequalities in 2-D
- showing the solution to more than one inequality
- using inequalities to solve problems

Inequalities on number lines

$x > 3$

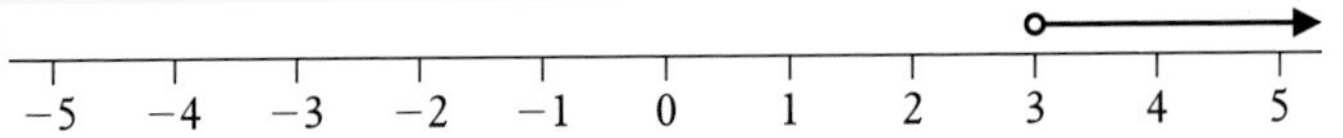

The open circle means that the end point is not included.

$x \leqslant 1$

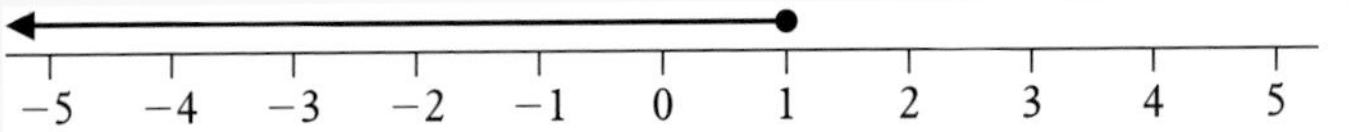

The solid circle means that the end point is included.

$-2 \leqslant x < 4$

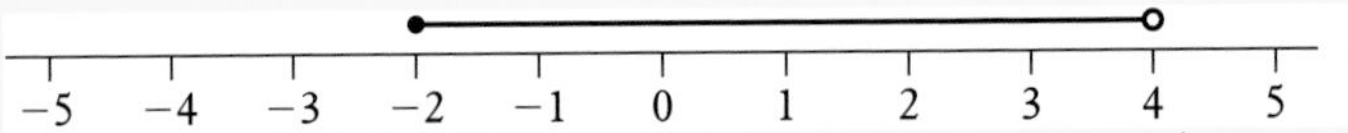

Solving inequalities using algebra

You solve inequalities in the same way that you solve equations.
The only exception to this occurs if you multiply or divide an inequality by a negative number. In this case you must reverse the inequality sign.

Examples

Solve:	$5x - 4 > 6$	Solve:	$20 - 4x \leqslant 6 + 3x$
Add 4 to both sides:	$5x > 10$	*Add 4x to both sides:*	$20 \leqslant 6 + 7x$
Divide both sides by 5:	$x > 2$	*Subtract 6 from both sides:*	$14 \leqslant 7x$
		Divide both sides by 7:	$2 \leqslant x$
		So:	$x \geqslant 2$
Solve:	$6 - 2x > 18$	Write down the smallest integer solution to $3x > 10$.	
Take 6 from both sides:	$-2x > 12$	An integer is a whole number.	
Divide both sides by −2 and reverse the inequality	$x < -6$	$x = 4$ is the smallest integer solution.	

Double inequalities

A double inequality is one with two inequality signs, like $13 < 3x + 4 \leqslant 22$.
This is actually the two separate inequalities $13 < 3x + 4$ and $3x + 4 \leqslant 22$.
Solve the two parts separately and then write the answer as a double inequality if you can

$13 < 3x + 4$	$3x + 4 \leqslant 22$
$9 < 3x$	$3x \leqslant 18$
$3 < x$	$x \leqslant 6$
$x > 3$	

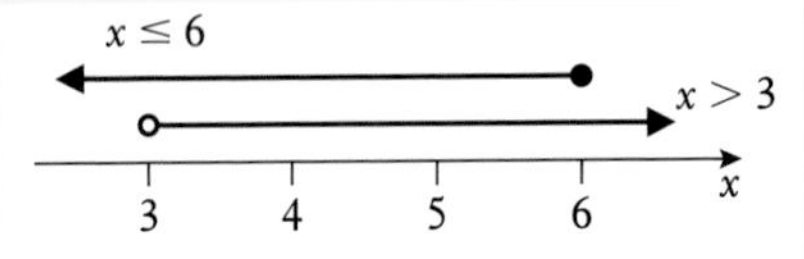

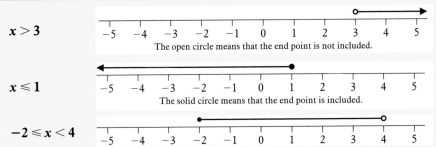

So $3 < x \leqslant 6$ is the solution.

Quadratic inequalities Quadratic inequalities involve an x^2 term. They will give you two possible solutions.

If $x^2 < 49$, then $x > -7$ and $x < 7$ or, written as a double inequality, $-7 < x < 7$.
If $x^2 \geqslant 25$, then $x \leqslant -5$ and $x \geqslant 5$. This 'split' solution should not be written as a double inequality. You may be tempted to write $5 \leqslant x \leqslant -5$ but this is nonsense as it says that 5 is less than something which is less than -5 : 5 cannot be less than -5!

Regions You can also show the solution to inequalities as regions.

$x \geqslant 3$

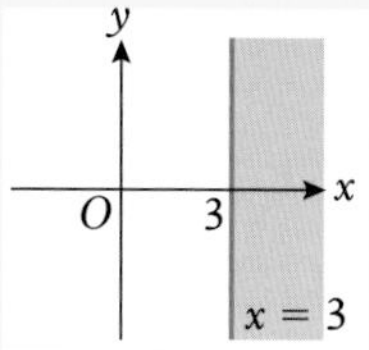

$y < -2$

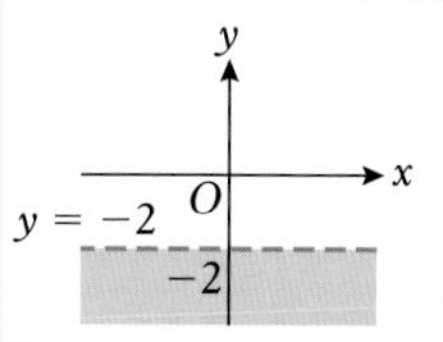

The line is $x = 3$.
All the points in the shaded region have an x co-ordinate greater than 3.
The solid line shows that the boundary **is** included.

The line is $y < -2$
All the points in the shaded region have a y co-ordinate less than -2.
The dashed line shows that the boundary is **not** included.

For inequalities that involve both x and y you need to draw the boundary line first. Then pick a point above and below the line and substitute the co-ordinates into the inequality to see which region satisfies the inequality.

Example Show the region for the inequality $x + y > 6$.
The boundary line is $x + y = 6$.
This passes through (0, 6) and (6, 0).
The point (3, 5) shown in blue gives $x + y = 8$, which is greater than 6 and so this is in the region you want.

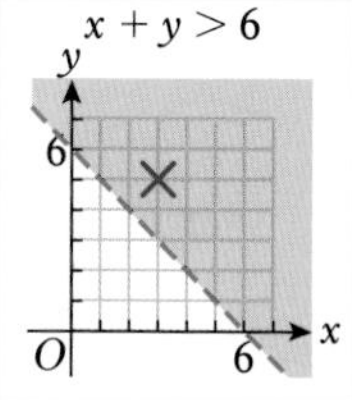

Showing the solution to more than one inequality

It is a good idea to cross out the regions that you don't want when you are showing the solution of several inequalities.
The region that remains unshaded is the solution to the question.
The region that satisfies the inequalities $x \geqslant 0, y \geqslant 0, x + y < 6$ and $x + 3y > 6$ is shown unshaded here.

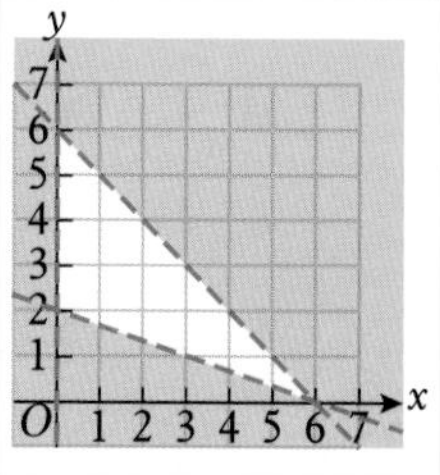

Solving problems

You will sometimes face problems in which information is given to you that requires you to form inequalities and then show their solution as a region.
Then you will often need to find a particular point in the region. This is usually to find how to maximise or minimise one of the quantities in the question, e.g. finding a maximum profit. The point will be near one of the vertices of the solution region and you need to test points to find the particular solution needed.

1 Draw a number line to show each of these inequalities.

a $x \geqslant -2$ **b** $x < 4$ **c** $-5 < x \leqslant 1$

2 Write down the inequality shown on each of these number lines.

a

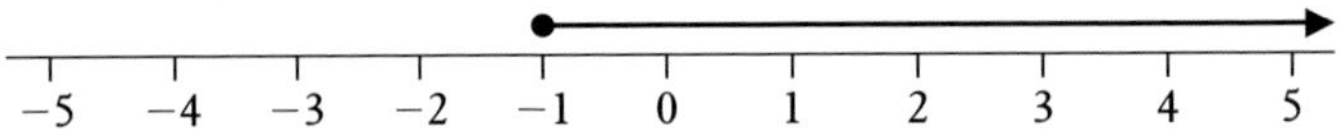

b

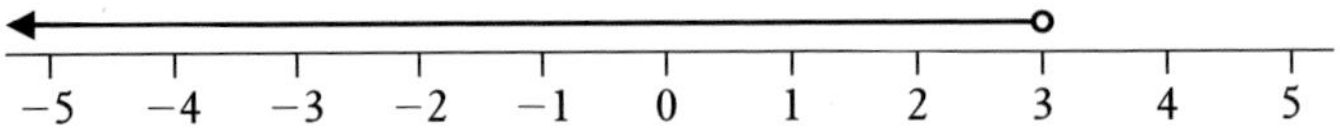

c

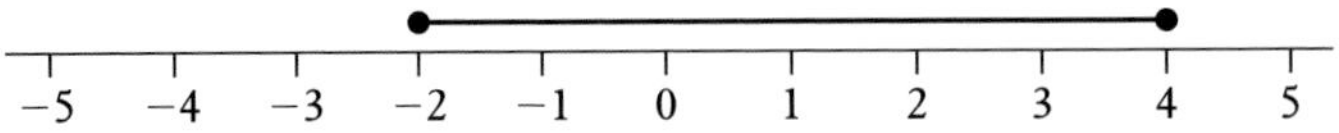

3 List the integers included in each of these inequalities.

a $-2 \leqslant x \leqslant 1$

b $-4 < x \leqslant 2$

c $4 < x < 10$

d $-7 \leqslant x < -1$

4 Solve each of these inequalities.

a $3x + 4 < 13$

b $2x - 1 \geqslant 23$

c $3 - 4x \leqslant 7$

d $4 - 5x > 9$

e $4x + 9 < 8$

f $3x - 4 \geqslant 1$

g $\frac{x}{4} \geqslant 7$

h $\frac{3x}{4} + 3 \leqslant 5$

i $\frac{2x}{5} - 4 > 1$

j $3 - \frac{2x}{3} < 5$

k $5 < 2x + 6 \leqslant 7$

l $2 < \frac{x}{4} + 1 \leqslant 6$

5 Solve each of these double inequalities.

a $1 < \frac{x}{5} \leqslant 3$

b $-3 \leqslant \frac{x}{3} + 1 \leqslant 4$

c $-2 < 1 - 3x < 4$

d $-7 \leqslant 3 - 4x < 3$

6 Write down the inequalities that describe the red region in each of these:

a

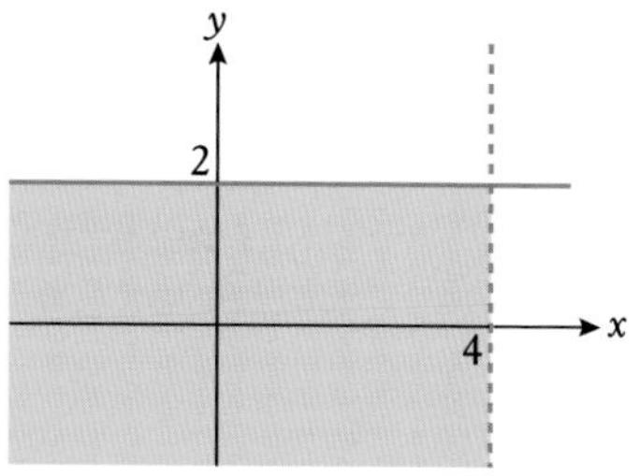

b

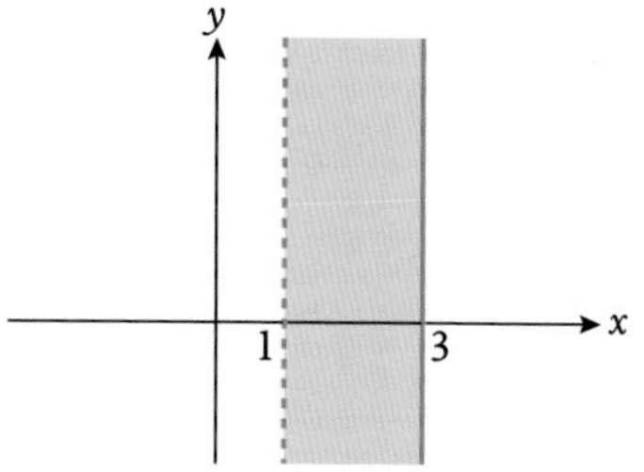

c

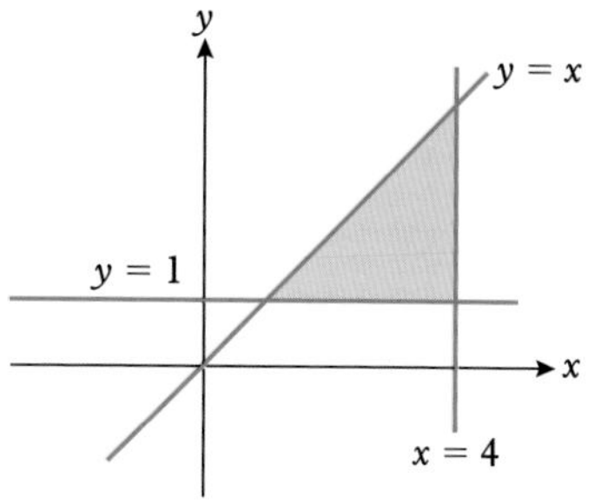

d

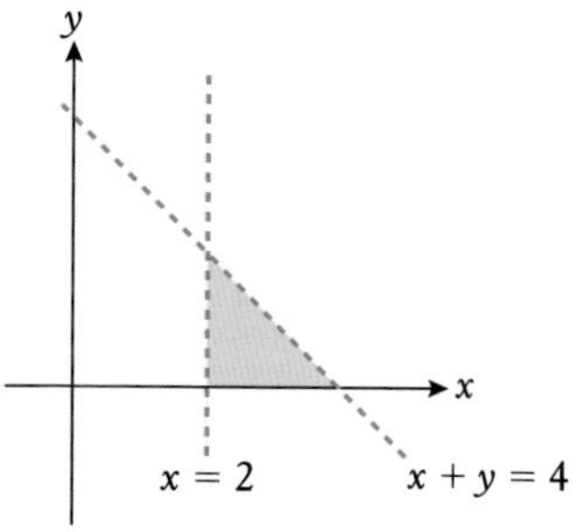

7 Show the region, R, that satisfies the inequalities.

a $x \geqslant 0, y \geqslant 0, x < 6, y < 5$ and $x + y > 4$

b $x \geqslant 0, x + y < 5$ and $y > x + 1$

8 Cath is buying refill cartridges for colour printers. She needs to buy both colour and black and white refills. A colour refill costs £30 and a black and white refill costs £20.
She wants at least 5 colour refills.
She wants at least 3 black and white refills.
She has £300.
She buys x colour, and y black and white refills.

a Show that $3x + 2y \leqslant 30$.

b Write down two more inequalities that describe this problem.

c Draw an x axis from 0 to 10 and a y axis from 0 to 16 using 1 cm per unit in both directions. Show the region that satisfies all three inequalities on your graph.

d Cath wants to buy the most refills that she can. She would also like to have some money left over if possible.
Use your graph to decide how many of each type of refill Cath should buy to satisfy all of her requirements.

1 List all possible integer values of n, where

a $1 < n \leqslant 7$ **(2 marks)** **b** $-3 < 2n < 12$ **(2 marks)**

a

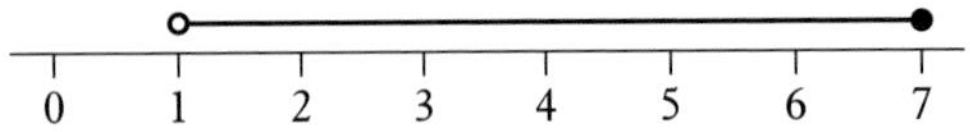

$n = 2, 3, 4, 5, 6, 7$

1 mark for 2, 7
1 mark for 3, 4, 5, 6

b *First get n and not 2n, ÷ 2 to give*

$-\frac{3}{2} \leqslant n < 6$

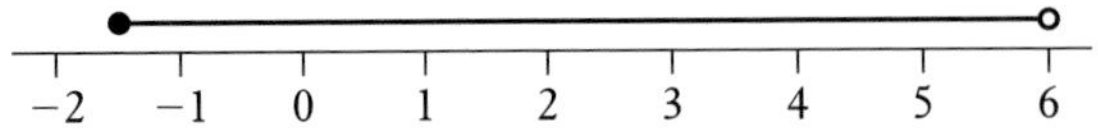

$n = -1, 0, 1, 2, 3, 4, 5$

1 mark for −1, 5
1 mark for 0, 1, 2, 3, 4

2 On a grid, indicate the region that satisfies the inequalities.

$y \leqslant 12 - x$ $x > 2$ $y \geqslant 5$ **(4 marks)**

Draw the boundary lines:
$y = 12 - x$ is shown as a solid line since the inequality is $\leqslant$.
$y = 5$ is shown as a solid line since the inequality is $\geqslant$.
$y = 2$ is shown as a dashed line since the inequality is $>$.

3 marks for boundary lines

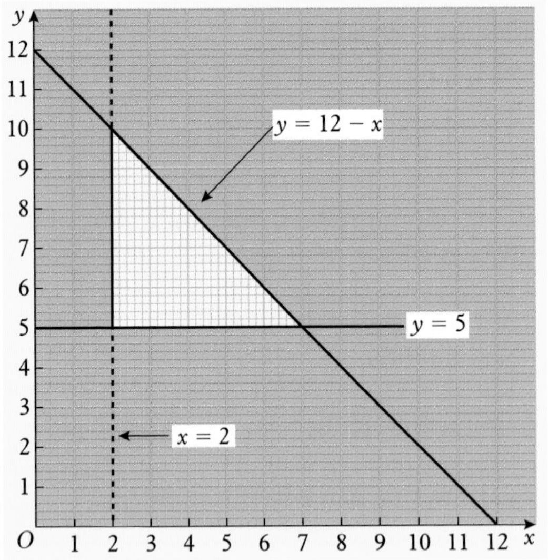

Now cross out the regions you don't want.
$y \leqslant 12 - x$ means you cross out above the line.
$x > 2$, so cross out to the left of the line.
$y \geqslant 5$, so cross out below the line.
The unshaded region is the answer. **1 mark**

3 A community centre is organising a trip for its family members.
There are up to 200 places on the trip.
There must be no more than 4 children per adult.
The trip price is £16 per adult, £10 per child.
The booking cost of the trip is £1600, which will be cancelled if this cost is not covered.

Find: **a** the minimum number of adults required to go on the trip
b the maximum number of children that can go on the trip. **(8 marks)**

This is an unstructured problem: you need to organise the information.
Start by writing expressions for each of the above conditions.
You will pick up marks by doing so.

Make sure you explain what x and y stand for.

Let x be the number of children and y be the number of adults.

200 places for the trip: $x + y \leqslant 200$ **1 mark**

Maximum of 4 children per adult: $\frac{x}{4} \leqslant y$ **1 mark**

£1600 booking cost: $1600 \leqslant 10x + 16y$ **1 mark**

Total amount collected for children and adults.

Next draw the graphs for these expressions.
Shade out those areas not needed.

3 × 1 mark
(1 mark per line)

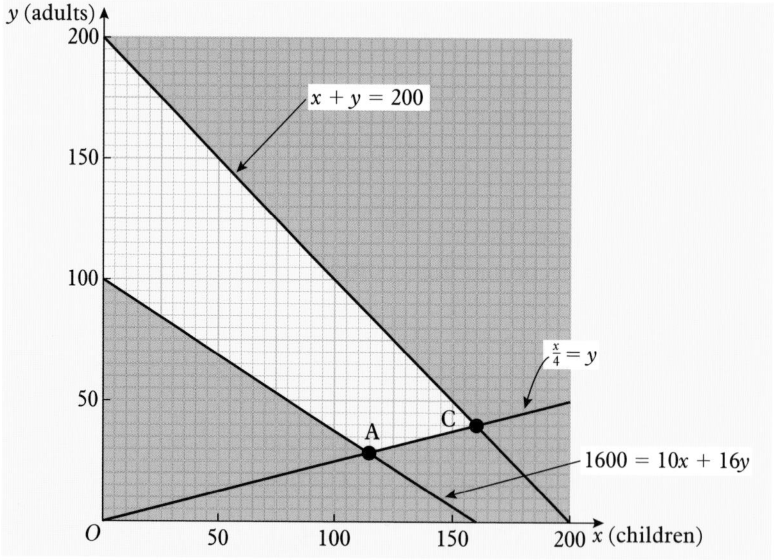

The unshaded area contains all points that meet the conditions of the trip.

a The minimum number of adults is shown at point A: 29 adults.
(This is the point with the least y value in the region.) **1 mark**

b The maximum number of children is shown at point C: 160 children
(This is the point with the biggest x value in the region.) **1 mark**

1 **a** Solve the inequality
$4x - 3 > 2.$ **(2 marks)**
b Write down the smallest whole number that is a solution of the inequality
$4x - 3 > 2.$ **(1 mark)**
[N1999 P5 Q1]

2 **(1)** Solve the inequality $3n > -8$.
(2) Write down the smallest integer which satisfies the inequality
$3n > -8.$ **(2 marks)**
[N1998 P5 Q3]

3 n is a whole number such that
$6 < 2n < 13.$

List all the possible values of n. **(3 marks)**
[N2000 P6 Q5]

4 Solve the inequality
$7y > 2y - 3.$ **(2 marks)**
[S1997 P5 Q9]

5 x is an integer.

Write down the greatest value of x for which $2x < 7$. **(1 mark)**
[S1997 P5 Q2]

6 A company makes compact discs.
The total cost, P pounds, of making n compact discs is given by the formula

$P = a + bn$

where a and b are constants.

The cost of making 1000 compact discs is £58 000.
The cost of making 2000 compact discs is £64 000.

a Calculate the values of a and b. **(4 marks)**

The company sells the compact discs at £10 each.
The company does not want to make a loss.

b Work out the minimum number of compact discs the company must sell. **(3 marks)**
[S1998 P5 Q17]

7 Mr Watt wants to tile his kitchen.
Plain tiles cost 20p each.
Patterned tiles cost £1 each.

Mr Watt can spend a maximum of £300.
He buys x plain tiles and y patterned tiles.

a Show that $x + 5y \leqslant 1500$. **(2 marks)**

He will use at least one patterned tile to every 5 plain tiles.

b Express this condition as an inequality which x and y must satisfy. **(1 mark)**

c By drawing straight lines and shading, indicate the region within which x and y must lie to satisfy both the inequalities.

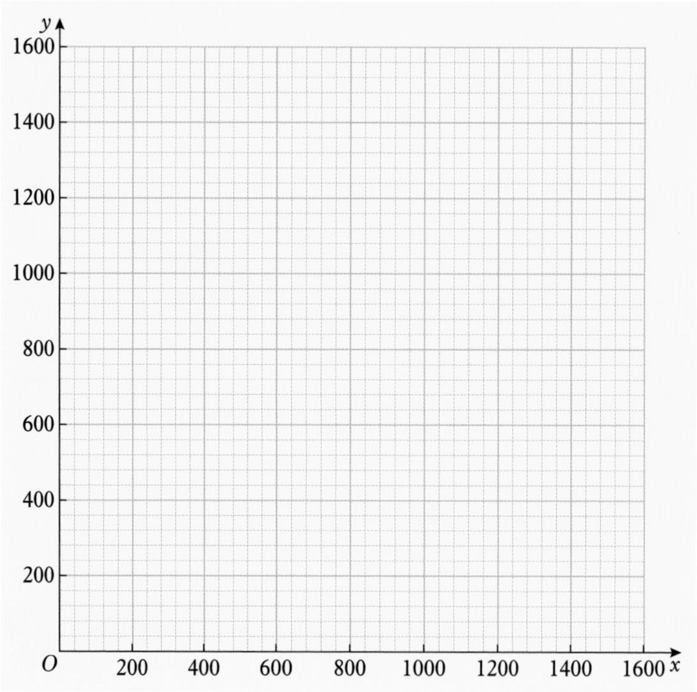

(3 marks)

d Use your diagram to find the greatest number of tiles he could buy. **(1 mark)**

Mr Watt buys a total of 600 tiles.

e Use your diagram to find the lowest price he could pay for these tiles. **(4 marks)**

He pays the lowest price.

f Write down the number of patterned tiles he buys. **(1 mark)**

[S1997 P6 Q11]

8 **a** Make y the subject of the equation $x + 2y = 6$. **(2 marks)**

b On the grid draw the line with equation

$$x + 2y = 6$$

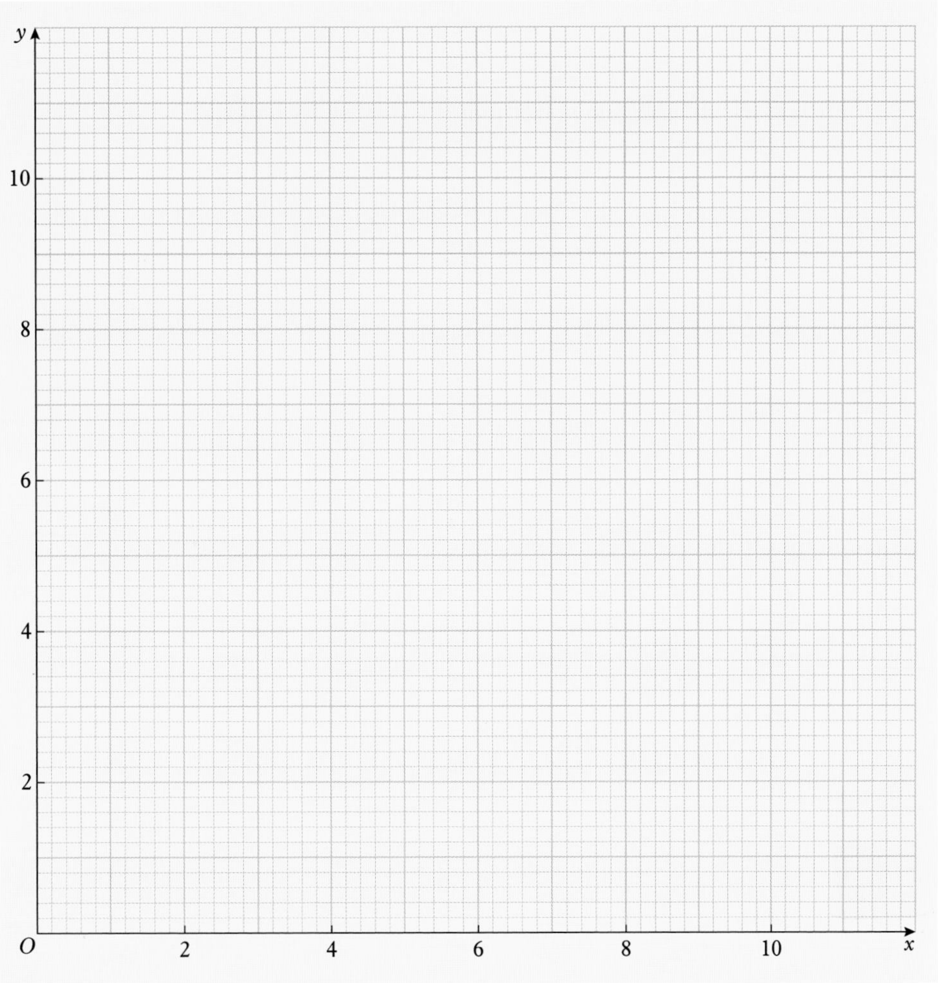

c On the grid shade the region for which

$x + 2y \leqslant 6$, $0 \leqslant x \leqslant 4$ and $y \geqslant 0$. **(2 marks)**

d When

$$4y = 3x$$

what is the greatest value of x in the shaded region? **(2 marks)**

[N1997 P5 Q7]

9 The ingredients for one Empire Cake include 200 g of self-raising flour and one egg.
The ingredients for one Fruit Cake include 100 g of self-raising flour and three eggs.
Alphonso has 1200 g of self-raising flour and 12 eggs.
He makes x Empire Cakes and y Fruit Cakes.

Using the weight of self-raising flour available means that x and y must satisfy the inequality $2x + y \leqslant 12$.

a Write down another inequality which x and y must satisfy, other than $x \geqslant 0$ and $y \geqslant 0$. **(1 mark)**

Use the grid below for part **b**.

b By drawing straight lines and shading, indicate the region within which x and y must lie to satisfy all the inequalities.

The line $2x + y = 12$ has been drawn for you. **(3 marks)**

c Use your diagran to find the greatest number of cakes that Alphonso can make. **(2 marks)**

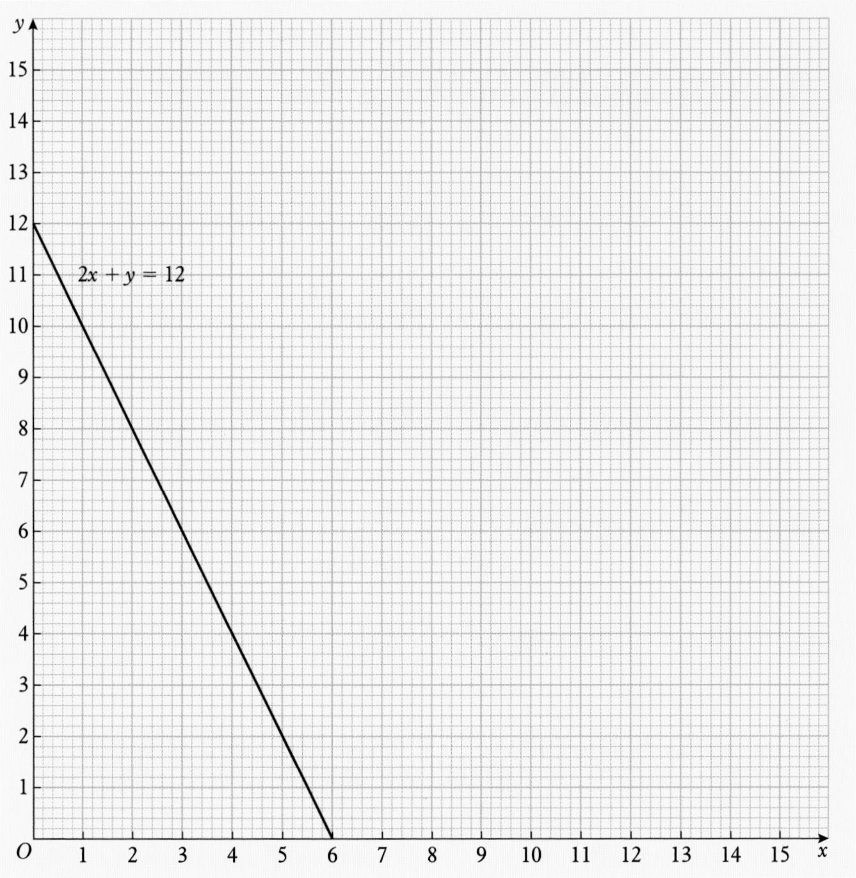

19 Trigonometry: making waves

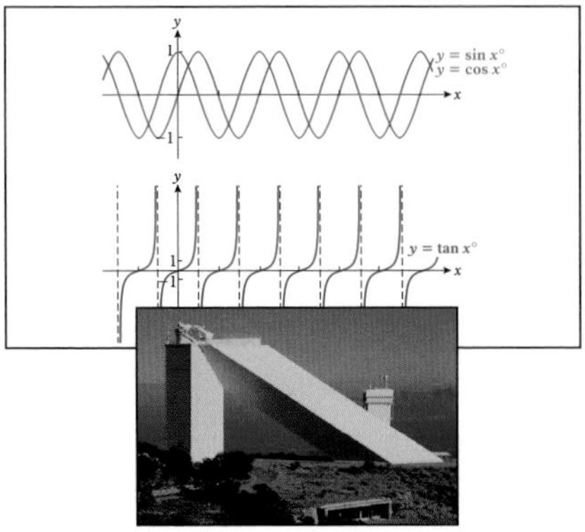

You need to know about:
- solving trigonometric equations for angles of any size
- the graphs of sin x, cos x and tan x for $0 \leqslant x \leqslant 360°$
- transforming trigonometric graphs
- the sine and cosine rules
- the area of a triangle
- bearings problems involving the sine and cosine rules
- 3-D problems involving the sine and cosine rules

Solving trigonometric equations for angles of any size

Find the first solution using your calculator.
Then you can use circle diagrams to find the other solutions in the range you need.
If you have one solution of an equation of the form $\tan x = k$ then you can find all of the others by adding or subtracting multiples of 180°.

Trigonometric graphs

These are the graphs of sin x, cos x, and tan x for $0 \leqslant x \leqslant 360°$.

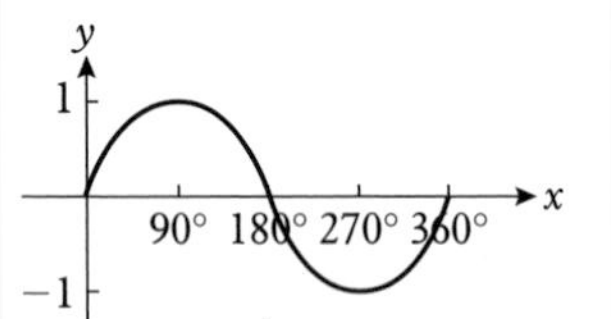

$y = \sin x$

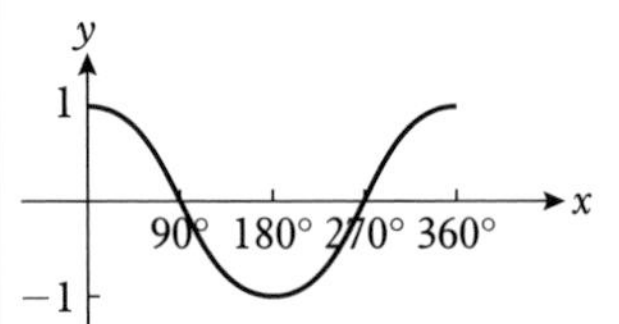

$y = \cos x$

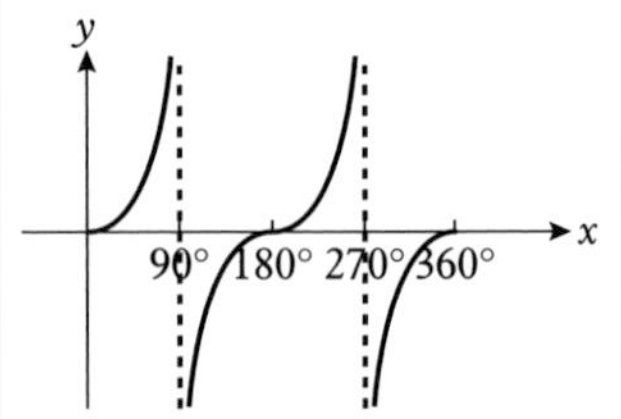

$y = \tan x$

Transforming trigonometric graphs

To sketch the graph of $y = a \sin bx$, you stretch the graph of $y = \sin x$ using a factor of a in the y direction and a factor of $\frac{1}{b}$ in the x direction.
You treat $y = a \cos bx$ in the same way starting from the graph of $y = \cos x$ and similarly $y = a \tan bx$ from the graph of $y = \tan x$.

$y = 2\sin x$

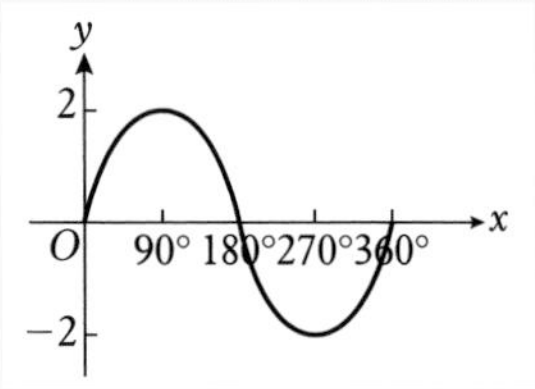

$y = \sin 2x$

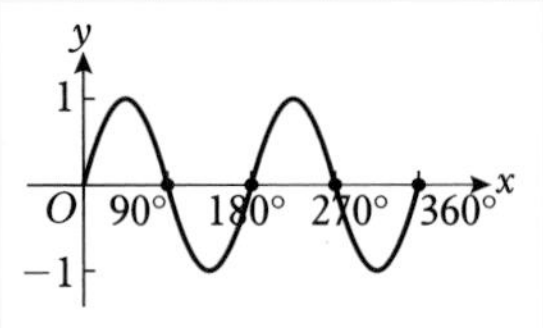

$y = \cos \frac{x}{2}$

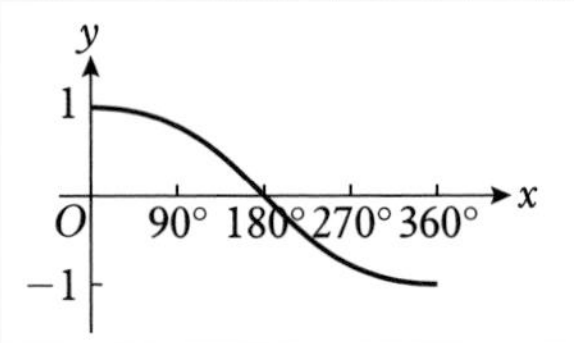

Sine and cosine rules

These rules allow you to do trigonometry in triangles that don't have a right angle. Label triangle ABC like this.

The angles are labelled with capital letters and the sides are labelled with lower case letters.
Each angle has the same letter as the opposite side.
This makes it easier to write some important rules in a way that you can remember.

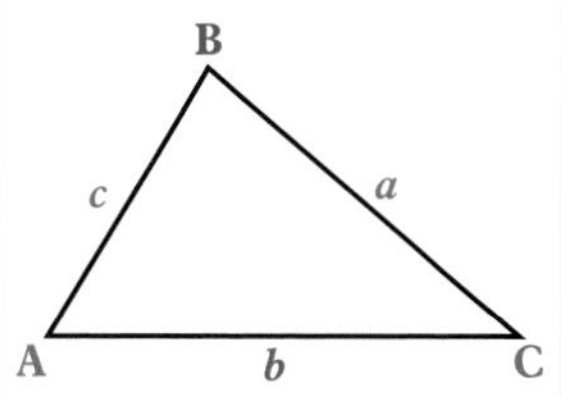

Sine rule

Use $\dfrac{a}{\sin A} = \dfrac{b}{\sin B} = \dfrac{c}{\sin C}$ to find a length or $\dfrac{\sin A}{a} = \dfrac{\sin B}{b} = \dfrac{\sin C}{c}$ to find an angle.

Cosine rule

Use $a^2 = b^2 + c^2 - 2bc \cos A$ to find a length or $\cos A = \dfrac{b^2 + c^2 - a^2}{2bc}$ to find an angle.

The area of a triangle

Area of a triangle = $\mathbf{\frac{1}{2}}\boldsymbol{ab}\,\mathbf{\sin C}$

You can think of this as:

Area $= \dfrac{1}{2} \times$ the product of any two sides $\times$ the sine of the angle between them.

Bearings

You have seen simple bearings questions in chapter 11.
Draw a diagram that shows the information you are given.
Decide on which triangle you need to work in.
You will need to use the sine or cosine rules to work out the lengths and angles that you need if the triangle is not right-angled.
Keep the exact value of any length or angle that you find in your calculator in case you need it in another part.
Remember to write down the answer to the actual question at the end of each part!

3-D problems

The sine and cosine rules can also be used in more complicated 3-D problems.
You still need to draw the triangle that you need to work in.
You did this in chapter 11 but the triangle was always right-angled then. You will still get some right-angled triangles here too and then you can use Pythagoras' theorem and ordinary trigonometry.
If the triangle is not right-angled, then use the sine or cosine rules to work out the lengths and angles that you need.
Keep the exact value of any length or angle that you find in your calculator in case you need it in another part.

1 **a** Sketch the graph of $y = \cos x$ for $0 \leqslant x \leqslant 360°$.
b Find a solution to $\cos x = 0.73$ using your calculator. Round your answer to 1 dp.
c Draw the line $y = 0.73$ on your graph in **a** and show the solution that you found in **b**.
d Use your graph to write down the other solution to the equation $\cos x = 0.73$ for $0 \leqslant x \leqslant 360°$.

2 **a** Sketch the graph of $y = \sin x$ for $0 \leqslant x \leqslant 360°$.
b Find a solution to $\sin x = 0.35$ using your calculator. Round your answer to 1 dp.
c Draw the line $y = 0.35$ on your graph in **a** and show the solution that you found in **b**.
d Use your graph to write down the other solution to the equation $\sin x = 0.35$ for $0 \leqslant x \leqslant 360°$.

3 Explain why the equation $\sin x = -1$ only has one solution for $0 \leqslant x \leqslant 360°$. Write down this solution.

4 Solve the equation $\cos x = -0.44$ for $0 \leqslant x \leqslant 360°$. Round your answers to 1 dp.

5 Solve these equations for $0 \leqslant x \leqslant 360°$.

a $\sin x = \frac{1}{2}$ **c** $\sin x = -\frac{1}{\sqrt{2}}$

b $\cos x = -\frac{\sqrt{3}}{2}$ **d** $\cos x = \frac{1}{\sqrt{2}}$

6 Sketch the graphs of these equations for $0° \leqslant x < 360°$.

a $y = 2\sin x$ **d** $y = \sin 2x$ **g** $y = \sin 3x$ **j** $y = \tan \frac{x}{2}$

b $y = \frac{1}{2}\cos x$ **e** $y = \cos 3x$ **h** $y = \cos \frac{x}{2}$ **k** $y = \cos \frac{x}{3}$

c $y = \tan 2x$ **f** $y = \tan 3x$ **i** $y = \sin \frac{x}{2}$ **l** $y = \frac{1}{2}\sin x$

7 **a** What is the maximum possible value of $4 \sin 2x$?
b What is the minimum possible value of $4 \sin 2x$?
c Sketch the graph of $y = 4 \sin 2x$ for $0° \leqslant x < 360°$.

8 **a** What is the maximum possible value of $5 \cos 3x$?
b What is the minimum possible value of $5 \cos 3x$?
c Sketch the graph of $y = 5 \cos 3x$ for $0° \leqslant x < 360°$.

9 **a** Sketch the graph of $y = \tan x$ for $0 \leqslant x \leqslant 360°$.
One solution of the equation $\tan x = 1.5$ is $x = 56.3°$.
b Draw the line $y = 1.5$ on your graph in **a** and show the solution given above.
c Use the sketch of $y = \tan x$ to find another solution of $\tan x = 1.5$ for $0 \leqslant x \leqslant 360°$ and show this solution on your graph.
d Use the sketch to solve the equation $\tan x = -1.5$ for $0° \leqslant x \leqslant 360°$.

10 Use the sine rule to find the marked side or angle in each of these.
Give your answers to 1 dp.

a

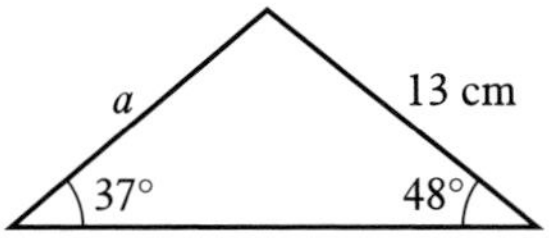

b

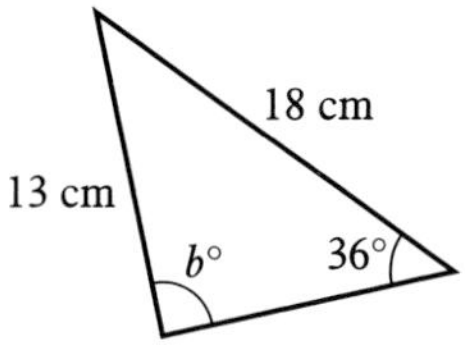

11 Use the cosine rule to find the marked side or angle in each of these.
Give your answers to 1 dp.

a

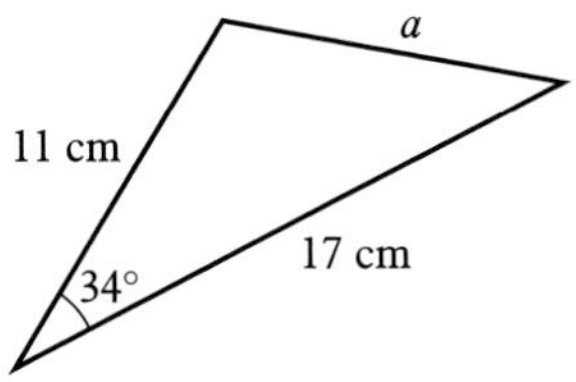

b

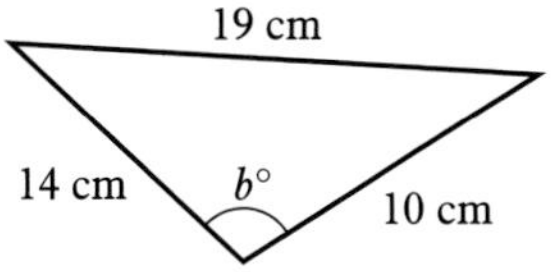

12 Find the area of this triangle.

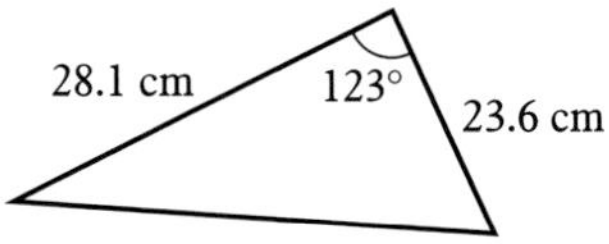

13 A helicopter flies 7 miles from Accrington to Blackburn on a bearing of 268°.
It then flies 8 miles on a bearing of 217° to Chorley.
a Draw a diagram to show this information.
b Find the direct distance from Accrington to Chorley.
c Find the bearing of Chorley from Accrington.

14 WXYZ is a tetrahedron.
Z is vertically above W.
WX = 12 cm, XY = 14 cm,
WZ = 5 cm, ∠WXY = 130° and
∠ZXY is obtuse. Find:

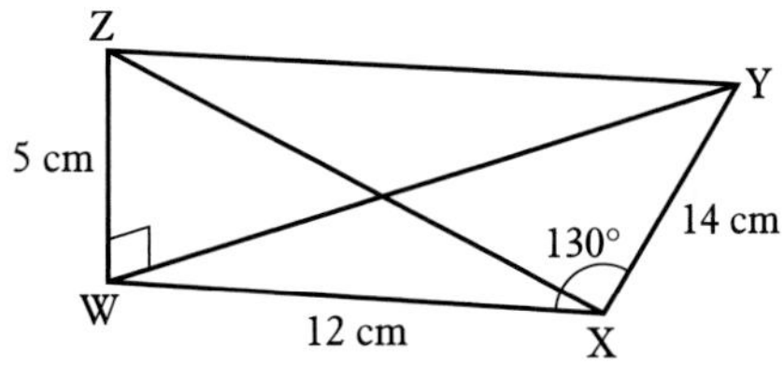

a WY
b ∠WYX
c YZ
d XZ
e ∠XZY
f ∠ZXY

1 The distance between two radio masts, A and B, is 35 km.
B is east of A.
A third mast, C, is on a bearing of 047° from A, and on a bearing of 302° from B.
Find the distances of A and B from C, correct to 3 significant figures.

(5 marks)

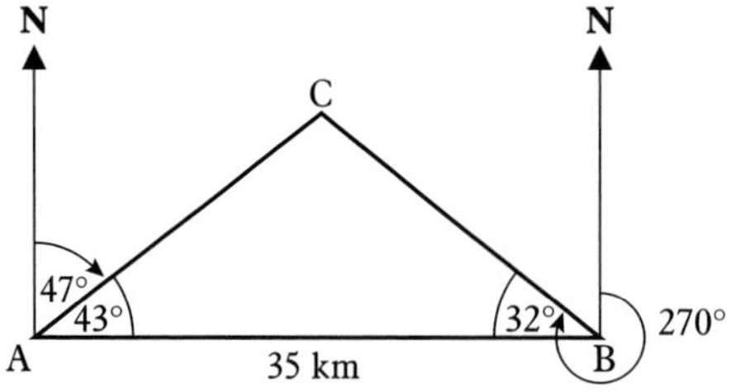

Add the missing angles to your diagram.

Angle C = 180° − 43° − 32° = 105°

Using the sine rule, put in what you know already.

$$\frac{BC}{\sin 43°} = \frac{AC}{\sin 32°} = \frac{35}{\sin 105°}$$

You need to show this first step, as this is where the first mark is earned. **1 mark**

The missing sides can now be found: *Show the next step.*

$$\frac{BC}{\sin 43°} = \frac{35}{\sin 105°} \qquad \frac{AC}{\sin 32°} = \frac{35}{\sin 105°}$$

$$BC = \frac{35 \sin 43°}{\sin 105°}$$ **1 mark**

$$AC = \frac{35 \sin 32°}{\sin 105°}$$ **1 mark**

$$BC = \frac{35 \times 0.682\ldots}{0.9659} \qquad AC = \frac{35 \times 0.5299\ldots}{0.9659}$$

Put numbers in only at the last stage.

BC = 24.711 98… AC = 19.201 448…

Remember to use the most accurate values you can, not the rounded decimals you may have written.

BC = 24.7 km **1 mark**

BC = 19.2 km **1 mark**

2 Find the two values of x for which $3 \cos(x - 20°) = 1.5$ and $0 \leqslant x \leqslant 360°$.

(3 marks)

The first step is to rearrange the equation:

$$\frac{\cancel{3} \cos(x - 20°)}{\cancel{3}} = \frac{1.5}{3}$$

Show the rearrangement.

$$\cos(x - 20°) = 0.5$$ **1 mark**

Press: **2nd F** **cos** **0** **·** **5** **=**
The calculator gives a single value of 60°.
Now use the graph of $y = \cos x$
to see the other solution is 300°.

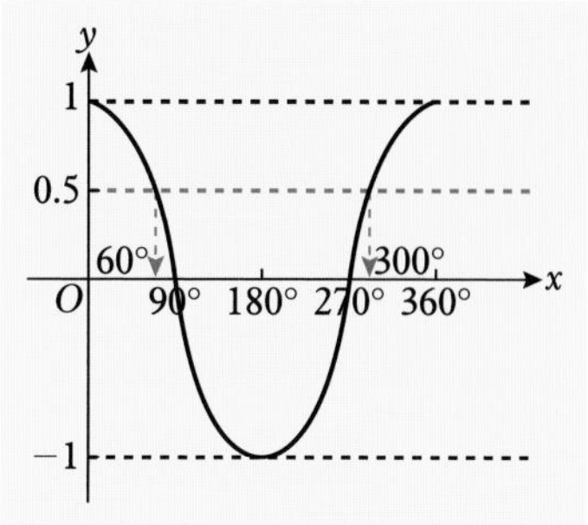

So $x - 20° = 60°$ and $x - 20° = 300°$.

So $x = 80°$ and $x = 320°$.

Draw a sketch graph to help you with the question. It is important to show where the 60° and the 300° come from.

2 marks: 1 each

3 Sketch the graph of $y = 2 \sin 3x$ for $0 \leqslant x \leqslant 180°$. **(3 marks)**
Start with graph of $y = \sin x$

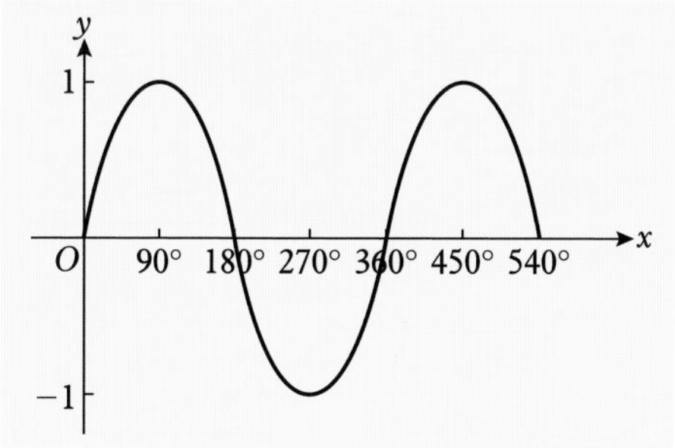

1 mark

Now draw the graph of $y = \sin 3x$:

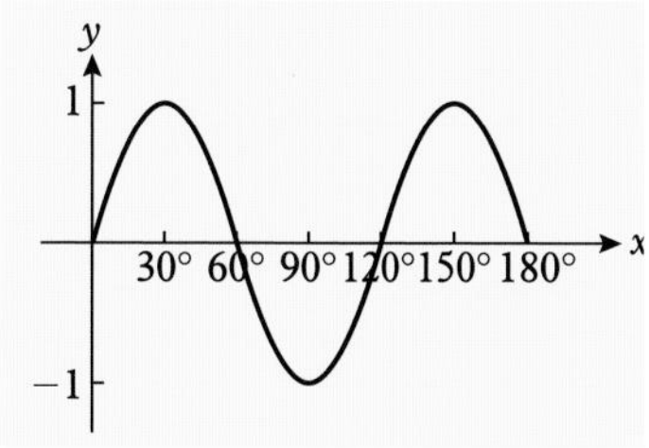

This is a stretch of factor $\frac{1}{3}$ in the x direction.
(360°, 0) will move to (120°, 0), (180°, 0) will move to (60°, 0), etc.
You need to draw as far as $x = 180°$, which comes from $x = 540°$ above.

1 mark

Now draw the graph of $y = 2 \sin 3x$:

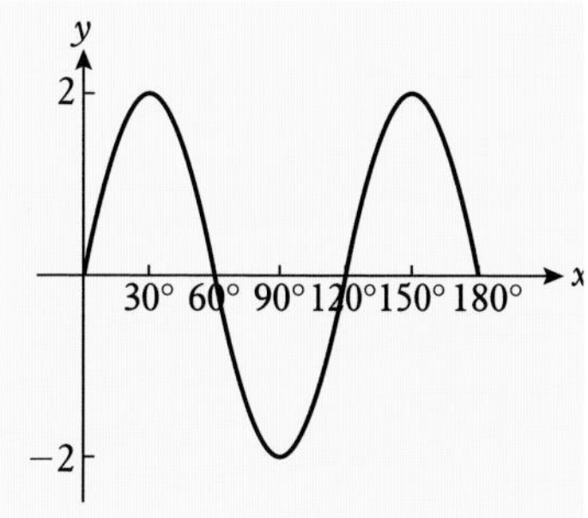

This is twice the size, a stretch of factor 2 in the y direction.

1 mark

1 This is a sketch of the graph of $y = \sin x°$ for values of x between 0 and 360.

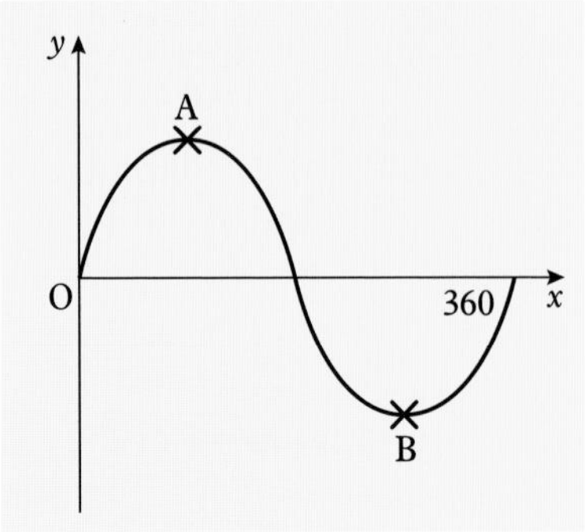

a Write down the co-ordinates of the points
(1) A, **(2)** B. **(2 marks)**

b On the same axes sketch the graph of $y = \sin 2x°$ for values of x between 0 and 360. **(2 marks)**

[S2001 P5 Q15]

2

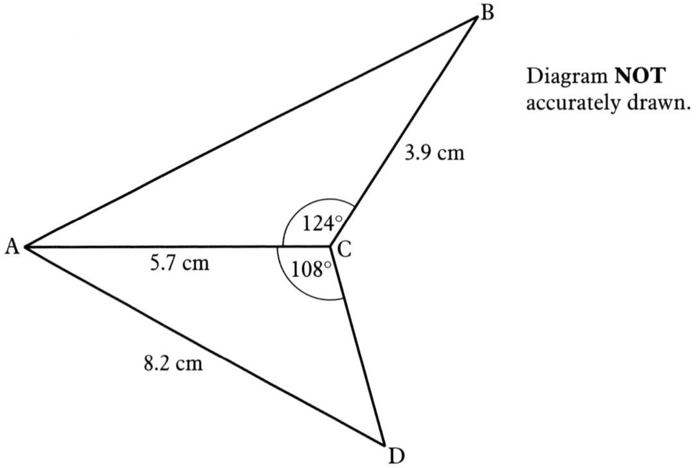

AC = 5.7 cm,
AD = 8.2 cm,
BC = 3.9 cm,
Angle ACD = 108°,
Angle ACB = 124°.

a Calculate the size of angle ADC.
Give you answer correct to 1 decimal place. **(3 marks)**

b Calculate the area of triangle ABC.
Give you answer, in cm^2, correct to 3 significant figures. **(2 marks)**

c Calculate the size of angle BAC.
Give you answer correct to 1 decimal place. **(5 marks)**

[N1999 P5 Q17]

3

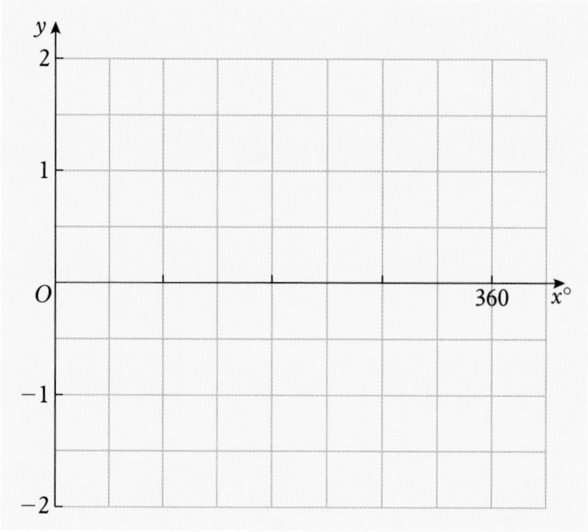

a On the grid above, sketch the graphs of
(1) $y = \sin x°$, **(2)** $y = \sin 2x°$,
for values of x between 0 and 360. Label each graph clearly. **(4 marks)**

b Calculate all the solutions to the equation

$2 \sin 2x° = -1$

between $x = 0$ and $x = 360$. **(3 marks)**

[S1998 P5 Q16]

4

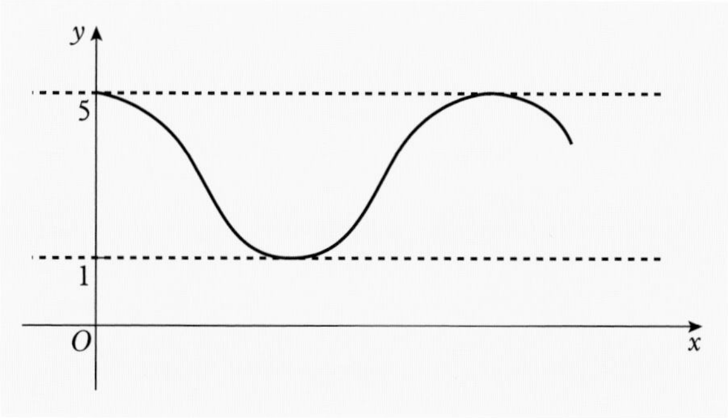

The diagram shows a sketch of part of the curve with equation

$y = p + q \cos x°$

where p and q are integers.

a Find the values of the integers p and q. **(2 marks)**

b Find the two values of x between 0 and 360 for which

$p + q \cos x° = 2$. **(3 marks)**

[N1999 P6 Q13]

5 The depth of water, d metres, at the entrance to a harbour is given by the formula

$d = 5 - 4\sin(30t)°$

where t is the time in hours after midnight on one day.

a On the axes below, draw the graph of d against t for $0 \leqslant t \leqslant 12$.

(3 marks)

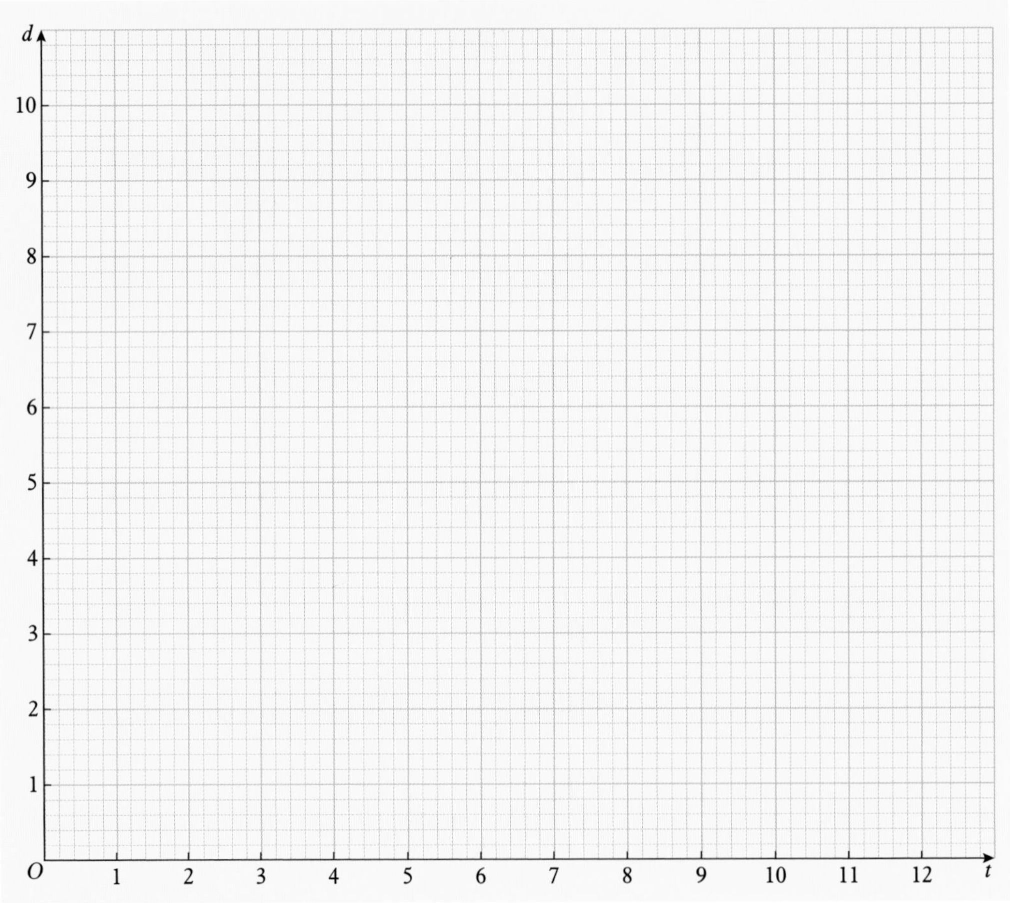

b Find the two values of t, where $0 \leqslant t \leqslant 24$, when the depth is least.

(1 mark)

[N2000 P6 Q15]

6

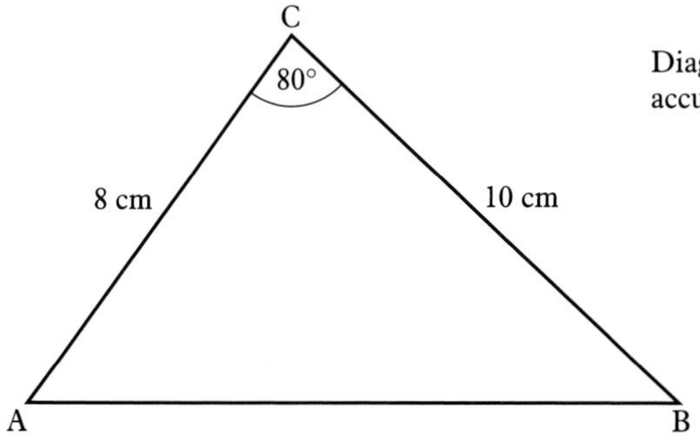

Diagram **NOT** accurately drawn.

a Calculate the length of AB.
Give your answer, in centimetres, correct to 3 significant figures.
(3 marks)

b Calculate the size of angle ABC.
Give you answer correct to 3 significant figures. **(3 marks)**
[S2001 P6 Q14]

7

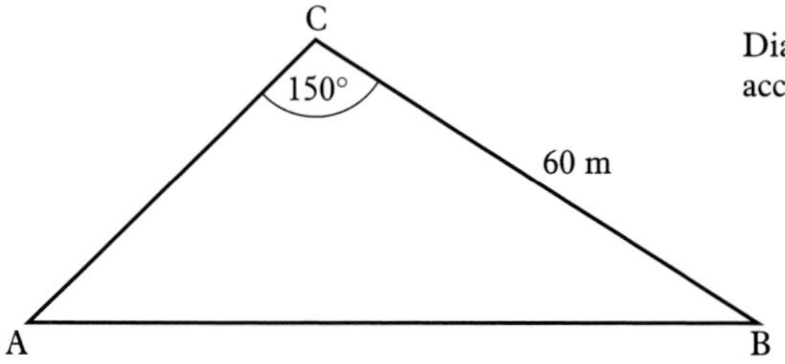

Diagram **NOT** accurately drawn.

Angle ACB = 150°.
BC = 60 m.
The area of triangle ABC is 450 m^2.
Calculate the perimeter of triangle ABC.
Give your answer correct to 3 significant figures. **(5 marks)**
[N2000 P6 Q20]

8 In the quadrilateral ABCD,
AB = 6 cm,
BC = 7 cm,
AD = 12 cm,
angle ABC = 120°,
angle ACD = 70°.

Calculate the size of angle ADC.
Give your answer correct to 3 significant figures.

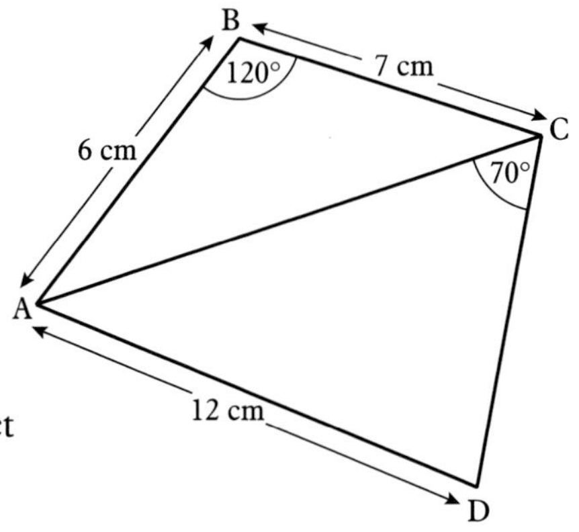

Diagram **NOT** accurately drawn.

(6 marks)
[N1998 P6 Q20]

20 Graphs: for real

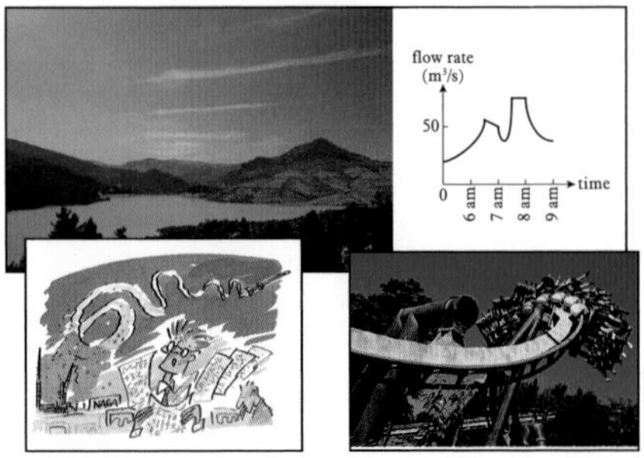

You need to know about:

- interpreting the gradient of straight line graphs
- sketching graphs that show real situations
- areas under straight line graphs
- modelling with exponential and trigonometric graphs

The gradient of straight line graphs

You can use the gradient to measure how quickly the values on the vertical axis change compared with the values on the horizontal axis.
The gradient of a distance–time graph is speed.
The gradient of a displacement–time graph is velocity.

Acceleration is the rate of change of velocity with respect to time. For velocity in m/s and time in s, acceleration is measured in m/s^2. The gradient of a velocity–time graph is the acceleration. For a straight line velocity–time graph, the acceleration is constant. It has the same value for the whole graph. Constant acceleration is also called uniform acceleration.

Sketching graphs to show real situations

This is a graph showing the height of water in the container as the container is filled at a constant rate.
For vertical sides the graph is a straight, angled line.
For angled sides the graph is curved. As the bottle narrows the rate at which the height changes will increase.

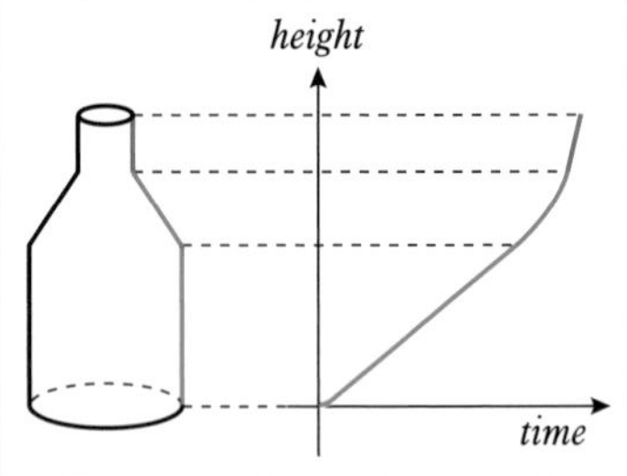

Area under graphs

The area under a speed–time graph is the distance travelled. Areas under other graphs have meaning too. To find what an area represents, multiply the units on each of the axes. So for a vertical axis showing flow of water in litres per minute, and a horizontal axis showing time in minutes, the area represents the volume of liquid in litres as $\frac{l}{\text{min}} \times \text{min} = l$.

Modelling using exponential and trigonometric functions

Exponent is another word for power. An **exponential function** is a function where the variable is the power or exponent.
The exponential function $f(x) = pq^x$ is very important because its behaviour models the way that changes occur in many situations in everyday life.
Once you know the values of p and q you can use the function to predict how the situation will change in the future.

Example The temperature T°C of an object is modelled by $T = pq^t$.
t stands for the time in minutes since the object started to cool.
p and q are constants.

a Use the information in the sketch to find the values of p and q.
b Find the temperature of the object when $t = 5$.

a From the sketch, $T = 100$ when $t = 0$.
so $100 = pq^0$
$q^0 = 1$, so $100 = p \times 1$
$p = 100$

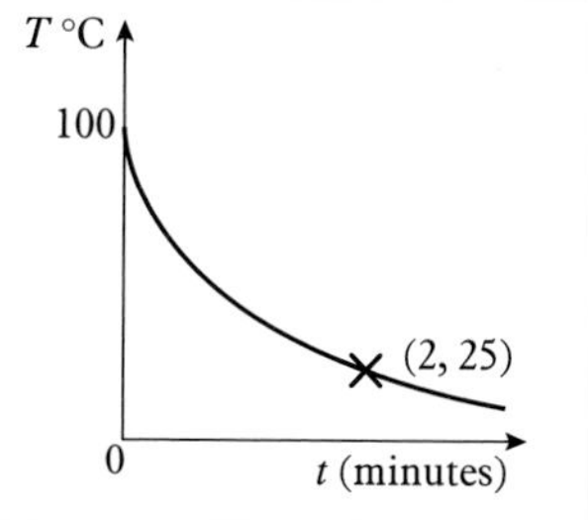

You can also see from the sketch that
$T = 25$ when $t = 2$, so $25 = 100q^2$
$q^2 = 0.25$
$q = 0.5$

b You can now use the formula $T = 100(0.5)^t$ to find T at any time.
When $t = 5$, $T = 100(0.5)^5$
$= 3.125$
So the temperature of the object when $t = 5$ is 3.1°C to 1 dp.

A different kind of function is needed to model behaviour that repeats itself in a cycle. You can use trigonometric functions to do this.

Example The depth of water in a harbour changes with the tide.
The depth d m is modelled by $d = 2\cos(30t)° + 3.5$ where t is the number of hours after high tide.

a How deep is the water at high tide?
b How deep is the water at low tide?
c High tide is at 12 noon. Find the depth of water at 5 pm.

a The largest possible value of $\cos(30t)°$ is 1.
So at high tide $d = 2 \times 1 + 3.5 = 5.5$ m.

b The least possible value of $\cos(30t)°$ is -1.
So at low tide $d = 2 \times -1 + 3.5 = 1.5$ m.

c At 5 pm $t = 5$. Using $d = 2\cos(30t)° + 3.5$
$d = 2\cos 150° + 3.5$
$= 1.767\ldots$
The depth of water at 5 pm is 1.8 m to 1 dp.

1 Glenda is out in her car one afternoon.
The graph shows how far away from home she is.

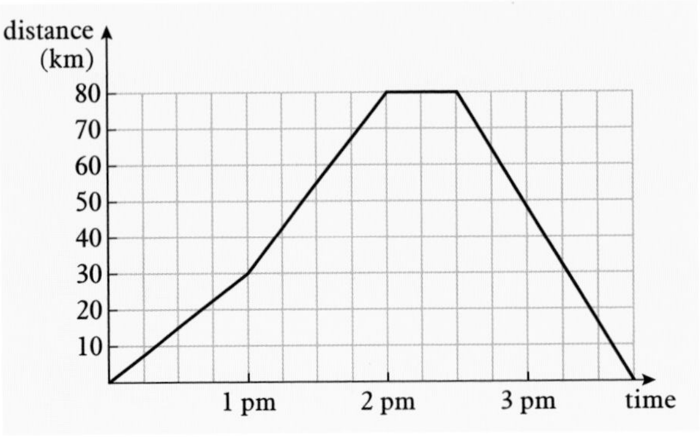

a How fast is Glenda travelling between noon and 1 pm? Give your answer in km/h.
b How fast is Glenda travelling between 1 pm and 2 pm? Give your answer in km/h.
c Explain what is happening between 2 pm and 2:30 pm.
d What is Glenda's speed on her return journey? Give your answer in km/h.

2 The graph shows how the flow rate of water from a pipe changes during a 5 minute interval.

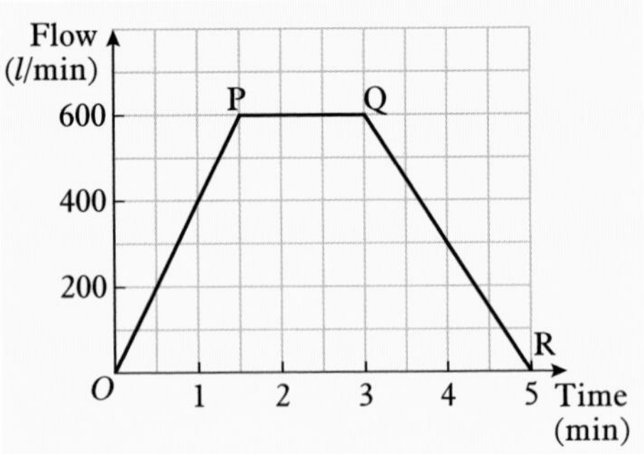

a Find the gradient of OP.
b What does the gradient represent? Include the units in your answer.
c What is happening between P and Q?
d Find the gradient of QR. What does the sign of the gradient show?
e Write down what the area under the graph represents.
f What is the value of each square of area on the graph?
g Find the total area under the graph.

3 Draw a graph to show how the height of water in each of these tanks changes as the tank is filled at a constant rate.

a

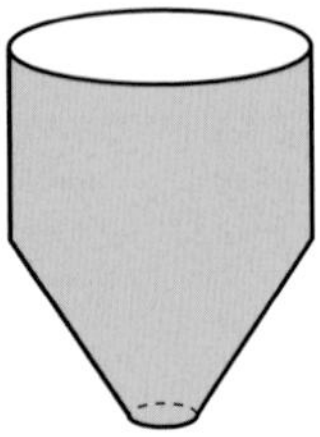

b

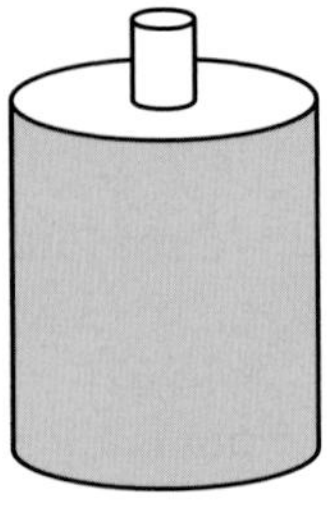

4 The number of bacteria in a colony doubles every hour.
Initially there are n bacteria.
Write down a formula for the number of bacteria in the colony after t hours.

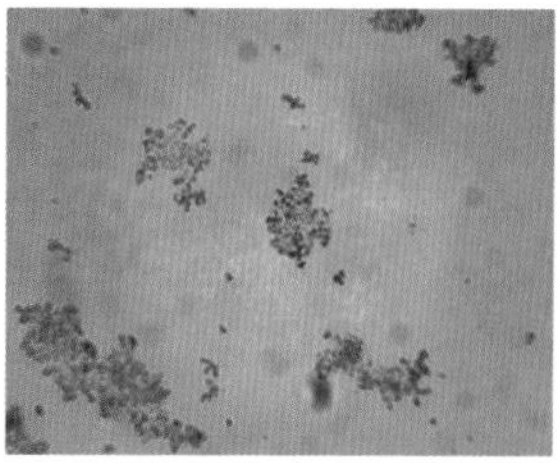

5 Uranium-238 is the most stable isotope of uranium. It has a half-life of about 10^9 years! This means that is takes 10^9 years for half of the amount of uranium that you start with to decay, then another 10^9 years for half of the remaining amount to decay and so on.

a How many half-lives are needed for a mass of 20 grams of uranium to decay to less than 1 gram?

b Write down a formula for the mass of uranium, m, remaining after t half-lives.

6 This is a diagram of a racetrack.
The start point is shown.
Cars travel round the circuit clockwise.

a Draw a graph of the speed of a car as it completes its first lap.

b Continue the graph to show the speed of the car on the second lap.

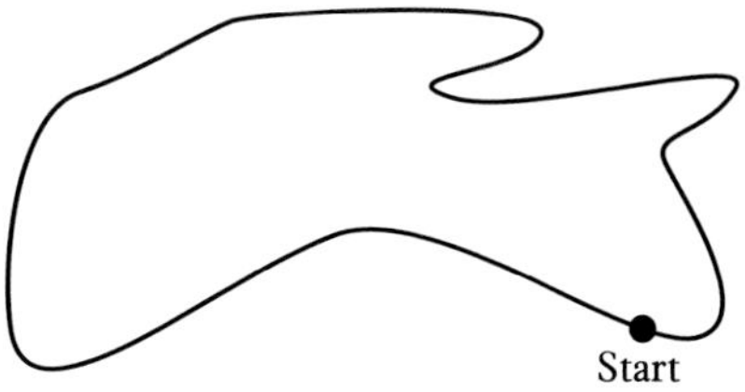

7 Phil puts money into an investment account.
The value, £v, of the account after t years is modelled by $v = pq^t$.
The values of p and q are constant.

a Use the sketch to find p and q.

b What does p represent?

c What is the value of the account after 6 years?

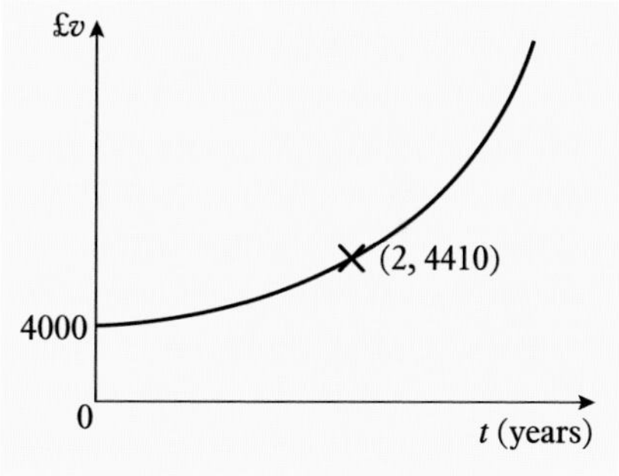

8 The depth of water, d m, in a harbour is given by the formula $d = 6 + 3\sin(30t)°$ where t is the number of hours since midnight.

a What is the depth of the water at midnight?

b What is the depth of the water at high tide?

c What is the depth of the water at low tide?

d What time is it when high tide first occurs?

e What time is it when high tide next occurs?

f What time is it when low tide first occurs?

g Find the depth of water at 9 pm.

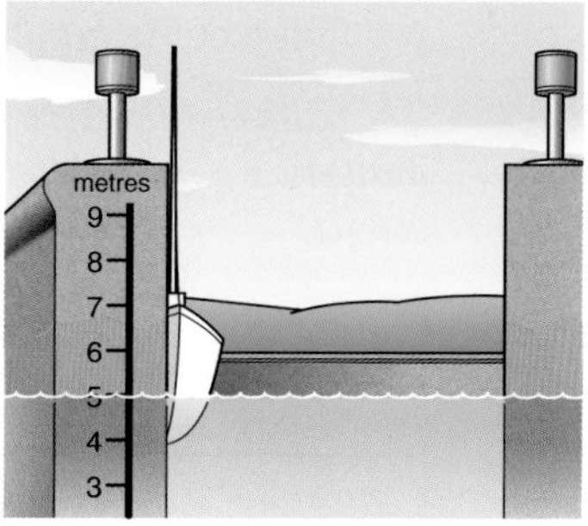

1

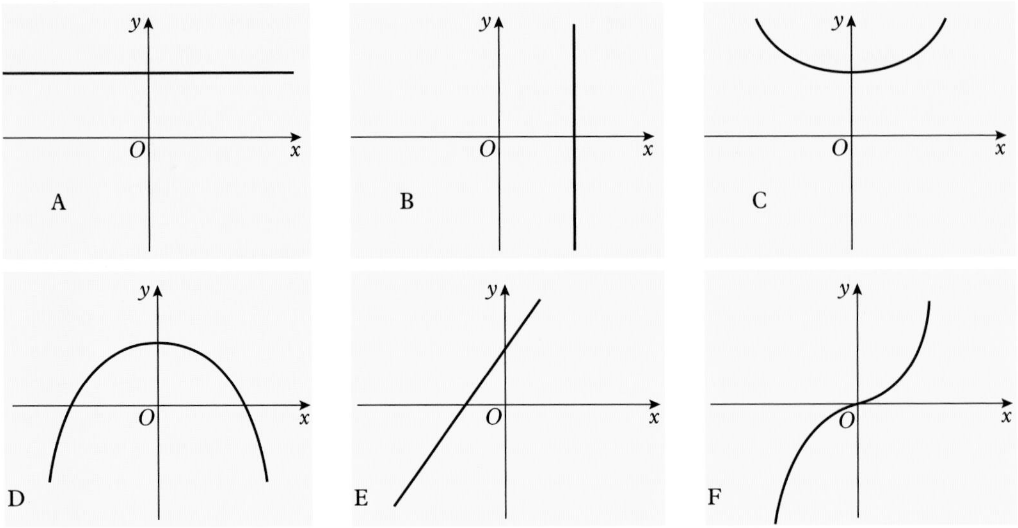

There are six sketches of graphs shown above.
There are six equations shown below.

$y = -x^2 + 2$		$x = 3$	
$y = 2$		$y = x^2 + 2$	
$y = x^3$		$y - 2x = 2$	

Next to each equation, write the letter of the graph for the appropriate function.

(4 marks)

First do the straight lines:
$x = 3$ crosses the x axis vertically, so graph B.
$y = 2$ crosses the y axis horizontally, so graph A.

$y - 2x = 2$ is the same as $y = 2x + 2$, a diagonal straight line crossing the y axis at (0, 2), with positive gradient, so graph E.

$y = x^2 + 2$ is a quadratic ($\cup$ shape) crossing at (0, 2), so graph C.
$y = -x^2 + 2$ is an inverted quadratic ($\cap$ shape) crossing at (0, 2), so graph D.

$y = x^3$ is a cubic graph, which is graph F.

The marks for this type of question are normally in proportion to the number of correct answers: e.g. 1 mark for 2 correct answers, 2 marks for 3, 3 marks for 4, 4 marks all correct.

$y = -x^2 + 2$	D	$x = 3$	B
$y = 2$	A	$y = x^2 + 2$	C
$y = x^3$	F	$y - 2x = 2$	E

4 marks

2 A radioactive substance decreases in mass with the emission of particles and rays. For a particular radioactive substance the following yearly results are shown:

Time, t years	0	1	2	3
Mass, m kg	30	27	24.3	21.87

a Calculate the percentage decrease per year. **(2 marks)**
b Write a formula for m in terms of t. **(3 marks)**
c Find the mass after 20 years. **(2 marks)**
d Find the year in which the mass becomes 15 kg. **(2 marks)**

a *Calculate the percentage decreases for each year:*

$$\frac{30-27}{30} \times 100 = 10\%$$ **1 mark**

$$\frac{27-24.3}{27} \times 100 = 10\%,$$

$$\frac{24.3-21.87}{24.3} \times 100 = 10\%$$ **1 mark**

b This is an exponential function of the form: $m = a \times b^t$

When $t = 0, m = 30$. $\quad 30 = a \times b^0$

Substitute in values. $\quad 30 = a \times 1$ so $a = 30$ **1 mark**

Show clearly what values you are substituting.

For $m = 30 \times b^t$

When $t = 1, m = 27$ so $27 = 30 \times b^1$

$$b = \frac{27}{30} = 0.9$$ **1 mark**

So the formula is $m = 30 \times 0.9^t$ **1 mark**

Always write down the complete formula after finding a and b.

c $m = 30 \times 0.9^{20}$ **1 mark**

$= 30 \times 0.121\,576\ \ldots.$

$= 3.65$ kg **1 mark**

Work out 0.9^{20} first, then $\times 30$.

d $15 = 30 \times 0.9^t$

so $0.5 = 0.9^t$ **1 mark**

$0.9^1 = 0.9 \quad 0.9^2 = 0.81, \ldots$ Now use trial and improvement on the calculator until 0.9^t becomes 0.5 or less.

$0.9^5 = 0.590\,59$

$0.9^6 = 0.531\,441$

$0.9^7 = 0.478\,296\,9$

It is important that you show all trials, with answers.

So the mass becomes 15 kg in the seventh year. **1 mark**

1 The diagram shows a water tank.
The tank is a hollow cylinder joined to a hollow hemisphere at the top.
The tank has a circular base.

The empty tank is slowly filled with water.

a On the axes, sketch a graph to show the relation between the volume, V cm^3, of water in the tank and the depth, d cm, of water in the tank. **(3 marks)**

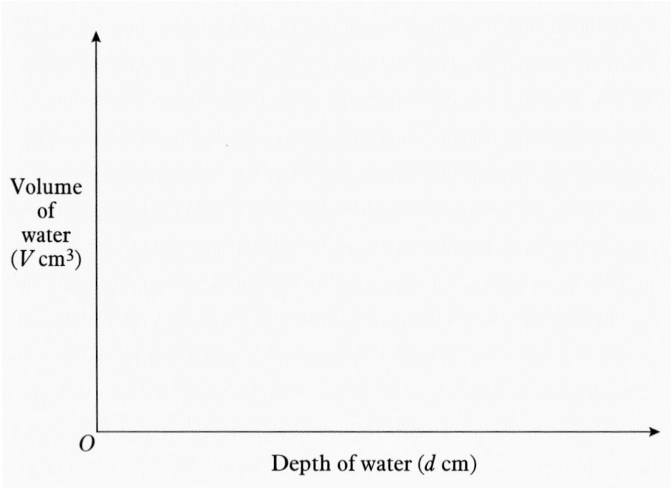

Both the cylinder and the hemisphere have a diameter of 46 cm.
The height of the tank is 90 cm.

b Work out the volume of water which the tank holds when it is full. Give your answer correct to 3 significant figures. **(4 marks)**

[S2000 P6 Q10]

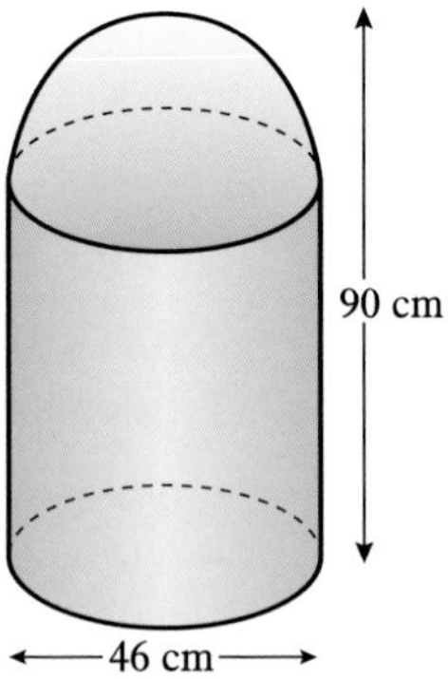

Diagram **NOT** accurately drawn.

2 **a** Simplify

(1) $pq \times q^2$ **(2)** $3^x \times 3^y$ **(2 marks)**

$2^c \times 8^{2c} = 2^k$

b Express k in terms of c. **(2 marks)**

The sketch graph shows a curve with equation $y = pq^x$ where $q > 0$.

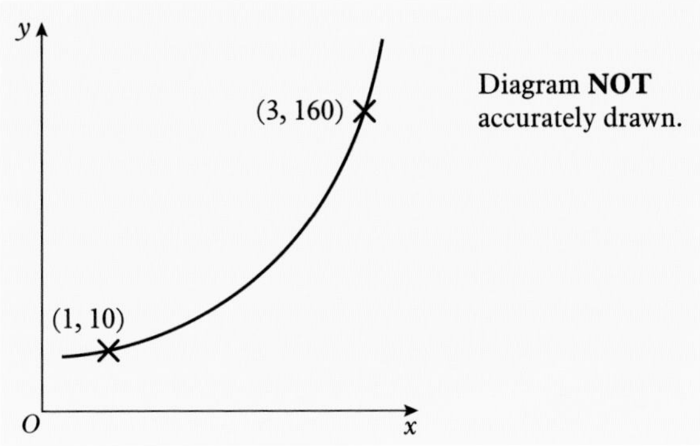

The curve passes through the points (1, 10) and (3, 160).

c Calculate the values of p and q. **(3 marks)**

[N2001 P6 Q14]

3 This sketch shows part of the graph with equation

$y = pq^x$,

where p and q are constants.
The points with coordinates (0, 8), (1, 18) and (1.5, k) lie on the graph.
Calculate the values of p, q and k.

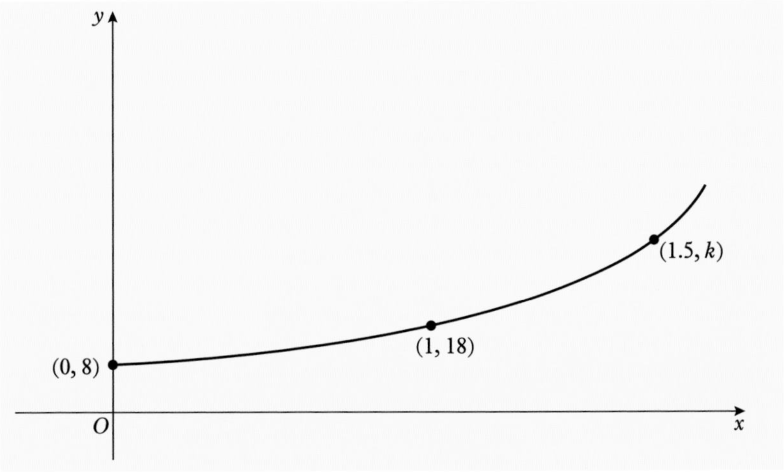

(6 marks)

[S1999 P5 Q17]

4 In the College games, Michael Jackson won the 200 metres race in a time of 20.32 seconds.

a Calculate his average speed in metres per second.
Give your answer correct to 1 decimal place. **(2 marks)**

b Change your answer to part a to kilometres per hour.
Give your answer correct to 1 decimal place. **(2 marks)**

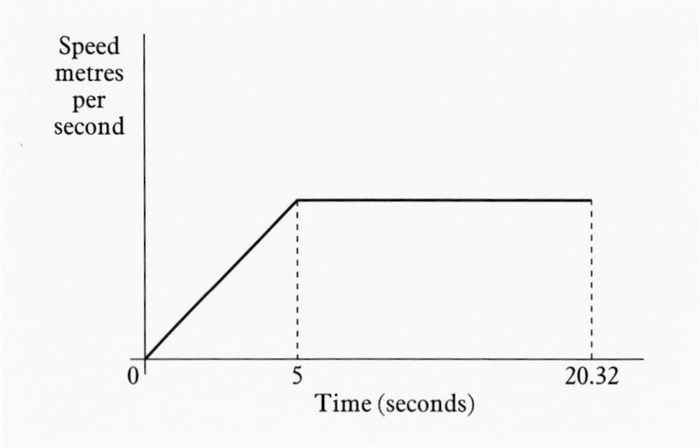

The diagram shows a sketch of the speed/time graph for Michael Jackson's race.

c Calculate his maximum speed in metres per second.
Give your answer correct to 1 decimal place. **(2 marks)**

d Calculate his acceleration over the first 5 seconds.
State the units in your answer.
Give your answer correct to 2 significant figures. **(2 marks)**
[S1998 P6 Q11]

CHAPTER 1

Questions

1 **a** $\begin{pmatrix} 3 \\ -3 \end{pmatrix}$ **c** $\begin{pmatrix} -5 \\ 0 \end{pmatrix}$

b $\begin{pmatrix} 4 \\ 4 \end{pmatrix}$ **d** $\begin{pmatrix} 0 \\ -4 \end{pmatrix}$

2

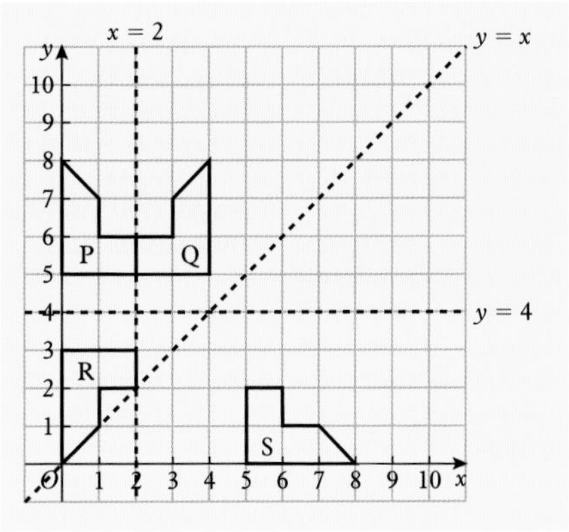

3

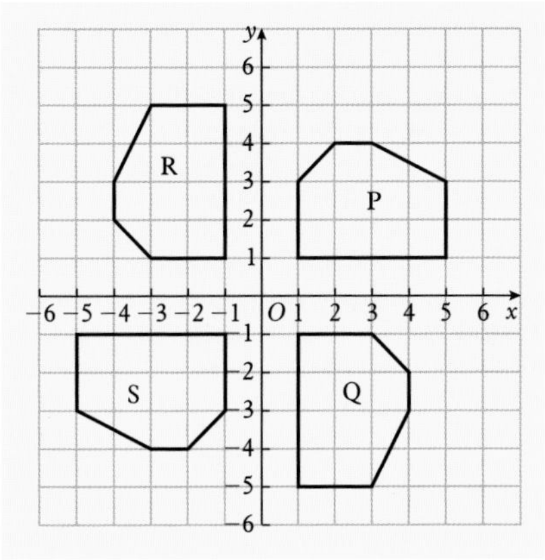

4 **a, b, c**

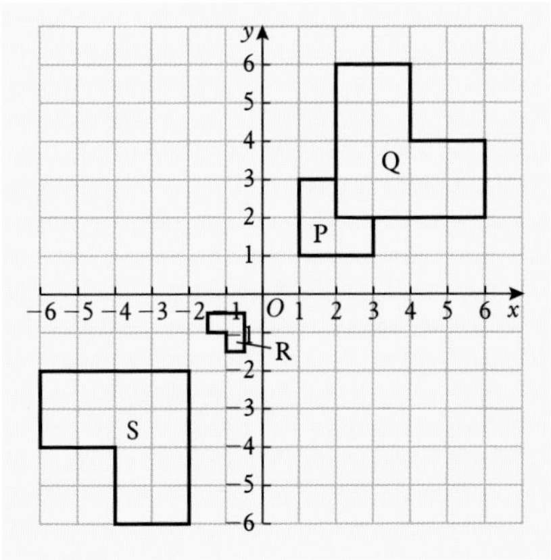

5 **a, b, c**

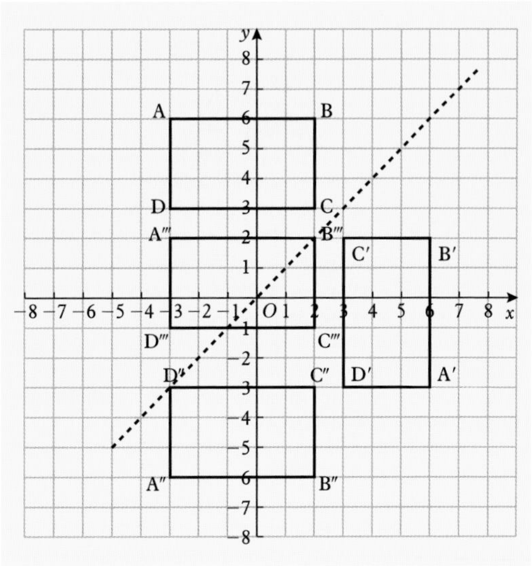

d Translation of $\begin{pmatrix} 0 \\ 4 \end{pmatrix}$

6 **a, b, c**

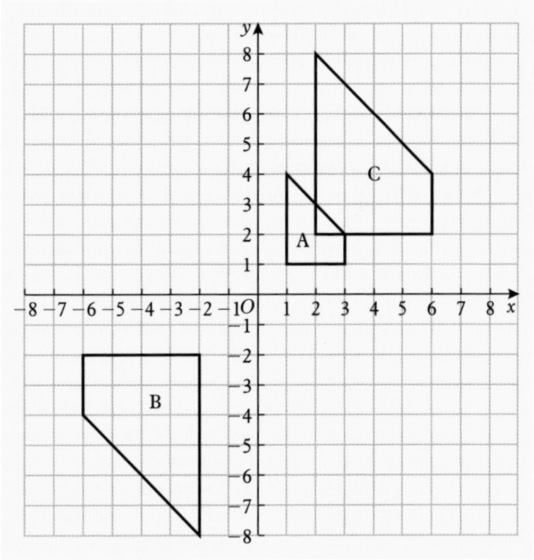

c Enlargement with scale factor $\frac{1}{2}$, centre O.

Exam questions

1 **a** 180° rotation, centre (0, 0)

b

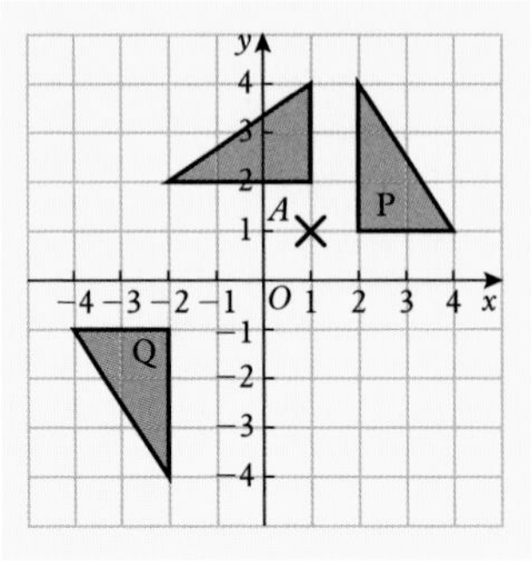

2 Enlargement with scale factor $\frac{1}{3}$, centre $(-5, 0)$

3 **a**

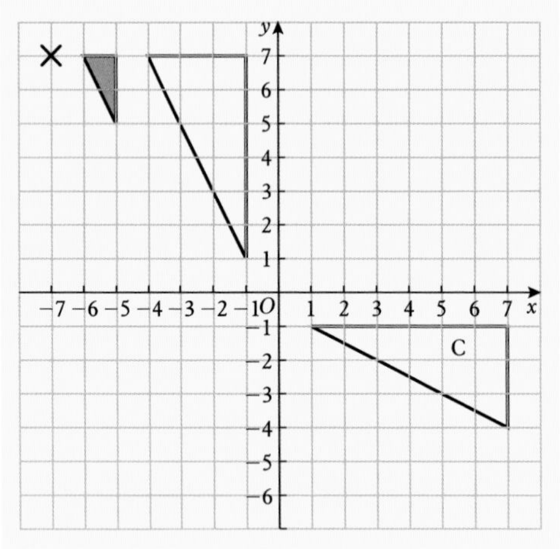

b Reflection in $y = x$

4 **a,b**

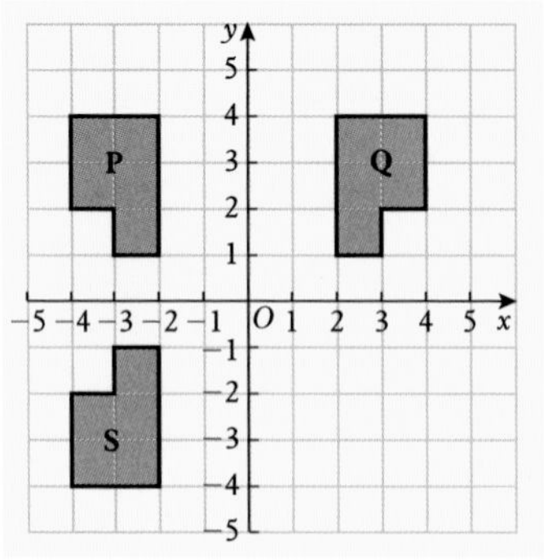

c Reflection in the x axis

5 **a,b**

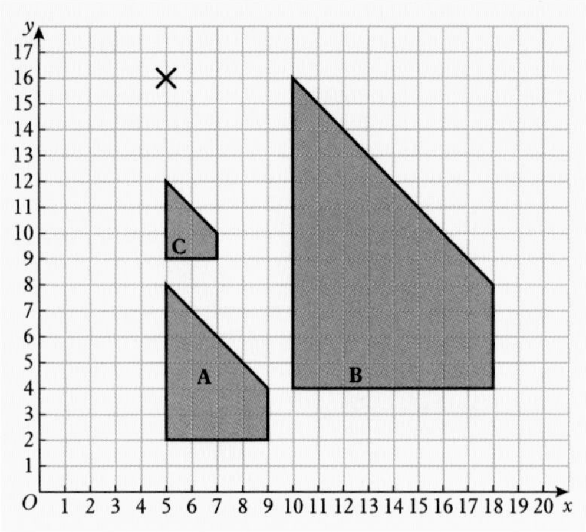

6 **a** $\frac{1}{3}$

b

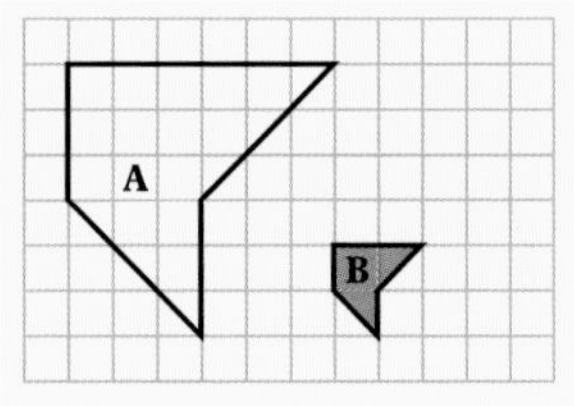

CHAPTER 2
Questions

1 **a** 3, −7, 0, 16, 5, 25, 15 **d** 3, 5
b 3, 16, 5, 25, 15 **e** 16, 25
c All except $\sqrt{5}$ **f** 3, 15

2 $30 = 1 \times 30$ factors: 1, 2, 3, 5, 6, 10, 15, 30
2×15
3×10
5×6

3 **a** $36 = 2 \times 2 \times 3 \times 3$
b $42 = 2 \times 3 \times 7$
c HCF of 36 and 42 is $2 \times 3 = 6$
(2 and 3 are the common prime factors.)

4 7, 14, 21, 28, 35

5 **a** $24 = 2 \times 2 \times 2 \times 3$
b $14 = 2 \times 7$
c $\text{LCM} = 2 \times 2 \times 2 \times 3 \times 7$
$= 168$

6 **a** $\frac{3}{10}$ **c** $\frac{7}{9}$
b $\frac{184}{1000} = \frac{23}{125}$ **d** $\frac{36}{99} = \frac{4}{11}$

7 $\sqrt{\frac{25}{4}} = \frac{5}{2}$ so this is rational as it is a fraction.

8 **a** $\sqrt{9} = 3$
b $\sqrt{8.5}$, for example.

9 **a** irrational (irrational plus rational)
b $\sqrt{30}$ = irrational
c rational
d irrational (π is irrational, irrational − rational)
e $\sqrt{36} = 6$ rational
f $\sqrt{17}$ irrational
g irrational (rational × irrational)
h irrational (rational + irrational)

10 **a** $0.\dot{4}2857\dot{1}$ **c** $\frac{2}{10} = \frac{1}{5}$
b $0.0\dot{4}2857\dot{1}$ **d** $\frac{1}{5} + \frac{3}{70} = \frac{17}{70}$

11 **a** $\sqrt{20} = 2\sqrt{5}$ **e** $\dfrac{3\sqrt{5}}{5}$
b $4\sqrt{2}$ **f** $\sqrt{12} = 2\sqrt{3}$
c $2\sqrt{15}$ **g** $\dfrac{7\sqrt{13}}{13}$
d $\sqrt{\dfrac{9}{4}} = \dfrac{3}{2}$ **h** $\dfrac{2\sqrt{2}+2}{2} = \sqrt{2} + 1$

12 **a** $18 + 3\sqrt{5} - 6\sqrt{5} - 5 = 13 - 3\sqrt{5}$
b $9 + 3\sqrt{2} + 3\sqrt{2} + 2 = 11 + 6\sqrt{2}$
c $9 - 3\sqrt{2} + 3\sqrt{2} - 2 = 7$
d $4 + 2\sqrt{3} - 2\sqrt{3} - 3 = 1$

13 **a** $2\sqrt{2} + \sqrt{2} = 3\sqrt{2}$
b $\sqrt{3} + 3\sqrt{3} = 4\sqrt{3}$
c $3\sqrt{6} - \sqrt{6} = 2\sqrt{6}$
d $\sqrt{36 \times 5} - \sqrt{16 \times 5} = 6\sqrt{5} - 4\sqrt{5} = 2\sqrt{5}$

14 **a** No, irrational
b Yes, $\frac{2}{9} = 0.\dot{2}$
c No, irrational
d No, irrational
e No, irrational
f Yes, $\sqrt{5\dfrac{4}{9}} = \sqrt{\dfrac{49}{9}} = \dfrac{7}{3} = 2.\dot{3}$

15 **a** $2n$
b $2m - 1$
c $2n + 2m - 1$
$= 2(n + m) - 1$, which is double a number minus 1 so it is odd.
d $2n(2m - 1)$
$= 4nm - 2n$
$= 2(2nm - n)$, which is double a number so it is even.

16 **a** e.g. $2n$
b e.g. $2(n + 1) = 2n + 2$
c $2(n + 2) = 2n + 4$
d $2n + 2n + 2 + 2n + 4$
$= 6n + 6$
$= 6(n + 1)$, which is 6 times a number so it is a multiple of 6.

Exam questions

1 **a** e.g. $x = \sqrt{20}$ **b** $k = 3\frac{1}{2}$
2 **a** **i** $m = 3\frac{1}{2}$ **ii** $n = 1$
b $t = 3$
3 **a** A perfect square
b 4
c $\sqrt{20} - 4$
4 **a** 67×73 **b** $(2n + 1)^2 - (2n - 1)^2 = 8n$
5 **a** 1.5 **b** 4.5
6 **a** e.g. $k = \sqrt{2}$ or $n = 4$ **b** $2^{1.5}$
c $d = -\frac{1}{4}$

7 **a** e.g. $\sqrt{8}$ and $\sqrt{2}$ **b** e.g. $\frac{125}{8} = 15.625$
8 **a** $\sqrt{2\frac{1}{4}}$
b 2
c $\sqrt{7} - \sqrt{5} = \dfrac{2}{\sqrt{7} + \sqrt{5}}$, so cannot be written as a fraction.
9 $\dfrac{\sqrt{12}}{\sqrt{3}}$ and $\sqrt{\dfrac{1}{36}}$

CHAPTER 3

Questions

1 **a** Total weight = 3.4 g
$1\text{ g} = \frac{360}{3.4} = 105.9°$
Protein 10.9 = 11°
Carbohydrate 232.9 = 233°, adjust to 232°.
Fat 63.5 = 64°
Fibre 31.8 = 32°
Sodium 21.1 = 21°

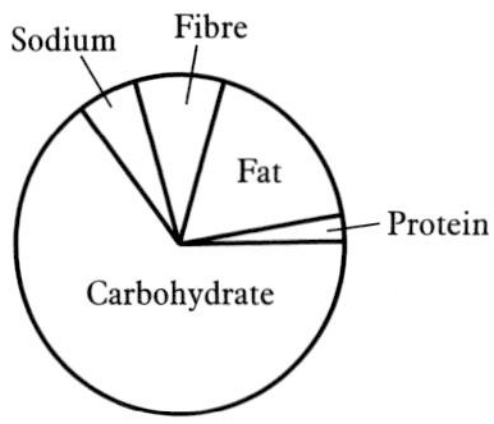

b As part (**a**) but split Fat section into 2 equal parts and Carbohydrate section has $\frac{0.5}{2.2} \times 232.9° = 53°$ for sugars.

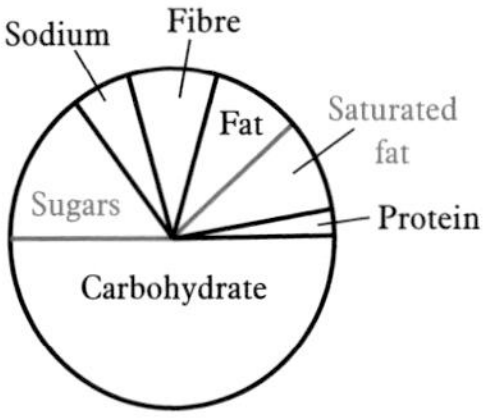

2

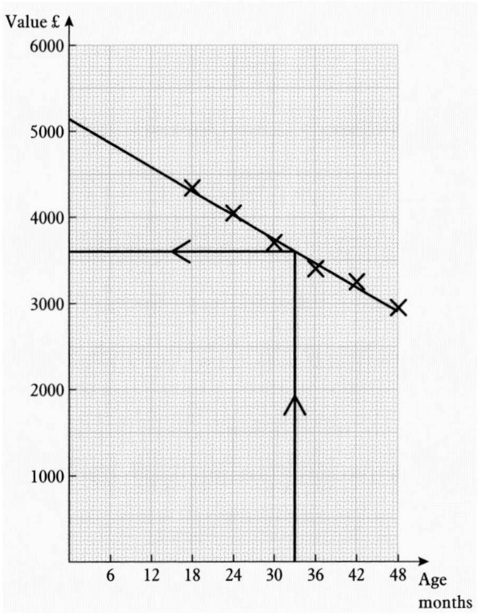

a Very strong negative correlation
b See graph
c £3600 – see graph
d £5150. A scatter diagram can only be used to read off values within the ranges given. A new car loses a lot of money very quickly, then its depreciation begins to follow the line of best fit shown.

3

	Time in seconds, t	$30 \leqslant t < 40$	$40 \leqslant t < 50$	$50 \leqslant t < 60$	$60 \leqslant t < 200$
a	No. of children	6	8	12	14
b	Class width	10	10	10	140
c	Frequency density	0.6	0.8	1.2	0.1

d

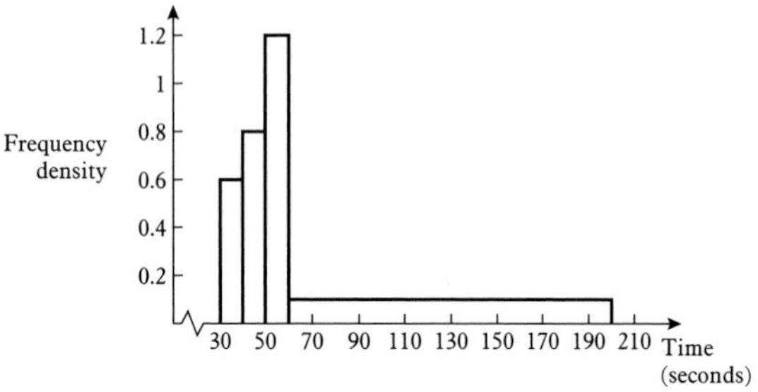

e One or two pupils probably took a long time (dropped their eggs a lot maybe!).

4 **a**

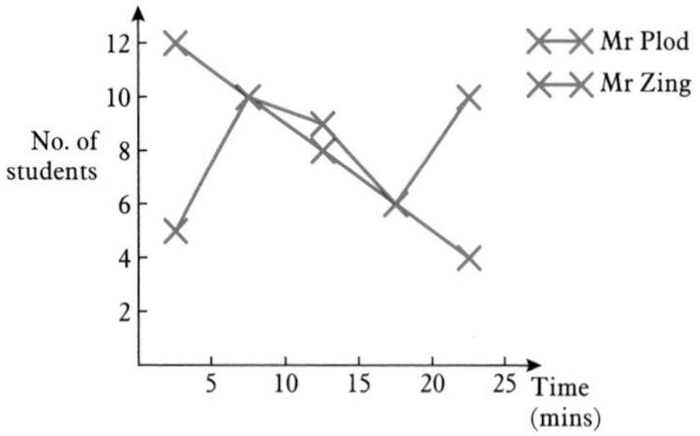

b That Mr Zing is more efficient, seeing more patients very quickly and fewer taking a long time.
c Increase the sample size. 40 is not a very large sample on which to base this judgement.

Exam questions

1 **a,c**

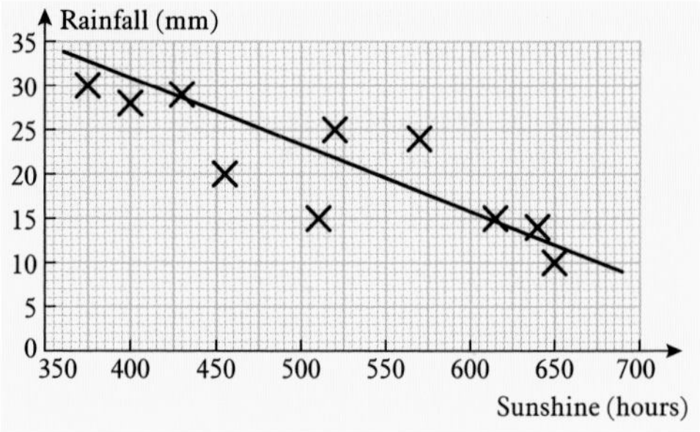

d **(1)** ≈25 mm **(2)** ≈550 h

2 **a**

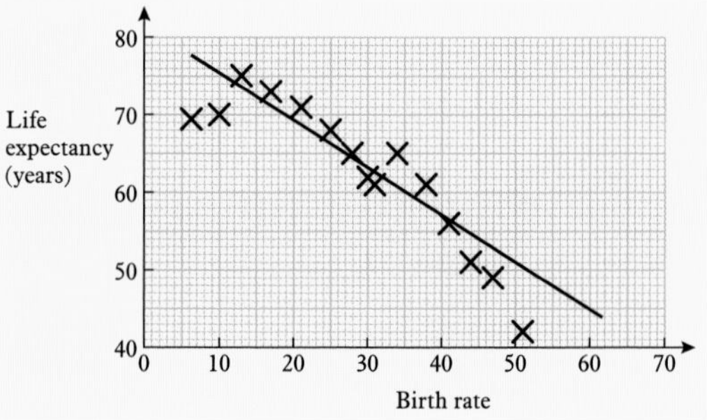

b 52–57 **c** 22–28

3 **a**

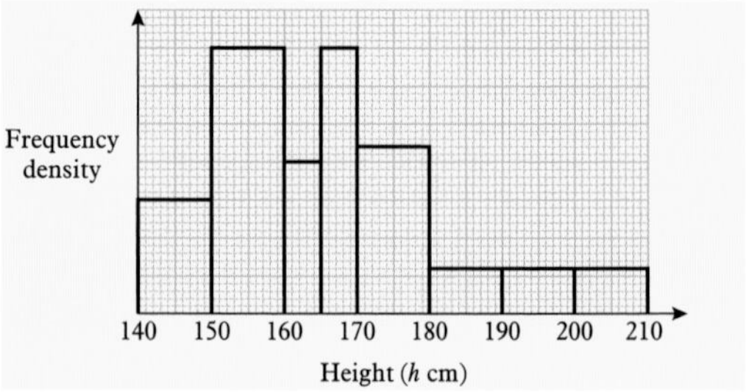

b

Height (h cm)	Frequency
$140 \leqslant h < 150$	15
$150 \leqslant h < 160$	35
$160 \leqslant h < 165$	20
$165 \leqslant h < 170$	18
$170 \leqslant h < 180$	22
$180 \leqslant h < 190$	12
$190 \leqslant h < 210$	12

4 **a**

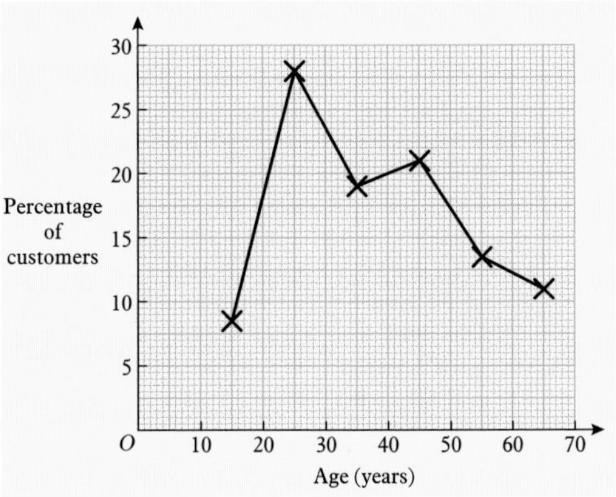

b 31–40

CHAPTER 4

Questions

1 **a** $7n - 2$

b $3n - 7$

2 **a** $2n - 5 = 45$

$2n = 50$

$n = 25$

b $2n - 5 = 39$

$2n = 44$

$n = 22$

3 **a**

1 8 17 28 41

7 9 11 13

2 2 2

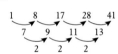

n^2	1	4	9	16	25
$4n - 4$	0	4	8	12	16

nth term $= n^2 + 4n - 4$

b

3 4 7 12 19

1 3 5 7

2 2 2

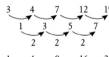

n^2	1	4	9	16	25
$-2n + 4$	2	0	-2	-4	-6

nth term $= n^2 - 2n + 4$

4 **a** 7 **b** $\frac{2}{3}$ **c** -2

5 **a** (0, 8) **b** (0, −3) **c** (0, 0)

6 $y = 2x - 4$ and $y = 2x + 2$

7 **a** $\frac{3}{5}$

b $y = \frac{3}{5}x + 3$

c $y = \frac{3}{5} \times -10 + 3 = -3$

d $12 = \frac{3}{5}x + 3$

$9 = \frac{3}{5}x$

$x = \frac{45}{3} = 15$

8 **a** $x = 8$

b $x = -12$

c $\frac{x}{2} = 11$

$x = 22$

d $\frac{x}{6} = 7$

$x = 42$

e $\frac{y}{2} = 7$

$y = 14$

f $\frac{y}{6} = 18$

$y = 108$

9 **a** $4x = 12$

$x = 3$

b $x = 12$

c $8x = 4$

$x = \frac{1}{2}$

d $-3 = 2x$

$x = -\frac{3}{2} = -1\frac{1}{2}$

10 **a** $\frac{2x}{3} = 8$

$x = 12$

b $\frac{3x}{5} = 3$

$x = 5$

c $2x + 6 = 11$

$2x = 5$

$x = \frac{5}{2} = 2\frac{1}{2}$

d $15x - 21 = 18$

$15x = 39$

$x = 2\frac{9}{15} = 2\frac{3}{5}$

11 **a** $8x$

b $6x + 4$

c $8x = 6x + 4$

$2x = 4 \quad x = 2$

Perimeter = 8 m

12 **a** $x = 5, y = 1$ **c** $x = -1, y = -2$

b $x = 3, y = -2$ **d** $x = 8, y = -8$

13 **a** $2n$ **b** $2m$

c $2n \times 2m = 4nm = 2(2nm)$ which is the $(2nm)$th term of the even numbers and so is even.

d $2p - 1$

e $2n \times (2p - 1)$

$= 4np - 2n$

$= 2(2np - n)$

which is the $(2np - n)$th term of the even numbers and so is even.

f $2q - 1$

g $(2p - 1)(2q - 1)$

$= 4pq - 2q - 2p + 1$

$= 2(2pq - q - p) + 1$

which is an even number plus 1, so is odd.

[OR $2(2pq - q - p + 1) - 2 + 1$

$= 2(2pq - q - p + 1) - 1$

which is the $(2pq - q - p + 1)$th odd number and so is odd.]

14 $3b + 2f = 3.55$

$4b + 3f = 4.95$

$b = 0.75, \quad f = 0.65$

burger 75p, fries 65p

Exam questions

1 $2n(n + 1)$ or $2n^2 + 2n$
2 $(n + 1)^2 - 1$
3 $20 - 3n$
4 **a** $(p + q)(p - q)$
b $n^2 - 1$
c $(n - 1)n(n + 1)(n + 2)$
5 **a** $n(n + 1)$ **b** $2(n + 1)$
6 **a** $(x - 2y)(x + 2y)$ **b** $x = 5, y = \frac{1}{2}$
7 $x = \frac{3}{4}, y = -\frac{1}{2}$
8 **a** $y = 2x - 1$ **b** $y = 2x$
9 -1.5
10 20 cm
11 **a** (2, 2), (1, 0), etc.
b $x = 1.5, y = 1$
c $3y = -2x + 7$
12 $x = 1.5, y = -2$
13 **a** $x^2 + 4x$
b $x = 3, y = \frac{1}{4}$

CHAPTER 5

Questions

1 **a** 36.1 cm **b** 22.2 cm
c 14.2 cm **d** 17.5 cm
2 **a** 12.4 cm **b** 26.2 cm
c 0.8 cm **d** 38.4 cm
3 **a** $\sqrt{7}$ cm **b** $\sqrt{200} = 10\sqrt{2}$ cm
4 $h^2 = 20^2 - 10^2$
$h = \sqrt{300}$ cm
$= 17.3$ cm (1 dp)

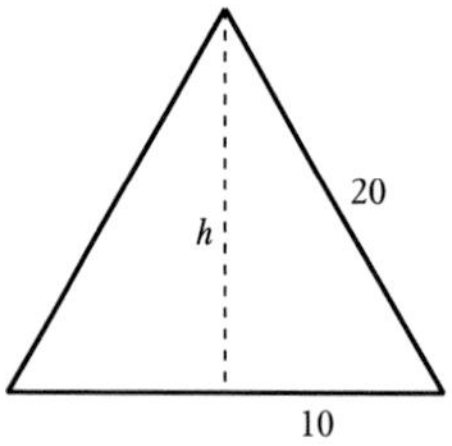

5

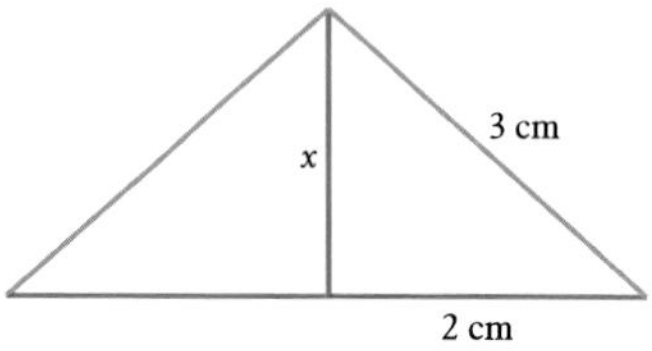

$x^2 = 3^2 - 2^2$
$x = \sqrt{5}$ cm

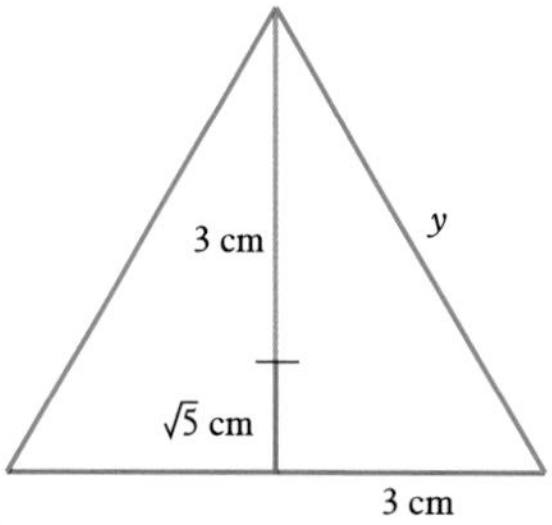

$y^2 = (3 + \sqrt{5})^2 + 3^2$
$y = 6.0$ cm

6
$$(x + 2)^2 = x^2 + 36$$
$$x^2 + 4x + 4 = x^2 + 36$$
$$4x = 32$$
$$x = 8 \text{ cm}$$

7 **a** $\cos a = \dfrac{3}{7}$
$a = 65°$
b $\sin b = \dfrac{15}{23.5}$
$b = 40°$
$c = 50°$

8 **a** $\tan 30° = \dfrac{a}{30}$
$a = 30 \tan 30°$
$a = 17.3$ cm
b $\cos 47° = \dfrac{b}{10}$
$b = 10 \cos 47°$
$b = 6.82$ m
c $\sin 78° = \dfrac{24}{c}$
$c = \dfrac{24}{\sin 78°}$
$c = 24.5$ cm
d $\cos 25° = \dfrac{33.2}{d}$
$d = \dfrac{33.2}{\cos 25°}$
$d = 36.6$ cm

Exam questions

1 17.7 cm
2 7.21 cm
3 28.3 cm
4 **a** 116 cm
b 147.5, 120.5, 15.5
5 6.52 cm
6 1.38 m
7 **a** 16.6 m
b 30.4°
8 **a** 5.29 cm
b 41.4°
c 12.4 cm
9 **a** 28.3 km
b 298°
10 **a** 5.26 m
b 46.9 m

CHAPTER 6

Questions

1 **a** 1.01 **c** 1.125
b 1.65 **d** 1.122
2 **a** 0.85 **c** 0.825
b 0.6 **d** 0.669
3 **a** £11.75 **c** £8518.75
b £145.70 **d** £27 025
4 **a** £160 **c** £158
b £110 **d** £2586

5 **a** £7035.50 **c** 40.7%

b £2035.50

6 **a** $12\,000 \times 0.88^5 = £6333$ **b** 47.2% (1 dp)

7 **a** $T = 4v$ **b** 12 **c** 25

8 $A = kr^2$ $706.9 = k15^2$

$k = 3.14$ (2 dp)

$A = 3.14r^2$

9 **a** $y = 2h^4$

b $162 = 2h^4$ $h^4 = 81$ $h = 3$

10 **a** $y = 8\sqrt{x}$

b $y = 16$

c $x = 16$

11 **a** $A = \dfrac{40}{B}$ **b** $A = 4$

12 **a** $v = \dfrac{200}{w^2}$ **c** $w = \pm 5$

b $v = 12.5$

13 39.8%

14 $\frac{2}{9} = \frac{12}{54}$ $\frac{5}{18} = \frac{15}{54}$ $\frac{7}{27} = \frac{14}{54}$

So, in order $\frac{2}{9}, \frac{7}{27}, \frac{5}{18}$.

15 **a** $\dfrac{7}{9}$ **c** $\dfrac{11}{2r}$

b $\dfrac{6}{p}$

16 **a** $\dfrac{y}{y^2} + \dfrac{2}{y^2} = \dfrac{y+2}{y^2}$

b $\dfrac{9x^2}{x} - \dfrac{1}{x} = \dfrac{9x^2 - 1}{x}$

c $\dfrac{5(x+1)}{x+1} - \dfrac{1}{x+1} = \dfrac{5(x+1) - 1}{x+1} = \dfrac{5x+4}{x+1}$

17 **a** $\dfrac{{}^{2}\cancel{8}}{\cancel{x}} \times \dfrac{x^{\cancel{2}}}{\cancel{12}_{3}} = \dfrac{2x}{3}$

b $\dfrac{{}^{4}\cancel{12p}q}{{}_{1}\cancel{5x}} \times \dfrac{{}^{2}\cancel{10x}}{{}_{3}\cancel{9p}} = \dfrac{8q}{3}$

c $\dfrac{2\cancel{(p+3)}}{\cancel{3}} \times \dfrac{\cancel{3}p}{\cancel{p+3}} = 2p$

18 **a** $\dfrac{3(x-2) - 6}{(x+1)(x-2)} = \dfrac{3x - 12}{(x+1)(x-2)} = \dfrac{3(x-4)}{(x+1)(x-2)}$

b $\dfrac{\cancel{p+2}}{\cancel{p}} \times \dfrac{\cancel{p}}{\cancel{2(p+2)}} \times \dfrac{\cancel{q}}{p+4} \times \dfrac{\cancel{2}p}{\cancel{q}} = \dfrac{p}{p+4}$

19 **a** $\dfrac{6x}{9} = \dfrac{2x}{3}$

b $\dfrac{{}^{2}\cancel{10}}{3\cancel{y}} \times \dfrac{y^{\cancel{2}}}{\cancel{5}_{1}} = \dfrac{2y}{3}$

c $\dfrac{{}^{1}\cancel{5t}}{\cancel{8pr}_{1}} \times \dfrac{{}^{2}\cancel{16}q\cancel{r}}{\cancel{15t}_{3}} = \dfrac{2q}{3p}$

Exam questions

1 **a** **(1)** £360 **(2)** £230.40 **b** £375

2 £48.50

3 **a** £360

b £288.26

4 **a** £61.80

b 1.23

5 Tracey £4000, Wayne £3200

6 **a** 75

b $37\frac{1}{2}\%$

7 **a** Ruth £100, Ben £80

b 60%

8 **a** $d = 0.000\,005\,9L^3$

b 136

9 **a** $8x^3$

b **(1)** $y = 1$ **(2)** $x = 1.5$

10 **a** $y = \dfrac{48}{x^2}$

b $y = 1.92$

11 **a** 11.2

b

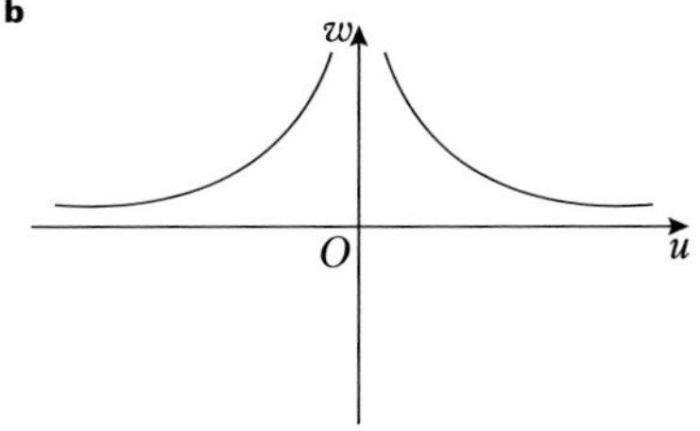

c $\frac{10}{3}$

12 $\dfrac{4x + 2}{(2x+3)(2x-1)}$

13 **a** $(x+1)(x+2)$ **b** $\dfrac{6}{x+2}$

14 **a** 5.6 **b** $\frac{1}{5}$

15 **a** $\frac{5}{3}$ **b** 10.5

16 **a** $x^2 + x - 4 = 0$ **b** 1.56, −2.56

17 **a** £5755.11 **b** 1.158

CHAPTER 7

Questions

1 **a**

	1	2	3	4	5	6
H	H, 1	H, 2	H, 3	H, 4	H, 5	H, 6
T	T, 1	T, 2	T, 3	T, 4	T, 5	T, 6

b **(1)** $\frac{1}{12}$ **(3)** $\frac{3}{12} = \frac{1}{4}$

(2) $\frac{3}{12} = \frac{1}{4}$ **(4)** $\frac{2}{12} = \frac{1}{6}$

2 **a** $P(C') = 1 - P(C) = \frac{19}{20}$

$P(A) + P(B) = \frac{1}{5} + \frac{3}{4} = \frac{19}{20}$

So $P(C') = P(A) + P(B)$

b $P(\text{A and B}) = P(A) \times P(B)$ since A and B are independent

$= \frac{1}{5} \times \frac{3}{4} = \frac{3}{20}$

c $P(\text{A or B}) = P(A) + P(B)$ if A and B are mutually exclusive

$= \frac{1}{5} + \frac{3}{4} = \frac{19}{20}$

Since $P(\text{A or B}) = \frac{17}{20}$, A and B are not mutually exclusive.

3 **a**

	1	2	3	4	5	6
1	2	3	4	5	6	7
2	3	4	5	6	7	8
3	4	5	6	7	8	9
4	5	6	7	8	9	10
5	6	7	8	9	10	11
6	7	8	9	10	11	12

b $\frac{6}{36} = \frac{1}{6}$

c $\frac{6}{36} = \frac{1}{6}$

d 'Double' and '7' are mutually exclusive so probability $= \frac{6}{36} + \frac{6}{36} = \frac{1}{3}$

e $\frac{5}{36}$

f $\frac{10}{36} = \frac{5}{18}$ ("Double" and "8" are *not* mutually exclusive (4, 4) so this is *not* $\frac{6}{36} + \frac{5}{36}$.)

4 **a**

	3	3	4	4	5	5
1	4	4	5	5	6	6
3	6	6	7	7	8	8
3	6	6	7	7	8	8
4	7	7	8	8	9	9

b $\frac{14}{24} = \frac{7}{12}$

5 **a**

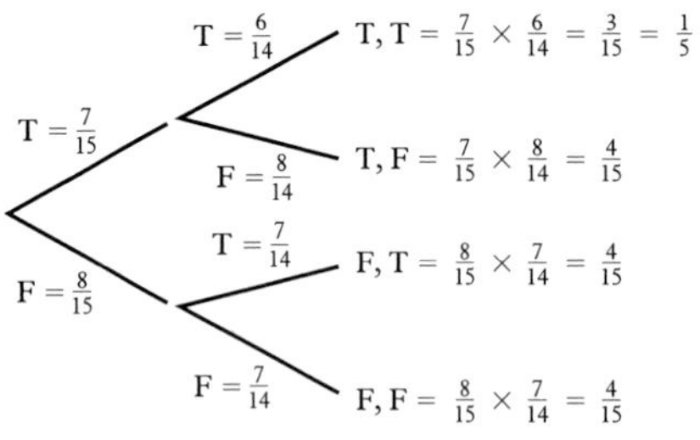

b **(1)** $\frac{1}{5}$ **(3)** $\frac{4}{15} + \frac{4}{15} = \frac{8}{15}$

(2) $\frac{4}{15}$ **(4)** $1 - P$ (no fruit creams) $= 1 - \frac{1}{5} = \frac{4}{5}$

6 **a** $\frac{12}{22} = \frac{6}{11}$

b $\frac{12}{22} \times \frac{11}{21} = \frac{2}{7}$

c (R, G) or (G, R) $= \frac{12}{22} \times \frac{10}{21} + \frac{10}{22} \times \frac{12}{21} = \frac{40}{77}$

d $\frac{12}{22} \times \frac{11}{21} \times \frac{10}{20} = \frac{1}{7}$

e $1 - P$ (no reds) $= 1 - P$ (GGG) $= 1 - \frac{10}{22} \times \frac{9}{21} \times \frac{8}{20} = 1 - \frac{6}{77} = \frac{71}{77}$

7 **a**

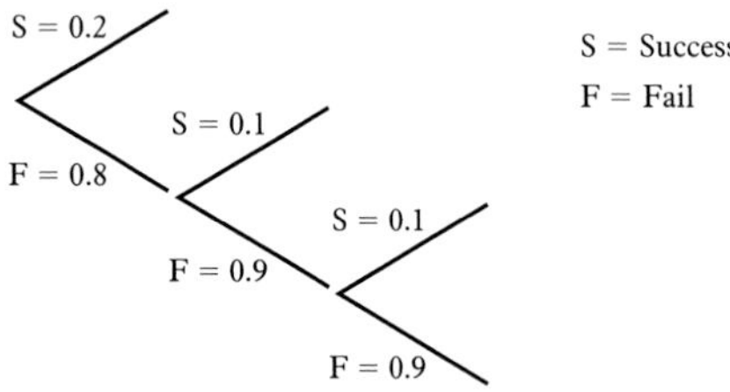

b F,S $= 0.8 \times 0.1 = 0.08$

c FFS $= 0.8 \times 0.9 \times 0.1 = 0.072$

d S or FS or FFS $= 0.2 + 0.08 + 0.072 = 0.352$

e $1 -$ **d** $= 0.648$ (or FFF $= 0.8 \times 0.9 \times 0.9$)

Exam questions

1 **a** 58 **b** $\frac{870}{11\,136}$ or 0.078 125 or $\frac{5}{64}$

2 **a** 0.51 **b** 0.34

3 **(1)** 0.0081 **(2)** 0.1638

4 **(1)** $\frac{1}{6}$ **(2)** $\frac{1}{4}$

5 **a** 0.0459 **b** 0.3941

6 **a** 0.05, 0.2, 0.2 **c** 0.23

b 0.76

7 **a** $\frac{5}{12}, \frac{7}{12}$

$\frac{5}{12}, \frac{7}{12}, \frac{5}{12}, \frac{7}{12}$

b **(1)** $\frac{25}{144}$ **(2)** $\frac{35}{72}$

8 **a** 0.4, 0.4, 0.6, 0.4 **c** 0.48

b 0.36

9 **(1)** 0.52 **(2)** 0.07

10 **a** $\frac{16}{100}$ **b** 4000 **c** 212

CHAPTER 8

Questions

1 **a** $7t = a + g$
$a = 7t - g$

b $2p = 4b + 7r$
$4b = 2p - 7r$
$b = \dfrac{2p - 7r}{4}$

c $ky = pc + 2t$
$c = \dfrac{ky - 2t}{p}$

d $4yk = ud + g$
$ud = 4yk - g$
$d = \dfrac{4yk - g}{u}$

e $w^2 = 5e - l$
$5e = w^2 + l$
$e = \dfrac{w^2 + l}{5}$

f $d + 5 = f^2$
$f = \sqrt{d + 5}$

g $t^2 = \dfrac{g}{5}$
$g = 5t^2$

h $5h = 4h + j$

i $pi = gi + 6j$
$pi - gi = 6j$
$i(p - g) = 6j$
$i = \dfrac{6j}{p - g}$

j $5aj = 4h + bj$
$5aj - bj = 4h$
$j(5a - b) = 4h$
$j = \dfrac{4h}{5a - b}$

k $uk = 4h - 3k$
$k(u + 3) = 4h$
$k = \dfrac{4h}{u + 3}$

l $5al = 4h + bl$
$l(5a - b) = 4h$
$l = \dfrac{4h}{5a - b}$

m $m^2 = 5m^2 - l$
$l = 4m^2$
$m^2 = \dfrac{l}{4}$
$m = \sqrt{\dfrac{l}{4}} = \dfrac{\sqrt{l}}{2}$

n $a^2n^2 = 6n^2 - 4g$
$n^2(a^2 - 6) = -4g$ or $4g = n^2(6 - a^2)$
$n^2 = \dfrac{-4g}{a^2 - 6}$ $n^2 = \dfrac{4g}{6 - a^2}$
$n = \sqrt{\dfrac{-4g}{a^2 - 6}}$ $n = \sqrt{\dfrac{4g}{6 - a^2}}$

2 **a**

x	−4	−3	−2	−1	0	1	2	3	4
x^2	16	9	4	1	0	1	4	9	16
$+3x$	−12	−9	−6	−3	0	3	6	9	12
-1	−1	−1	−1	−1	−1	−1	−1	−1	−1
y	3	−1	−3	−3	−1	3	9	17	27

b, c

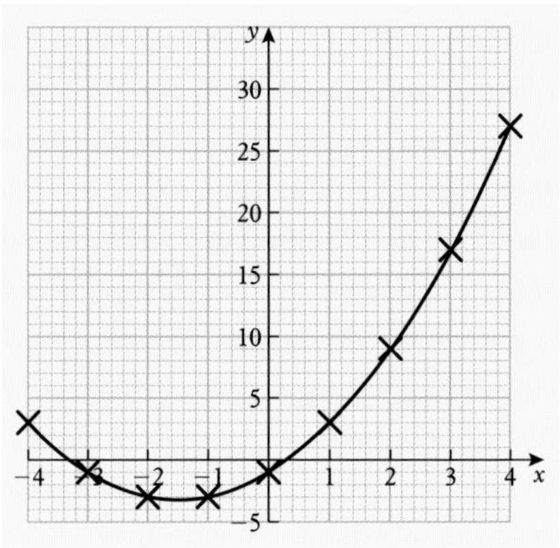

3 **a**

t	0	1	2	3	4	5
h	1.3	2.05	2.3	2.05	1.3	0.05

b

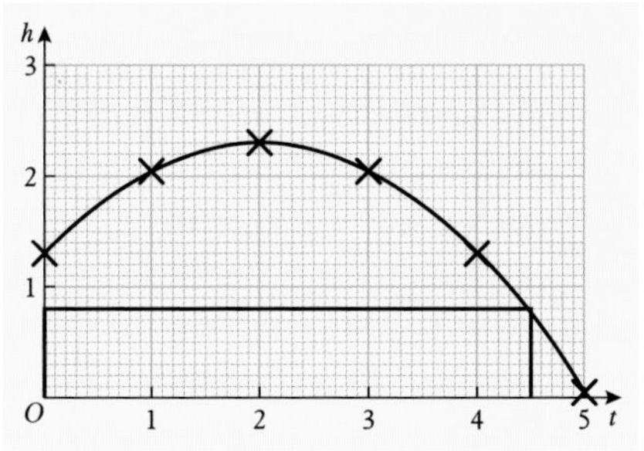

c 4.5 secs

4 **a** $x^2 + x - 6x - 6$
$= x^2 - 5x - 6$

b $10x^2 + 5x - 4x - 2$
$= 10x^2 + x - 2$

5 **a** $(x + 3)(x + 6)$

b $(3x - 2)(x + 2)$

c $(3x - 1)(x - 3)$

d $(3x + 4)(2x - 3)$

6 **a** $(x - 2)(x - 6) = 0$
$x = 2$ or $x = 6$

b $(2x + 1)(x + 1) = 0$
$x = -\frac{1}{2}$ or $x = -1$

c $(3x + 2)(x - 2) = 0$
$x = -\frac{2}{3}$ or $x = 2$

d $(3x - 1)(2x - 5) = 0$
$x = \frac{1}{3}$ or $x = \frac{5}{2} = 2\frac{1}{2}$

7 **a** $x = \dfrac{-9 \pm \sqrt{9^2 - 4 \times 1 \times 12}}{2}$
$x = -1.63$ or -7.37

b $x = \dfrac{-3 \pm \sqrt{3^2 - 4 \times 2 \times -1}}{4}$
$x = 0.28$ or -1.78

c $x = \dfrac{2 \pm \sqrt{(-2)^2 - 4 \times 3 \times -4}}{6}$
$x = 1.54$ or -0.87

d $x = \dfrac{10 \pm \sqrt{(-10)^2 - 4 \times 4 \times 3}}{8}$
$x = 2.15$ or 0.35

8 **a** $x(2x + 8)$

b $x + 4$

c $x + 5$

d $(x + 4)(x + 5)$

e $x(2x + 8) = (x + 4)(x + 5)$
$2x^2 + 8x = x^2 + 9x + 20$
$x^2 - x - 20 = 0$
$(x - 5)(x + 4) = 0$
So $x = 5$ ($x = -4$ not possible as x is a length)

9 **a** $x^2 + 4x - 1$
$(x + 2)^2 = x^2 + 4x + 4$
So $x^2 + 4x - 1 = (x + 2)^2 - 5$

b $3x^2 - 6x - 4 = 3(x^2 - 2x) - 4$
$3(x - 1)^2 = 3(x^2 - 2x + 1) = 3x^2 - 6x + 3$
So $3x^2 - 6x - 4 = 3(x - 1)^2 - 7$

10 $y = 2 - x$ $x^2 + (2 - x)^2 = 25$
$x^2 + 4 - 4x + x^2 = 25$
$2x^2 - 4x - 21 = 0$
$x = 4.391$ or $y = -2.391$
Solutions are $x = 4.391$ and $y = -2.391$
or $x = -2.391$ and $y = 4.391$

11 **a**

x	$x^3 + 15$	
5.6	190.616	too small
5.7	200.193	too big
5.65	195.36	too small

So $x = 5.7$ (1 dp)

b

x	$x^2 + \frac{3}{x}$	
−1.6	0.685	too small
−1.7	1.125	too big
−1.65	0.904	too small

So $x = -1.7$ (1 dp)

c

x	$2x^3 + x$	
3.3	75.174	too small
3.4	82.008	too big
3.35	78.541	too small

So $x = 3.4$ (1 dp)

d

x	$5x^2 - \frac{1}{x}$	
0.8	1.95	too small
0.9	2.9388	too big
0.85	2.436	too big

So $x = 0.8$ (1 dp)

Exam questions

1 $R = \dfrac{C - 2000}{180}$

2 **a** 0.94 or 0.937 **b** $\dfrac{8p}{tp - 8}$

3 **a** 1240 **b** $\dfrac{3v}{\pi(R^2 + Rr + r^2)}$ **c** 3.30

4 **a** **(1)** $10x + 4$ **(2)** $2x(3x + 2)$
b 112 cm^2

5 **(1)** $2x^2 + 5x - 12$ **(2)** $x^4 + 2x^2y^2 + y^4$ **(3)** $4xy$

6 **a** $2x^2 - 5x - 3$ **c** $2t(t + 2)$
b $6a^4b^2$ **d** $(3x + 2)(3x - 2)$

7 6.5, −3.5

8 **a** $2x^2 + x - 15$
b **(1)** $(x + 7)(x - 1)$ **(2)** $-7, 1$

9 **(1)** $(x - 4)(x - 2)$ **(2)** $x = 2$ or $x = 4$

10 **a** $(x + 4)(x - 3) = 78$
b **(1)** $x^2 + x - 90 = 0$ **(2)** 9, −10 **(3)** 13 cm, 6 cm

11 3.53

12 3.25

13 2.44

14 3.9

15 2.9

16 $x(x^2 - 8)$ or $x^3 - 8x$ (units cm²)

17 **a** $10x + 3$ **b** $6x^2 + 5xy - 4y^2$

18 $x = 3.2$

CHAPTER 9

Questions

1 **a** $\begin{pmatrix}8\\4\end{pmatrix} - \begin{pmatrix}9\\-1\end{pmatrix} = \begin{pmatrix}-1\\5\end{pmatrix}$

b $\begin{pmatrix}3\\10\end{pmatrix} + \begin{pmatrix}6\\-8\end{pmatrix} = \begin{pmatrix}9\\2\end{pmatrix}$

c $\begin{pmatrix}-7\\-5\end{pmatrix} - \begin{pmatrix}-6\\-2\end{pmatrix} = \begin{pmatrix}-1\\-3\end{pmatrix}$

d $\overrightarrow{AB} + \overrightarrow{BF} = \overrightarrow{AB}$

e $\overrightarrow{PB} + \overrightarrow{BQ} = \overrightarrow{PQ}$

f $\overrightarrow{GH} + \overrightarrow{HB} + \overrightarrow{BA} = \overrightarrow{GA}$

2 $\overrightarrow{AB} = \mathbf{p} + \mathbf{q} - 2\mathbf{p} + \mathbf{r}$
$= -\mathbf{p} + \mathbf{q} + \mathbf{r}$

3 **a** $\begin{pmatrix}13\\1\end{pmatrix}$ **d** $\begin{pmatrix}11\\0\end{pmatrix}$

b $\begin{pmatrix}16\\-4\end{pmatrix}$ **e** $\begin{pmatrix}-9\\6\end{pmatrix}$

c $\begin{pmatrix}1\\3\end{pmatrix}$ **f** $\begin{pmatrix}-3\\-4\end{pmatrix}$

4 **a** $\begin{pmatrix}24\\-3\end{pmatrix}$ **d** $\begin{pmatrix}3\\-2\end{pmatrix}$

b $\begin{pmatrix}12\\-8\end{pmatrix}$ **e** $\begin{pmatrix}16\\-2\end{pmatrix} - \begin{pmatrix}-18\\12\end{pmatrix} = \begin{pmatrix}34\\-14\end{pmatrix}$

c $\begin{pmatrix}4\\-\frac{1}{2}\end{pmatrix}$ **f** $\frac{1}{2}\begin{pmatrix}14\\-5\end{pmatrix} = \begin{pmatrix}7\\-2\frac{1}{2}\end{pmatrix}$

5 **a** $\sqrt{(-3)^2 + 4^2} = 5$
b $\sqrt{(-5)^2 + (-12)^2} = 13$
c $\sqrt{8^2 + (-15)^2} = 17$
d $\sqrt{3^2 + (-5)^2} = \sqrt{34} = 5.83$ (2 dp)
e $\sqrt{7^2 + (-1)^2} = \sqrt{50} = 5\sqrt{2} = 7.07$ (2 dp)
f $\sqrt{(-5)^2 + (-5)^2} = \sqrt{50} = 5\sqrt{2} = 7.07$ (2 dp)

6 **a** $\overrightarrow{OA} = \begin{pmatrix}5\\-2\end{pmatrix}$ $\overrightarrow{OB} = \begin{pmatrix}7\\2\end{pmatrix}$

b $\overrightarrow{AB} = \overrightarrow{OB} - \overrightarrow{OA} = \begin{pmatrix}2\\4\end{pmatrix}$

c $\frac{1}{2}\begin{pmatrix}12\\0\end{pmatrix} = \begin{pmatrix}6\\0\end{pmatrix}$

7 $\overrightarrow{KM} = 2\overrightarrow{KL}$
So KM is parallel to KL. But K is a common point so K, L and M are collinear.

8 **a** $\overrightarrow{RS} = \overrightarrow{OS} - \overrightarrow{OR} = 6\mathbf{a} + 2\mathbf{b}$
$\overrightarrow{ST} = \overrightarrow{OT} - \overrightarrow{OS} = 3\mathbf{a} + \mathbf{b}$
So $\overrightarrow{RS} = 2\overrightarrow{ST}$. Hence RS is parallel to ST. But S is a common point so R, S and T are collinear.
b RS is twice as long as ST, so S is $\frac{2}{3}$ of way from R to T. Hence S divides RT in ratio 2 : 1.

9 **a** **(1)** **a** **(3)** **b**
(2) **b** **(4)** 2**b**

b **(1)** $\overrightarrow{ED} = -\mathbf{b} - \mathbf{a} + 2\mathbf{b}$
$= \mathbf{b} - \mathbf{a}$

(2) $\overrightarrow{BA} = -\overrightarrow{AB} = -\overrightarrow{ED}$
$= \mathbf{a} - \mathbf{b}$

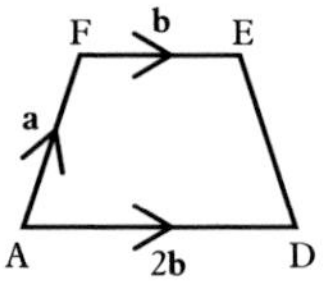

10 **a** $\sqrt{0.5^2 + 3^2}$
$= 3.04$ m/s
b $\tan \alpha = \frac{3}{0.5}$
$\alpha = 80.5°$

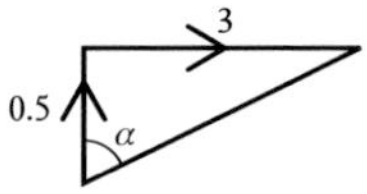

So Aimee travels at 80.5° to the perpendicular.
c 50 metres at 0.5 m/s (using perpendicular direction) gives
time $= \frac{50}{0.5} = 100$ secs

11 **a** $20\,000 \cos 45° + 30\,000 \cos 60°$
$= 29\,142$ N
b $30\,000 \cos 30° - 20\,000 \cos 45°$
$= 11\,839$ N

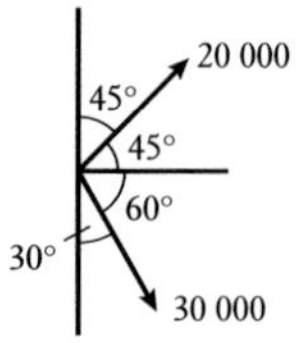

c 31 455 N at a bearing of 112.1°.

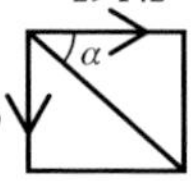

$\sqrt{29142^2 + 11839^2}$
$= 31455$
$\tan \alpha = \frac{11\,839}{29\,142}$
$\alpha = 22.1°$

Exam questions

1 **a** **(1)** $\begin{pmatrix}5\\0\end{pmatrix}$ **(2)** $\begin{pmatrix}0\\5\end{pmatrix}$
b $c = 9, d = -3$

2 **a** (3, 6) **b** $\begin{pmatrix}-3\\-4\end{pmatrix}$ **c** 5

3 **a** $\begin{pmatrix}-4\\-3\end{pmatrix}$ **b** (2, 9) **c** $\begin{pmatrix}0\\1.5\end{pmatrix}$

4 **a** **(1)** $\begin{pmatrix}-4\\3\end{pmatrix}$ **(2)** 5
b $\begin{pmatrix}-2\\6\end{pmatrix}$ **c** (−6, 3)

5 $\frac{3}{2}(\mathbf{b} - \mathbf{a})$

6 **a** **(1)** **b** − **a** **(2)** 2**a** **(3)** 2**b** − 2**a**
b Either QT and RS parallel or RS = 2QT.

7 **a** **(1)** 2**a** **(2)** **a** − **b** **(3)** −**a** −**b**
b **(1)** **a** − **b**
(2) AX = BX so AX is parallel to BA.
A is on AX and BA, so B, A and X lie on the same straight line.

8 **(1)** $a + \frac{1}{2}b$ **(2)** $\frac{1}{2}b - \frac{2}{3}a$

9 **a** **(1)** $a + b$ **(2)** $2a - b$

b $2a + \frac{1}{2}b$

c $a + b$

CHAPTER 10

Questions

1 **a** $a = 67°$ alternate angles

b $b = 123°$ corresponding angles

c $c = 124°$ alternate angles

$d = 44°$ alternate angles

$e = 180° - 124° - 44°$

$= 12°$ angles in a triangle

d $f = 122°$ corresponding angles

$g = 122°$ corresponding angles

$h = 180° - 68° - 65°$

$= 47°$ angles on a straight line and angles in a triangle

2 **a** $\frac{360°}{10} = 36°$ **b** $8 \times 180° = 1440°$

3 Exterior angle $= 20°$

$\therefore$ Sides $= \frac{360}{20} = 18$

4 **a**

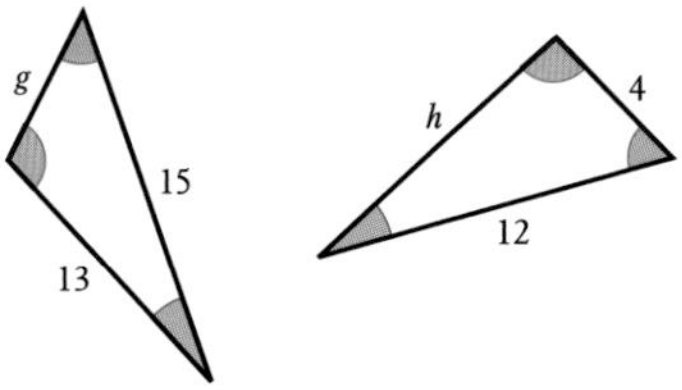

b $\frac{12}{15} = \frac{4}{5}$ $h = 13 \times \frac{4}{5} = 10.4$ cm

$g = 4 \div \frac{4}{5} = 5$ cm

5 BP = RD (P and R mid-points of equal sides)

BQ = SD (Q and S mid-points of equal sides)

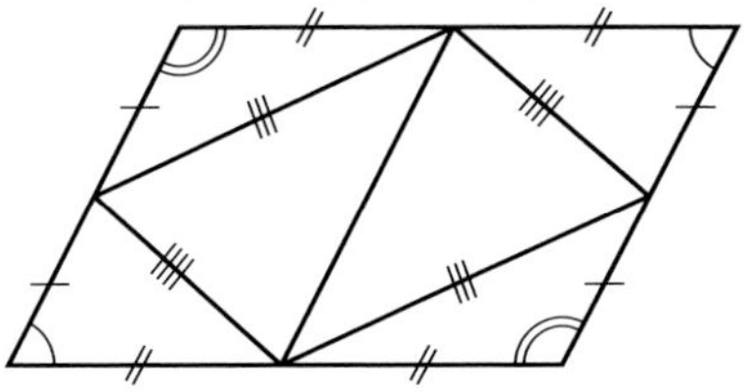

$P\hat{B}Q = R\hat{D}S$ (opposite angles in a parallelogram)

$\therefore \triangle$s PBQ and RDS are congruent (SAS).

Hence PQ = RS.

Similarly, using $\triangle$s APS and CRQ

PS = RQ

and QS = QS (common)

So $\triangle$s PQS and RSQ are congruent (SSS).

(There are other proofs too!)

6

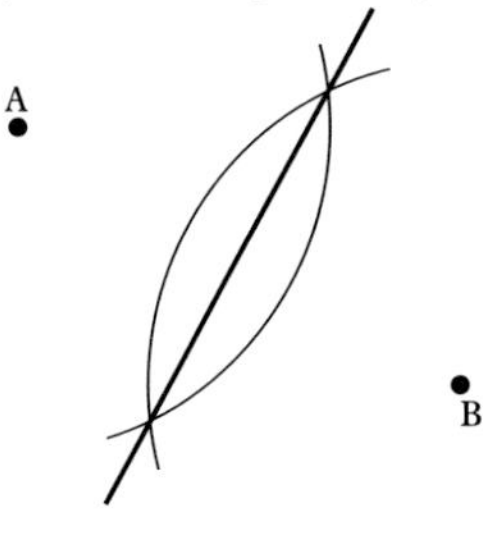

7 **a** $a = 134°$ (angle at centre $= 2\times$ angle at circumference)

b $b = 48°$ (angles in same segment)

$c = 90°$ (angle in a semi-circle)

c $d = 29°$ (angles in same segment)

$e = 29°$ (base angles in an isosceles triangle)

$f = 29°$ (base angles in an isosceles triangle)

$g = 58°$ (angle at centre $= 2\times$ angle at circumference)

d

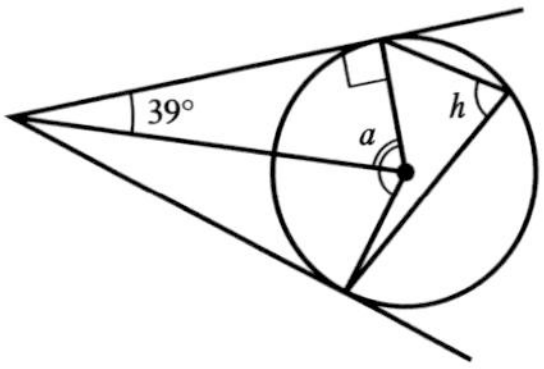

$a = 51°$ (angles in a triangle)

So $b = 102°$

$h = 51°$ (angle at circumference $= \frac{1}{2}$ of angle at centre)

$i = 129°$ (opposite angles in a cyclic quadrilateral)

8 **a** In $\triangle$ ABC, $\angle$ACB $= 90°$ (angle in a semi-circle)

So $\angle$BAC $= 180° - 90° - p$ (angles in a triangle)

$= 90° - p$

Now $\angle$TAO $= 90°$ (tangent perpendicular to radius)

So $a = 90° - (90° - p)$

$a = p$

b The angle between a tangent and a chord drawn to the point of contact is equal to the angle in the alternate segment. (Alternate Segment Theorem.)

Exam questions

1 135°

2 **(1)** 9.6 cm **(2)** 3 cm

3 8.25 cm

4

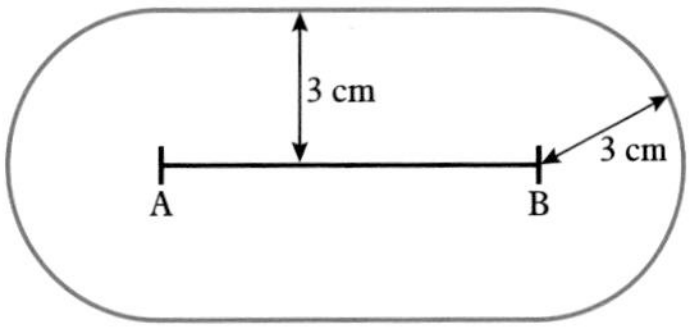

5

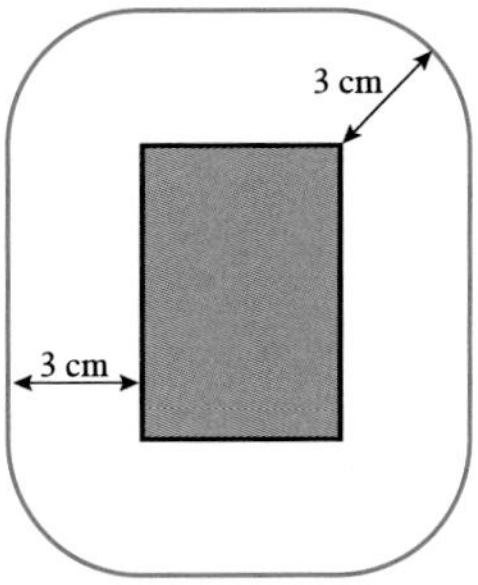

6 Side, side, angle

7 **a** 132° Cyclic quad

b 24° Isoscles triangle, alternate angle theorem

8 **a** 27° Angles in triangle, angle at centre $= 2\times$ angle at circumference

b 153° Cyclic quad

9 **a** 40° **b** 12° **c** Explain Angle EAD $\neq 90°$

CHAPTER 11

Questions

1 **a** **(1)** 078° **(2)** 258°
b **(1)** 237° **(2)** 057°
c **(1)** 314° **(2)** 134°

2 **a** 045°
b 270°
c 045°
d 315°

3 **a**

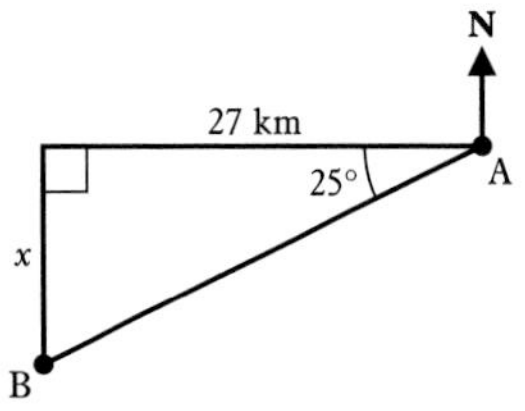

b $\tan 25° = \dfrac{x}{27}$ $x = 12.6$ km (3 sf)

4

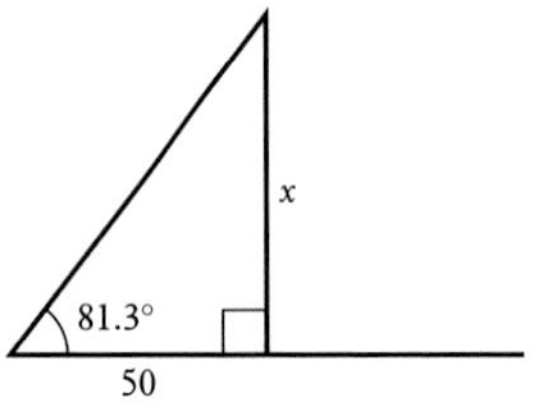

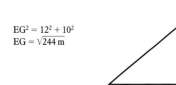

$\tan 81.3° = \dfrac{x}{50}$ $x = 327$ ft (nearest foot)

5

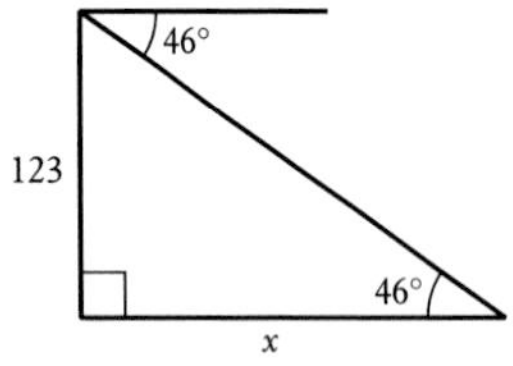

$\tan 46° = \dfrac{123}{x}$ $x = 118.8$ m (1 dp)

6

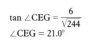

$EG^2 = 12^2 + 10^2$
$EG = \sqrt{244}$ m

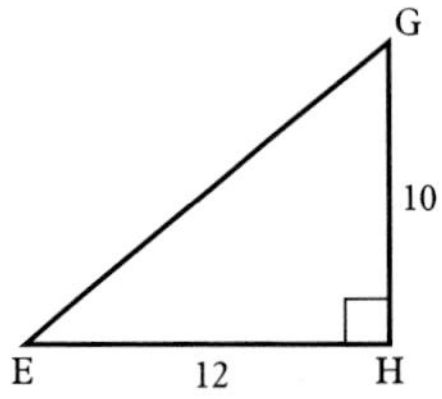

$\tan \angle CEG = \dfrac{6}{\sqrt{244}}$
$\angle CEG = 21.0°$

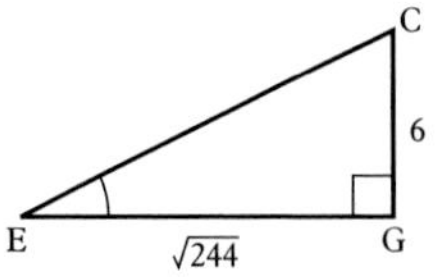

7 Base diagonal AC,
$AC^2 = 14^2 + 18^2$
$AC = \sqrt{520}$
$\tan \angle ECA = \dfrac{7}{\sqrt{520}}$
$\angle ECA = 17.1°$

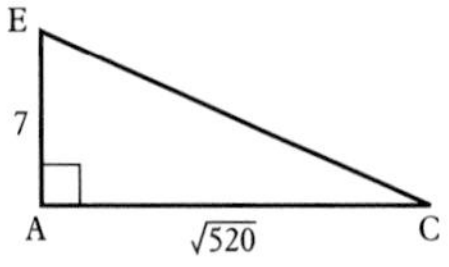

8 **a**
$QM^2 = 10^2 - 5^2$
$QM = \sqrt{75}$ cm (= 8.7 cm)

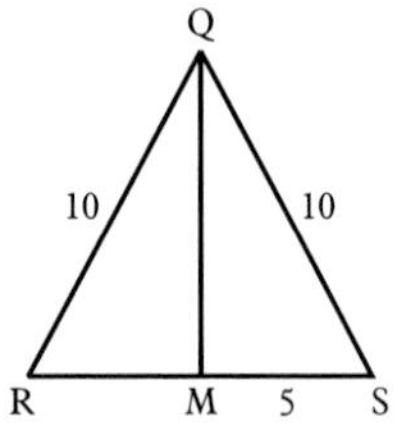

b $\sqrt{75}$ cm (= 8.7 cm)
c $\cos \angle PMX$
$= \dfrac{\sqrt{75}}{3\sqrt{75}} = \dfrac{1}{3}$
$\angle PMX = 70.5°$ (3 sf)

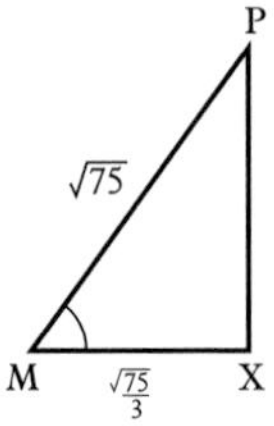

d $\cos \angle PQX$
$= \dfrac{2\sqrt{75}}{3 \times 10}$
$\angle PQX = 54.7°$ (3 sf)

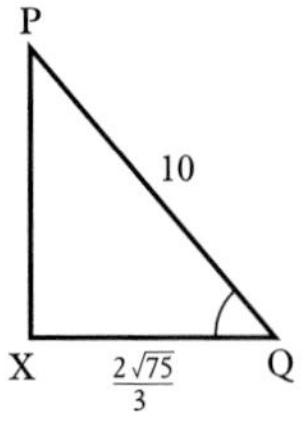

9 **a** $AC^2 = 12^2 + 15^2$
$AC = \sqrt{369}$
$\tan \angle FAC = \dfrac{5}{\sqrt{369}}$
$\angle FAC = 14.6°$ (3 sf)

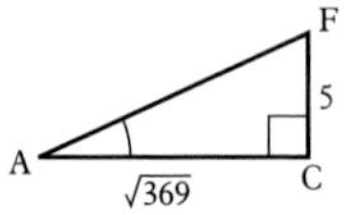

b $DF^2 = 15^2 + 5^2$
$DF = \sqrt{250}$
$\tan \angle AFD = \dfrac{12}{\sqrt{250}}$
$\angle AFD = 37.2°$

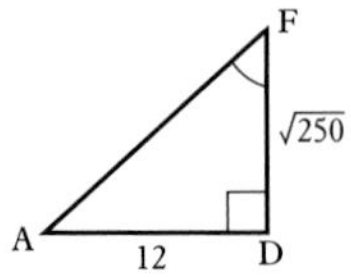

Exam questions

1 **a** 120° **b** 300°
2 **a** 0.5 or $\frac{1}{2}$ **b** 30, 60
3 **a** 6.24 km
b **(1)** 51.3° **(2)** 321.3° **(3)** 141.3°
4 213°
5 **a** 20.6° **b** 111°
6 **a** 6.18 m **b** 27.3°
7 **a** 2.63 m **b** 66.4°

CHAPTER 12

Questions

1 **a** 64
b 1
c $\dfrac{1}{2}$
d $\dfrac{1}{18}$
e $\dfrac{1}{5^2} = \dfrac{1}{25}$
f 1
g $\dfrac{1}{10^3} = \dfrac{1}{1000}$
h 6

i 2

j $\frac{1}{(\sqrt[3]{125})^2} = \frac{1}{5^2} = \frac{1}{25}$

k $\frac{1}{(\sqrt[8]{256})^3} = \frac{1}{2^3} = \frac{1}{8}$

l $\frac{1}{(\sqrt[5]{243})^3} = \frac{1}{3^3} = \frac{1}{27}$

2 **a** $\frac{1}{6}$ **b** $\frac{1}{20}$ **c** $\frac{5}{4} = 1\frac{1}{4}$ **d** $\frac{4}{15}$

3 **a** a^7b^7
b $6a^8b^{-2}$
c $8x^{-1}y$
d $6x^3y^{-6}$
e x^6y^{12}
f $p^{10}q^{-6}$
g $27x^6y^{-9}$
h $256x^{-8}y^{-12}$
i a^3b^2
j $2a^{-8}b^2$
k $2x^7y^2$
l $\frac{1}{2}x^3y^{-3}$

4 **a** $x = 3$ **b** $x = 4$ **c** $y = -2$ **d** $y = -3$

5 **a** 4.2×10^7 **b** 6.03×10^2 **c** 4.1×10^{-5} **d** 2.34×10^{-3}

6 **a** 31 200 **b** 698 000 000 **c** 0.0017 **d** 0.000 158 2

7 **a** Mercury
b $1.5 \times 10^8 - 5.8 \times 10^7 = 9.2 \times 10^7$
c $(5.92 \times 10^9) \div (1.5 \times 10^8) = 39$ (nearest whole number)

8 **a** $2 \times \pi \times 5.790\,92 \times 10^7 = 3.64 \times 10^8$ km (3 sf)
b a $\div 87.9686 \div 24 = 172\,000$ km/h (nearest 1000).

9 $3 \times 10^8 \times 60 \times 60 \times 24 \times 365$
$= 9.5 \times 10^{15}$ km (2 sf)

10 **a** 6×10^{10}
b $15 \times 10^{13} = 1.5 \times 10^{14}$
c $28 \times 10^5 = 2.8 \times 10^6$
d $16 \times 10^{-6} = 1.6 \times 10^{-5}$
e 3×10^7
f 2×10^4
g 3×10^3
h 2.5×10^{-11}

11 **a** 2×10^{100}
b $155 \times 10^{100} = 1.55 \times 10^{102}$
c $1000 \times 10^{100} = 1 \times 10^{103}$
d $0.5 \times 10^{100} = 5 \times 10^{99}$

12 **a** Chlamydia 1×10^{-6} m
Spirochaete $0.5 \div 1000 = 5 \times 10^{-4}$ m
b $(5 \times 10^{-4}) \div (1 \times 10^{-6}) = 500$

13 Volume of Earth $= \frac{4}{3} \times \pi \times \left(\frac{12\,756}{2}\right)^3$

Volume of Moon $= \frac{4}{3} \times \pi \times \left(\frac{3476}{2}\right)^3$

$\frac{\text{Volume of Earth}}{\text{Volume of Moon}} = 49.420\,16\ldots$

So $n = 49$ to 2 sf.

Exam questions

1 **a** x^8 **b** y^4 **c** $4w^2$

2 **a** **(1)** p^4 **(2)** q^4 **(3)** $16x^6$
b **(1)** $6a^4b^3$ **(2)** $\frac{1}{8a^6}$
c $3xy(3x - 2y^2)$

3 **a** $8x + 1$
b $2x^2 + 5x - 3$
c $3a^2(2a - 3)$
d **(1)** $\frac{1}{25}$ **(2)** 4 **(3)** $\frac{1}{7}$

4 **a** $81x^4y^{12}$ **b** $x = 20y^2 + 4$

5 **a** 3^4
b **(1)** $2^3 \times 5^{-2}$ **(2)** $2^{-9} \times 5^6$

6 **a** $7a^4$
b $\frac{c^6}{d}$
c p^8
d $5m^3$
e 9, −3
f $(2y - 3)(y - 4)$

7 **a** 1.6×10^8 **b** 28 000

8 **a** **(1)** 50 100 **(2)** 9×10^{-4}
b 2.4×10^9

9 **a** 8.4×10^7 **b** 2.1×10^{-5}

10 **a** 9.1×10^{-25} **b** 4.55×10^{-18} g

11 **a** 7.6×10^3 **b** $\frac{Rv^2}{G}$

12 **a** 1.4×10^{10} **b** 1.32×10^{23}

13 **a** 3.796×10^2 **b** 9.94×10^4

14 **a** 3×10^{-8} m **b** 3×10^{-10} m

15 $2p^2q^2(p - 2q)$

16 **a** 1×10^{297} **b** 4×10^{304}

17 7.5×10^4

18 **a** 3.62×10^8 **b** 5.12×10^8 km **c** 71%

CHAPTER 13

Questions

1 **a**

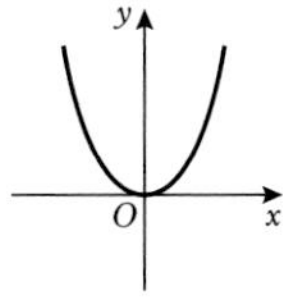

e

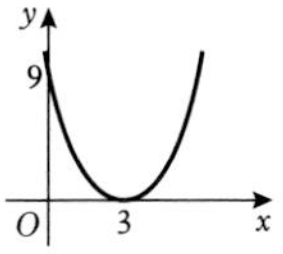

b

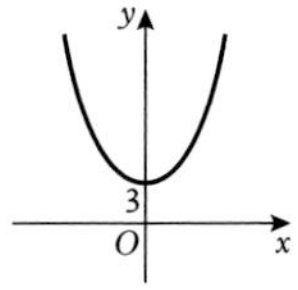

f

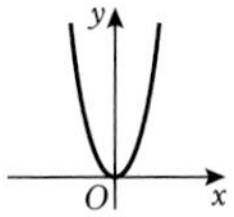

c

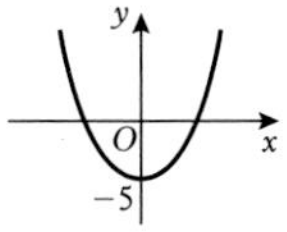

g

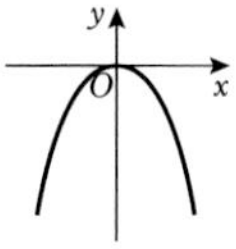

d

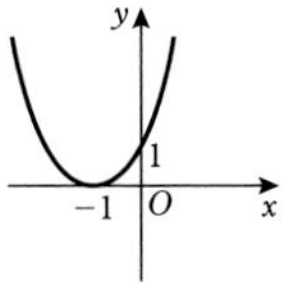

h

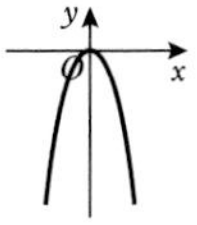

Parts **a–d**, **e** and **g** are all the same steepness (all $y = x^2$ shape)
Part **f** twice as steep
Part **h** three times as steep

2 **a** Translation $\begin{pmatrix} 1 \\ 0 \end{pmatrix}$, Stretch factor 3 in y direction.

b Translation $\begin{pmatrix} -4 \\ 0 \end{pmatrix}$, Stretch factor 2 in y direction,

Translation $\begin{pmatrix} 0 \\ -3 \end{pmatrix}$.

c Translation $\begin{pmatrix}1\\0\end{pmatrix}$, Stretch factor 2 in y direction,

Translation $\begin{pmatrix}0\\4\end{pmatrix}$.

d Translation $\begin{pmatrix}5\\0\end{pmatrix}$, Stretch factor 3 in y direction,

Translation $\begin{pmatrix}0\\-2\end{pmatrix}$.

3 **a**

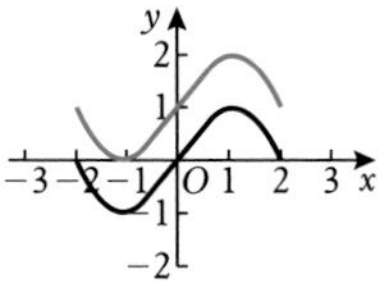

d

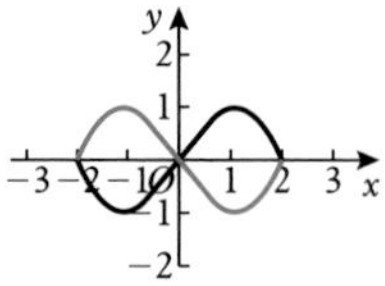

b

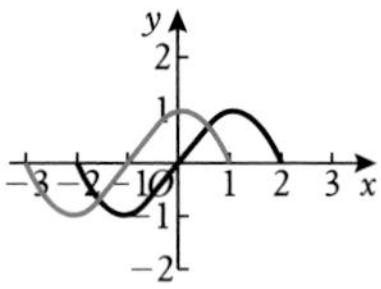

e

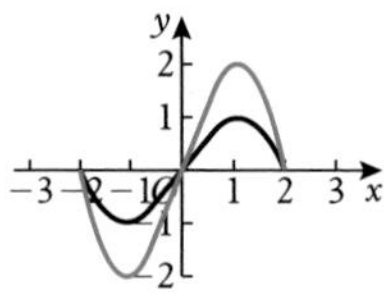

c

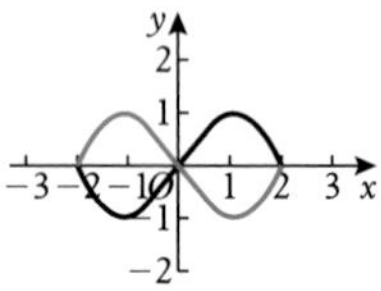

4 **a** $2^2 - 4 \times 2 + 3 = -1$
b $(-1)^2 - 4(-1) + 3 = 8$
c $(-3)^2 - 4(-3) + 3 = 24$
d $a^2 - 4a + 3$
e $(-p)^2 - 4(-p) + 3 = p^2 + 4p + 3$
f $(2a)^2 - 4(2a) + 3 = 4a^2 - 8a + 3$

5 $x^2 - x + 2 = 4$
$x^2 - x - 2 = 0$
$(x - 2)(x + 1) = 0$
$x = -1$ or 2

6 **a** $2x^2 - 8x + 3 = x - 4$
$2x^2 - 9x + 7 = 0$
$(2x - 7)(x - 1) = 0$
$x = 1$ or $\frac{7}{2}$
b $f(1) = 1 - 4 = -3$
$f(\frac{7}{2}) = \frac{7}{2} - 4 = -\frac{1}{2}$
c $(1, -3)$ and $(\frac{7}{2}, -\frac{1}{2})$

7 **a** $x = -1.2$ (Intersection with x axis)
b $x = 1.4$ (Intersection with $y = 3$)
c $x^2 - x + 2 = 0$
$x^2 - x + 1 + 1 = 0$
$x^2 - x + 1 = -1$
$x = -1.3$ (Intersection with $y = -1$)
d $x^3 - 2x + 4 = 0$
$x^3 - x + 1 - x + 3 = 0$
$x^3 - x + 1 = x - 3$
$x = -1.75$ (Intersection with $y = x - 3$)
e $-3 = x^3 - x$
$x^3 - x + 3 = 0$
$x^3 - x + 1 + 2 = 0$
$x^3 - x + 1 = -2$
$x = -1.4$ (Intersection with $y = x - 2$)

8 You need to plot T against $\sqrt{l}$ to see if you put a straight line.

l	0.2	0.4	0.6	0.8	1	1.2	1.4
$\sqrt{l}$	0.45	0.63	0.77	0.89	1	1.10	1.18
T	0.89	1.26	1.55	1.79	2.00	2.19	2.37

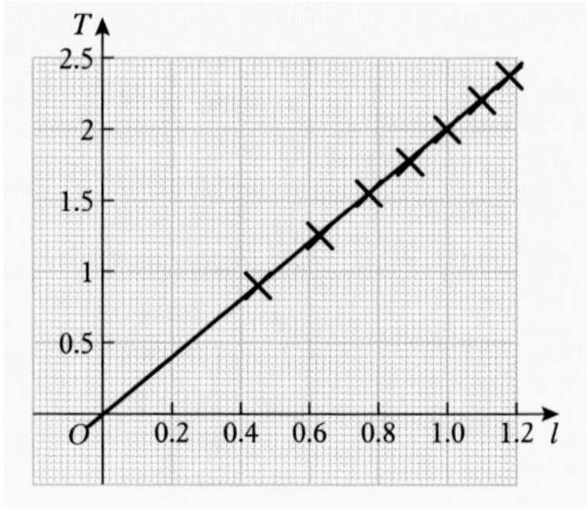

Since you get a straight line the relationship is shown to be valid.
a = gradient $= \frac{2}{1} = 2$ $b = y$ intercept $= 0$ So $T = 2\sqrt{l}$

Exam questions

1 **a** 1 and 5

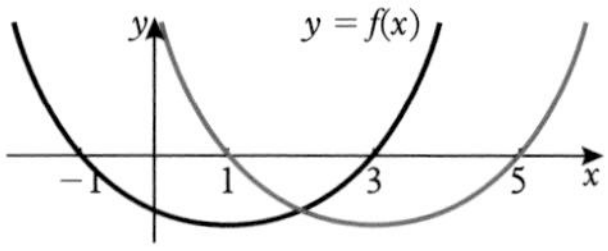

b $1 + \dfrac{a}{2}$
c $(x + 3)(x - 1)$

2

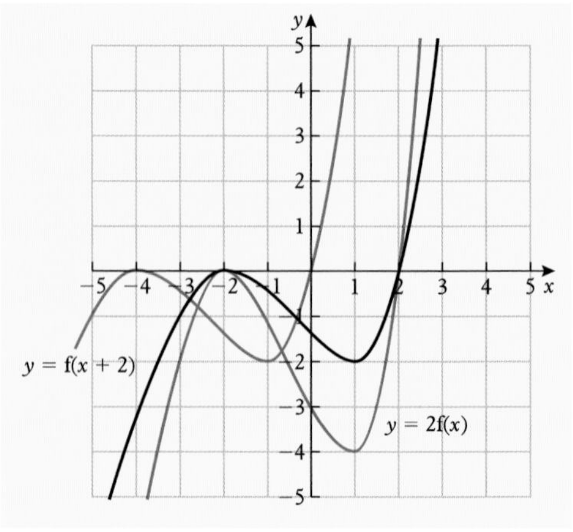

3 **a** Reflection in the x axis
b $p = 2, q = -4$
c Translation $\begin{pmatrix}-2\\-4\end{pmatrix}$

4 **a** $(2, 18)$ **b** $(-1, 12)$ **c** $(-2, 12)$ **d** $(\frac{1}{2}, 12)$

5 **a** −3.5, 0, 3.5

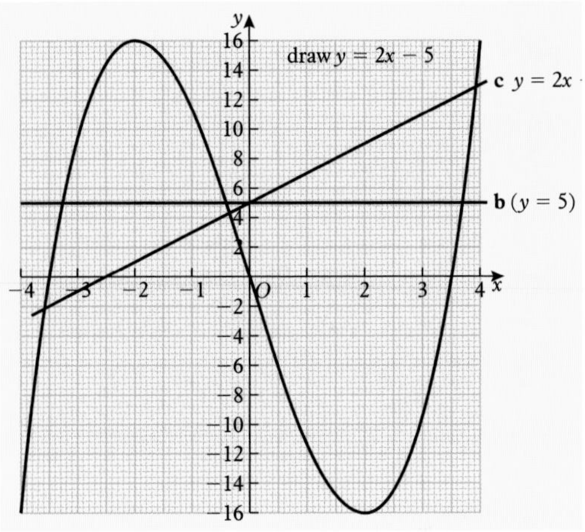

b −3.2, −0.4, 3.7 **c** −3.9, 0.4, 3.5

6 **a**

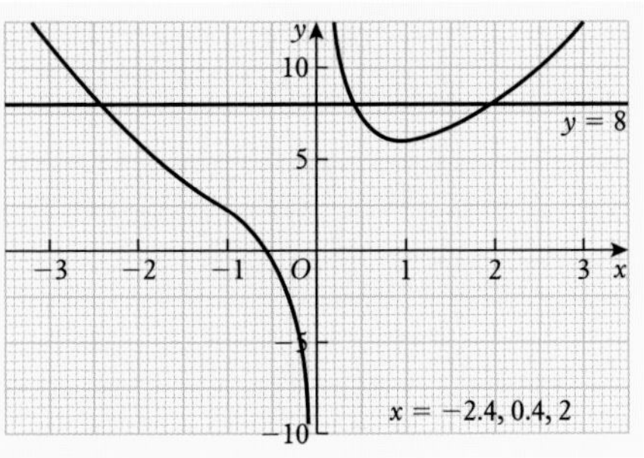

b $y = 3x + 3$

CHAPTER 14

Questions

1 $P(7\text{ sixes}) = P(6) \times P(6) \times P(6) \times P(6) \times P(6)$
$\times P(6) \times P(6) = \left(\frac{1}{6}\right)^7 = \frac{1}{279\,936}$

2 **a** $P(\text{double}) = \frac{6}{36} = \frac{1}{6}$

b $100 \times \frac{1}{6} = 16\frac{2}{3}$ So 17 doubles expected.

3 **a** $1 - (\frac{1}{20} + \frac{7}{20} + \frac{23}{40}) = \frac{1}{40}$

b $\frac{1}{40} \times \frac{1}{40} = \frac{1}{1600}$

c $\frac{1}{20} \times \frac{1}{20} = \frac{1}{400}$

d $P(\text{R, B}) + P(\text{B, R})$
$= \frac{7}{20} \times \frac{1}{20} + \frac{1}{20} \times \frac{7}{20}$
$= \frac{14}{400} = \frac{7}{200}$

e $P(\text{miss, not miss}) + P(\text{not miss, miss})$
$= \frac{1}{40} \times \frac{39}{40} \times \frac{39}{40} \times \frac{1}{40} = \frac{78}{1600} = \frac{39}{800}$

f $1 - P(\text{miss, miss})$
$= 1 - \frac{1}{40} \times \frac{1}{40} = \frac{1599}{1600}$

4 **a** $\frac{6}{20} = \frac{3}{10}$

b It will approach the probability of a 6 which is $\frac{1}{8}$.

5 **a** **(1)** $\frac{15}{100} = \frac{3}{20}$

(2) $\frac{32}{100} = \frac{8}{25}$ (5 + 12 + 15)

(3) $\frac{19}{100}$ (4 + 6 + 9)

(4) $\frac{37}{100}$ (12 + 17 + 6 + 2)

b $\frac{3}{7}$ (7 red-haired people of whom 4 have straight hair)

6 Perform an experiment, spinning the spinner a large number of times, say 200 times. Count the number of green results and use this number divided by 200 as the estimate of probability.

7 **a**

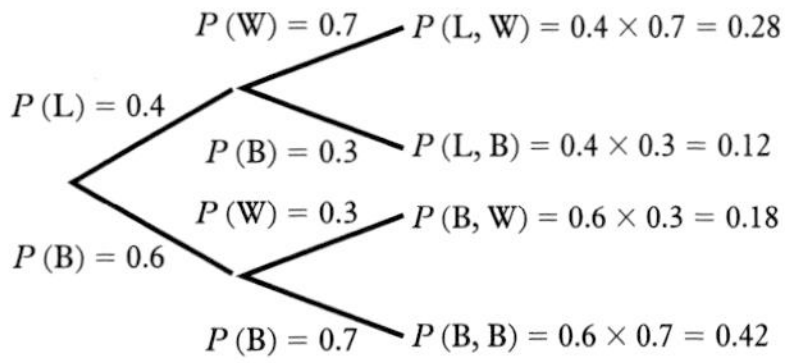

b **(1)** $P(\text{B, B}) = 0.42$

(2) $P(\text{L, B}) + P(\text{B, B}) = 0.12 + 0.42 = 0.54$

(3) $1 - P(\text{L, W}) = 1 - 0.28 = 0.72$

Exam questions

1 **a** **(1)** 0.125 **(2)** 0
b 0.42
c **(1)** 0.441 **(2)** 0.847

2 **a** $\frac{7}{15}$ **b** $\frac{1}{5}$

3 $\frac{41}{95}$

4 $\frac{21}{50}$

5 **(1)** 0.06 **(2)** 0.38

6 $n = 7$

7 **(1)** $\frac{20}{72}$ **(2)** $\frac{18}{72}$

8 **(1)** $\frac{1}{16}$ **(2)** $\frac{1}{6}$

9 **a** **(1)** 0.28 **(2)** 0.88
b 0.48

10 **a** $\frac{3}{8}$

11 **a** 700 **c** 0.18
b 0.42 **d** 0.91

CHAPTER 15

Questions

1 **a** 4905.6

b 956 4855 4992 5267 5422 5434 5445 5506 5523 5656

Median $= \frac{5422 + 5434}{2} = 5428$

c The median.
The mean is badly affected by the one very small value.

2 $\frac{x + x + 5 + x + 4}{3} = 12$

$\frac{3x + 9}{3} = 12$

$x + 3 = 12$

$x = 9$

3 Total score of 32 pupils = 68.5×32
$= 2192$
Total score of 33 pupils = 69×33
$= 2277$
$\therefore$ Bushra scored $2277 - 2192$
$= 85$ marks

4 **a** $n - 1, n, n + 1$

b $\frac{n - 1 + n + n + 1}{3} = \frac{3n}{3} = n$

c $(n - 1)^2 = n^2 - 2n + 1$
n^2
$(n + 1)^2 = n^2 + 2n + 1$

d n^2

e $\frac{n^2 - 2n + 1 + n^2 + n^2 + 2n + 1}{3} = \frac{3n^2 + 2}{3}$
$= n^2 + \frac{2}{3}$

5 **a, c, d**

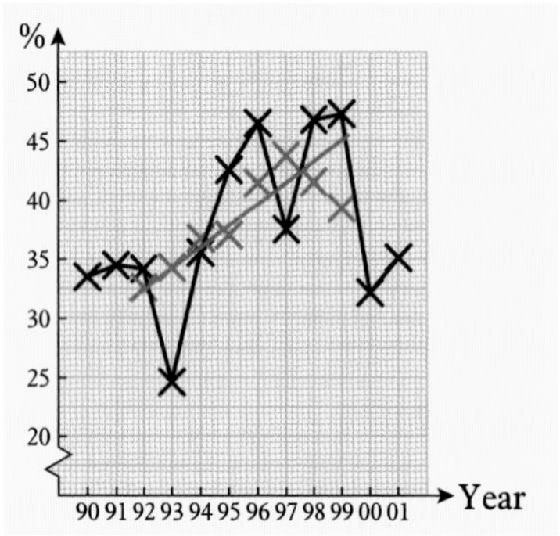

		5 point moving average	Plot at
b	90–94	32.46	92
	91–95	34.28	93
	92–96	36.68	94
	93–97	37	95
	94–98	41.44	96
	95–99	43.8	97
	96–100	41.7	98
	97–101	39.42	99

d Results have steadily improved, although the recent results show a downward trend. The head can certainly justify her claim over the first 10 or 11 years but the last moving average is well below the trend line.

6 **a**
Tel's Taxis
2 4 8 9 9 | 10 10 10 11 11 | 11 12 12 12 13 | 13 14 14 20 22

Median = 11 Range = 20
Lower quartile = 9.5 Upper quartile = 13

Speedy Steve's
1 1 1 2 2 | 2 4 5 6 11 | 13 14 15 18 20 | 22 23 24 26 33

Median = 12 Range = 32
Lower quartile = 2 Upper quartile = 21

b

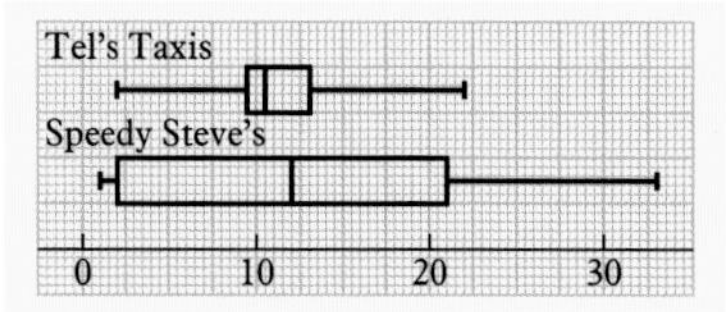

c The two companies have similar medians.
Tel's times are more consistent with a smaller range and Steve's have several very high values.
Steve also has more low values though.

d Either answer allowed (but don't give both).
Either Tel's because his times are more consistent.
Or Steve's because some of his times are very quick and I would take a chance on a very quick response.

7 **a** $40 < A \leqslant 60$

b $\frac{10 \times 4 + 30 \times 14 + 50 \times 41 + 80 \times 35 + 150 \times 6}{100}$
$= £62.10$

c Because you have not used the actual values but the mid-values of each group.

d The mean is greater than the upper end of the modal group. There are some high values that are causing the mean to be larger than the other averages.

e

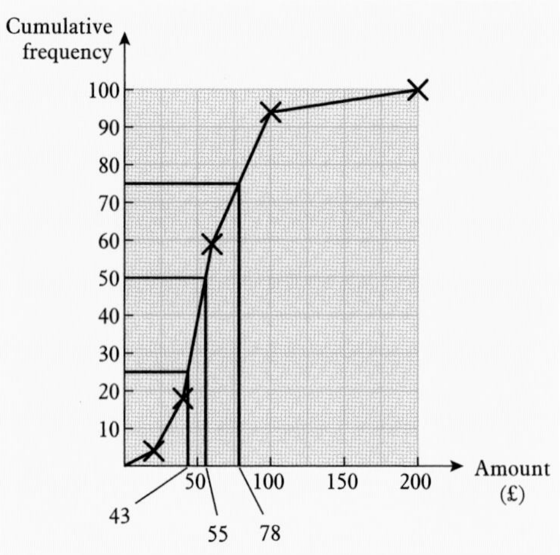

Amount	Cumulative frequency
$\leqslant 20$	4
$\leqslant 40$	18
$\leqslant 60$	59
$\leqslant 100$	94
$\leqslant 200$	100

f £55

g The median is lower than the mean.
The few high values are distorting the mean upwards.

h $78 - 43 = £25$

Exam questions

1 **a** 11 to 15
b 16.4

2 **a** 16.2
b 20

3 £380

4 **a** Use random numbers, for example.
b 2, 5, 10, 4, 3

5 8

6 **a** 7.9 years
b 41, 67, 87, 97, 100

c

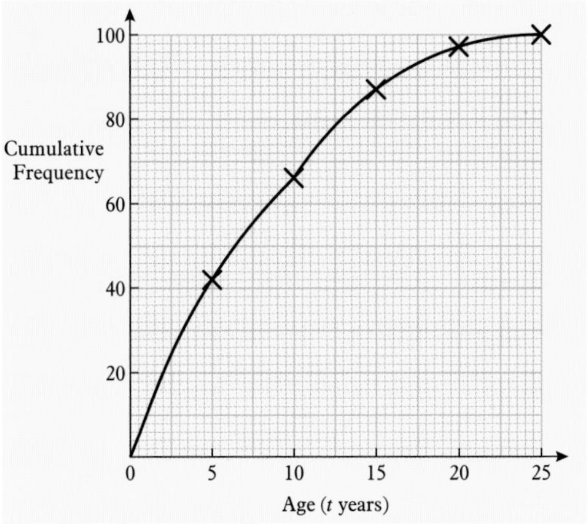

d 11.5–12.0

7 $n^2 + 4n + 6 - (n^2 + 4n + 4) = 2$

CHAPTER 16

Questions

1 **a** **(1)** $S = \frac{80}{360} \times 2 \times \pi \times 13$

$S = 18.2$ cm

(2) $A = \frac{80}{360} \times 2 \times \pi \times 13^2$

$A = 118.0$ cm^2

b **(1)** $S = \frac{54}{360} \times 2 \times \pi \times 23$

$S = 21.7$ mm

(2) $A = \frac{54}{360} \times \pi \times 23^2$

$A = 249.3$ mm^2

c **(1)** $S = \frac{284}{360} \times 2 \times \pi \times 5$

$S = 24.8$ cm

(2) $A = \frac{284}{360} \times \pi \times 5^2$

$A = 62.0$ cm^2

2 $\frac{75}{360} \times \pi \times 10^2 - \frac{1}{2} \times 10 \times 10 \times \sin 75°$

$= 17.2$ cm^2

3 $\frac{90}{360} \times \pi \times 20^2 - \frac{1}{2} \times 20 \times 20 \times \sin 90°$

(or just $\frac{1}{2} \times 20 \times 20$)

$= \frac{1}{4} \times 400\pi - 200$

$= 100\pi - 200$

$= 100(\pi - 2)$ cm^2

4 $\frac{1}{2} \times 19 \times 23 \times \sin(180° - 93° - 65°)$

$= 81.9$ cm^2

5 **a** **(1)** $4 \times \pi \times 4^2$

$= 64\pi$ cm^2

(2) $\frac{4}{3} \times \pi \times 4^3$

$= \frac{256\pi}{3}$ cm^3

b **(1)** $\pi \times 5 \times 13 + \pi \times 5^2$

$= 65\pi + 25\pi$

$= 90\pi$ cm^2

(2) $\frac{1}{3} \times \pi \times 5^2 \times 12$

$= 100\pi$ cm^3

c **(1)** $2 \times \pi \times 4^2 + 2 \times \pi \times 4 \times 15$

$= 32\pi + 120\pi$

$= 152\pi$ cm^2

(2) $\pi \times 4^2 \times 15$

$= 240\pi$ cm^3

6 **a** $8 \times 20 + \frac{1}{2} \times \pi \times 10^2$

$= 160 + 50\pi$

$= 317.1$ m^2

b a $\times$ 140 = 44 391.1

= 44 400 m^3 (3 sf)

7 **a** Volume scale factor = 8

∴ Length factor = $\sqrt[3]{8} = 2$

and Area factor = 4

∴ Base area = 4 × 46 = 184 cm^2

b 184 × 100 = 18 400 mm^2

8 Length scale factor = 0.25

∴ Area scale factor = $(0.25)^2 = 0.0625$

∴ Material needed = 0.0625 × 0.6

= 0.0375 m^2

= 375 cm^2

9 **a** volume

b area

c area + area = area

d volume + volume = volume

e volume + volume = volume

f length × (area + area) = volume

Exam questions

1 **a** 12.6 cm

b 240 cm^2

2 **a** $r\sqrt{2}$

b $\frac{AB}{2} = \sqrt{R^2 - r^2}$

$= \sqrt{r^2 = r}$

3 **a** $(2x - 3)(x + 11)$

b 33π cm^2

c 11 cm

4 8.9 cm

5 **a** 2.5 m

b 20 cm

6 **a** 262 cm^3

b 13.8 cm

7 $\pi r^2 l, 4\pi r^3, 3(a^2 + b^2)r$

8 50 kg

9 **a** 700 cm^3

b 13.51 kg

CHAPTER 17

Questions

1 **a** 350 **b** 10 **c** 23 300 **d** 1 232 000 **e** 1.9×10^5 **f** 6×10^{-7}

2
- **a** $(3 \times 10^5) \times (2 \times 10^4) = 6 \times 10^9$
- **b** $(1 \times 10^9) \times (5 \times 10^{-3}) = 5 \times 10^6$
- **c** $(7 \times 10^9) \times (4 \times 10^{-4}) = 28 \times 10^5 = 2.8 \times 10^6$
- **d** $(6 \times 10^8) \times (4 \times 10^{-3}) = 24 \times 10^5 = 2.4 \times 10^6$
- **e** $(6 \times 10^7) \div (2 \times 10^{-3}) = 3 \times 10^{10}$
- **f** $(8 \times 10^5) \div (2 \times 10^{-5}) = 4 \times 10^{10}$
- **g** $\dfrac{5 \times 10^6}{2 \times 10^{-3}} = 2.5 \times 10^9$
- **h** $\dfrac{7 \times 10^{-5}}{4 \times 10^7} = 1.75 \times 10^{-12}$

3
- **a** **(1)** $30 \times 800 = 24\,000$
  (2) 28 460.4708
  (3) $\dfrac{4460.4708}{28\,460.4708} \times 100 = 15.7\%$ (1 dp)
- **b** **(1)** $\dfrac{80 \times 4}{40} = 8$
  (2) 7.253 879 077
  (3) $\dfrac{0.746\,120\,922\,7}{7.253\,879\,077} \times 100 = 10.3\%$ (1 dp)
- **c** **(1)** $\dfrac{4 \times 6}{4} = 6$
  (2) 5.388 238 141
  (3) $\dfrac{0.611\,761\,858\,7}{5.388\,238\,141} \times 100 = 11.4\%$ (1 dp)

4
- **a** **(1)** $\dfrac{24 \times 8}{4} = 48$
  (2) 42.719 535 52
  (3) $\dfrac{(1) - (2)}{(2)} \times 100 = 12.4\%$ (1 dp)
- **b** **(1)** $\dfrac{\sqrt{100} \times 9}{18} = \dfrac{10}{2} = 5$
  (2) 5.262 790 852
  (3) $\dfrac{(2) - (1)}{(2)} \times 100 = 5.0\%$ (1 dp)
- **c** **(1)** $\dfrac{{}^{2}\cancel{12} \times \cancel{70}^{\,2}}{\cancel{35}_1 \times \cancel{6}_1} = 4$
  (2) 4.637 596 595
  (3) $\dfrac{(2) - (1)}{(2)} \times 100 = 13.7\%$ (1 dp)
- **d** **(1)** $\sqrt{\dfrac{60 \times 8}{6 \times 4}} = 20 \approx 4.5$
  ($\sqrt{16} = 4$ $\sqrt{25} = 5$ So $\sqrt{20} \approx 4.5$)
  (2) 4.552 120 958
  (3) $\dfrac{(2) - (1)}{(2)} \times 100 = 1.1\%$ (1 dp)

5
- **a** $\dfrac{12.3}{2}(8.6 + 17.1) = 158.055 \text{ cm}^2$
- **b** $\dfrac{10}{2}(9 + 20) = 5 \times 29 = 145 \text{ cm}^2$
- **c** $\dfrac{\text{a} - \text{b}}{\text{a}} \times 100 = 8.3\%$ (1 dp)
- **d** $\dfrac{12}{2}(9 + 17) = 6 \times 26 = 156 \text{ cm}^2$
- **e** $\dfrac{\text{a} - \text{d}}{\text{a}} \times 100 = 1.3\%$ (1 dp)

6 Journey via B = 290 m
Diagonal $= \sqrt{120^2 + 170^2} = 208.086\,520\,5 \text{ m} = 210 \text{ m}$
Distance shorter = 290 − 210 = 80 m
(Since the numbers in the question are only given to 2 sf [accurate to nearest 10 m] don't use more accuracy than 3 sf at the most. This would give 290 − 208 = 82 m.)

7

	Lower	Upper
a	$2.565 + 4.15 = 6.715$	$2.575 + 4.25 = 6.825$
b	$2.565 - 4.25 = -1.685$	$2.575 - 4.15 = -1.575$
c	$4.15 \times 0.65 = 2.6975$	$4.25 \times 0.75 = 3.1875$
d	$\dfrac{2.565}{0.75} = 3.42$	$\dfrac{2.575}{0.65} = 3.961\,538\,462$
e	$2 \times 2.565 + 3 \times 4.15 = 17.58$	$2 \times 2.575 + 3 \times 4.25 = 17.9$
f	$\dfrac{4.15 + 0.65}{2.575} = 1.864\,077\,67$	$\dfrac{4.25 + 0.75}{2.565} = 1.949\,317\,739$

8

	Lower	Upper
a	1.705×10^7	1.715×10^7
b	1.475×10^8	1.485×10^8
c	$\dfrac{1.475 \times 10^8}{1.715 \times 10^7} = 8.600\,583\,09 = 8.6$ (1 dp)	$\dfrac{1.485 \times 10^8}{1.705 \times 10^7} = 8.709\,677\,419 = 8.7$ (1 dp)
d	$\dfrac{1.705 \times 10^7}{70.5} = 241\,843.9716$	$\dfrac{1.715 \times 10^7}{69.5} = 246\,762.5899$
e	$\dfrac{1.475 \times 10^8}{2.55} = 57\,843\,137.25$	$\dfrac{1.485 \times 10^8}{2.45} = 60\,612\,244.9$

f Lower bound $= \dfrac{57\,843\,137.25}{246\,762.5899} = 234.408\,048\,9$
Upper bound $= \dfrac{60\,612\,244.9}{241\,843.9716} = 250.625\,411\,5$
So 234 < Population density < 251

9 Lower bound $= \dfrac{19.5}{24.5} = 0.796$ m/hour
Upper bound $= \dfrac{20.5}{23.5} = 0.872$ m/hour

Exam questions

1 41.2
2 **a** 46.416 376… **b** 46
3 $\frac{1}{80}$
4 **a** 40, 60, 300 **b** 280 < 300 or $42.8 \times 63.740 \times 60$
5 **a** **(1)** $a = 50, b = 30$ *or* 31
(2) 75, 77.5, 80
(3) 79.368 75
b 4.9×10^{11}
6 **(1)** $f = 9, g = 2$ **(2)** $\frac{89}{3}$ or $\frac{108}{4}$
7 **a** 4200 **b** **(1)** 98.5 **(2)** 97.5
8 **a** **(1)** 200.5 g **(2)** 199.5 g
b **(1)** 115 **(2)** 105
c 1.9095
9 1.865
10 **a** **(1)** 5.65 h **(2)** 0.045 cm
b 7.9 h
c 0.1 cm
11 28.5 cm²
12 11.14 ms^{-2}, 9.02 ms^{-2}

CHAPTER 18

Questions

1 **a**

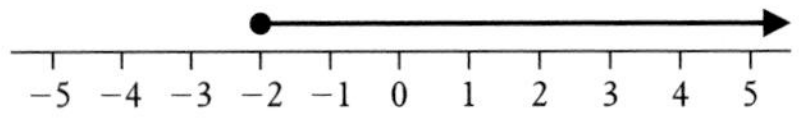

b

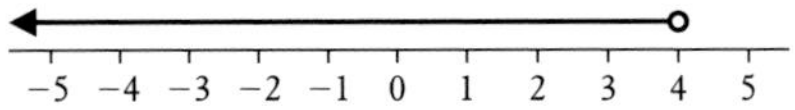

c

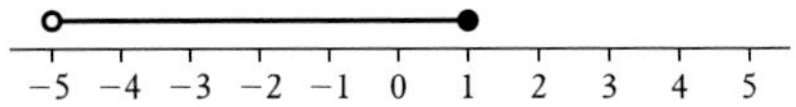

2 **a** $x \geqslant -1$ **c** $-2 \leqslant x \leqslant 4$

b $x < 3$

3 **a** $-2, -1, 0, 1$ **c** $5, 6, 7, 8, 9$

b $-3, -2, -1, 0, 1, 2$ **d** $-7, -6, -5, -4, -3, -2$

4 **a** $3x < 9$
$x < 3$

b $2x \geqslant 24$
$x \geqslant 12$

c $-4x \leqslant 4$
$x \geqslant -1$

d $-5x > 5$
$x < -1$

e $4x < -1$
$x < -\frac{1}{4}$

f $3x \geqslant 5$
$x \geqslant 1\frac{2}{3}$

g $\frac{x}{4} \geqslant 7$
$x \geqslant 28$

h $\frac{3x}{4} \leqslant 2$
$3x \leqslant 8$
$x \leqslant 2\frac{2}{3}$

i $\frac{2x}{5} > 5$
$2x > 25$
$x > 12\frac{1}{2}$

j $-\frac{2x}{3} < 2$
$-2x < 6$
$x > -3$

k $-1 < 2x \leqslant 1$
$-\frac{1}{2} < x \leqslant \frac{1}{2}$

l $1 < \frac{x}{4} \leqslant 5$
$4 < x \leqslant 20$

5 **a** $5 < x \leqslant 15$

b $-4 \leqslant \frac{x}{3} < 3,$
$-12 \leqslant x < 9$

c $-3 < -3x$ and $-3x < 3$
$x < 1$ and $x > -1$
So $-1 < x < 1$

d $-10 \leqslant -4x$ and $-4x < 0$
$x \leqslant 2.5$ and $x > 0$
So $0 < x \leqslant 2.5$

6 **a** $x < 4, y \leqslant 2$ **c** $x \leqslant 4, y \geqslant 1, y \leqslant x$

b $1 < x \leqslant 3$ **d** $x + y < 4, x > 2, y > 0$

7 **a**

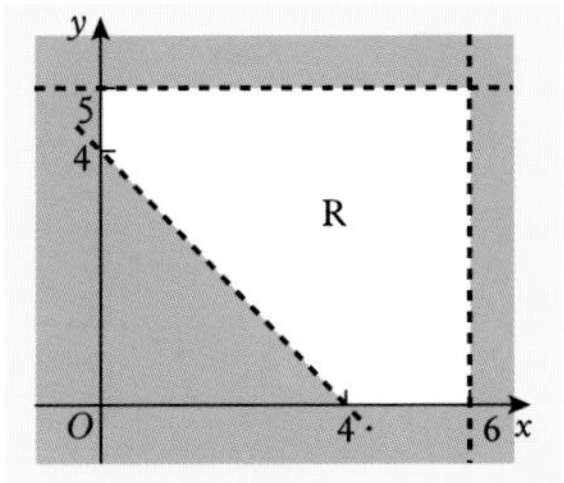

b

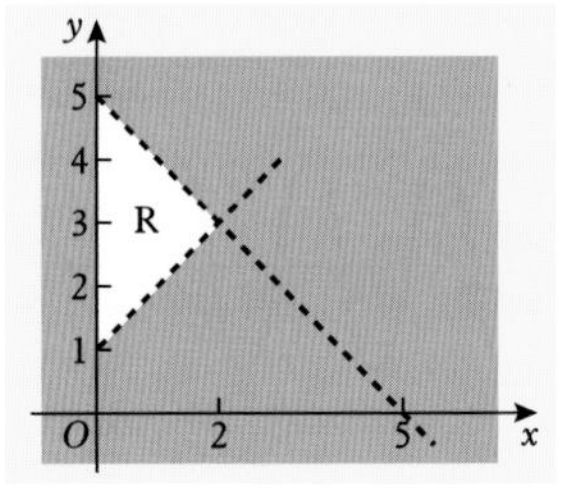

8 **a** Cost $30x + 20y \leqslant 300$
$3x + 2y \leqslant 30$

b $x \geqslant 5, y \geqslant 3$

c

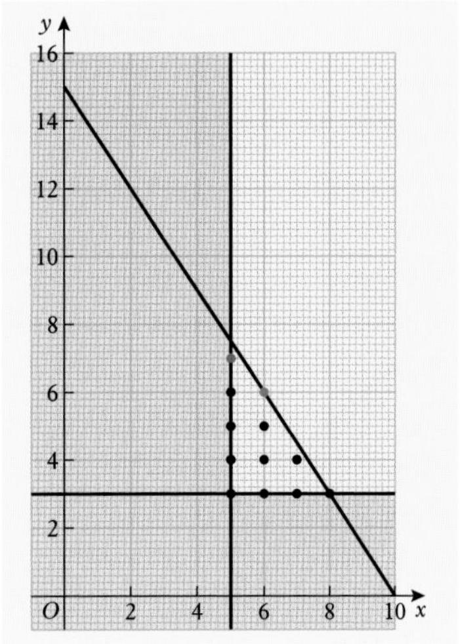

d The dots show possible solutions.
The red and blue dots both give 12 refills but the red dot gives change. So she should buy 5 colour and 7 black and white.

Exam questions

1 **a** $x > \frac{5}{4}$

b 2

2 **(1)** $n > -\frac{8}{3}$ **(2)** -2

3 4, 5, 6

4 $y > -\frac{3}{5}$

5 3

6 **a** $a = 52\,000, b = 6$ **b** 13 000

7 **a** $(0.2x + y \leqslant 300) \times 5$

b $y \geqslant \dfrac{x}{5}$ or $5y \geqslant x$

c

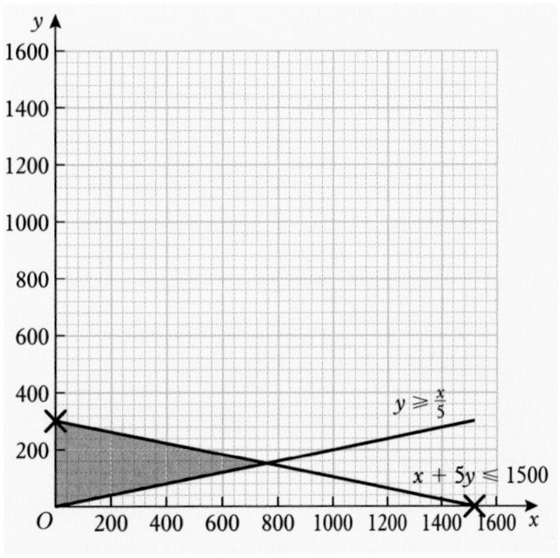

d 900 **e** £200 **f** 100

8 **a** $y = \dfrac{6-x}{2}$

b Line drawn, e.g. (0, 3), (6, 0), (2, 2)

c

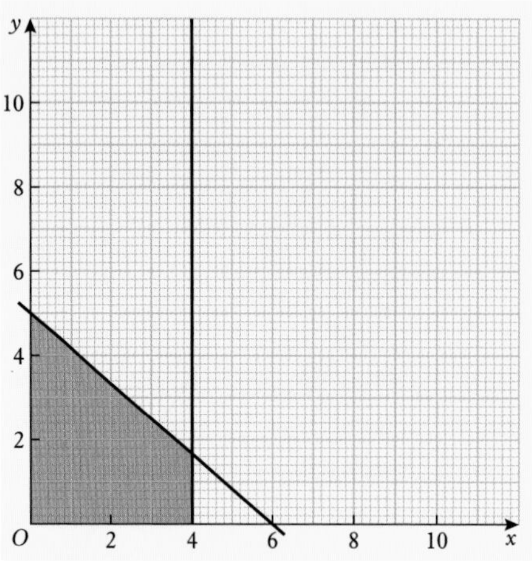

d $x = 2.4$

9 **a** $x + 3y \leqslant 12$

b

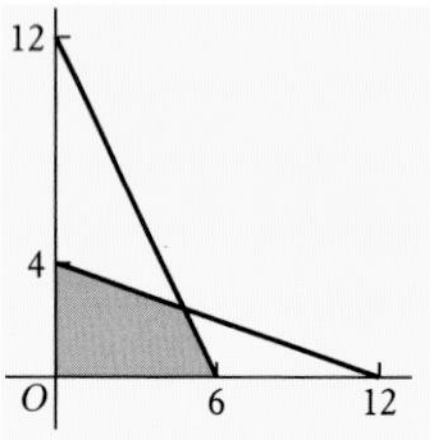

c 7

CHAPTER 19

Questions

1 **a**

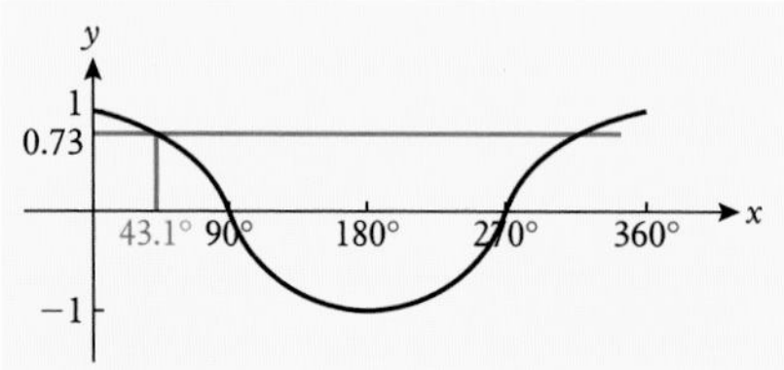

b 43.1°

c Red lines on diagram

d $360° - 43.1° = 316.9°$

2 **a**

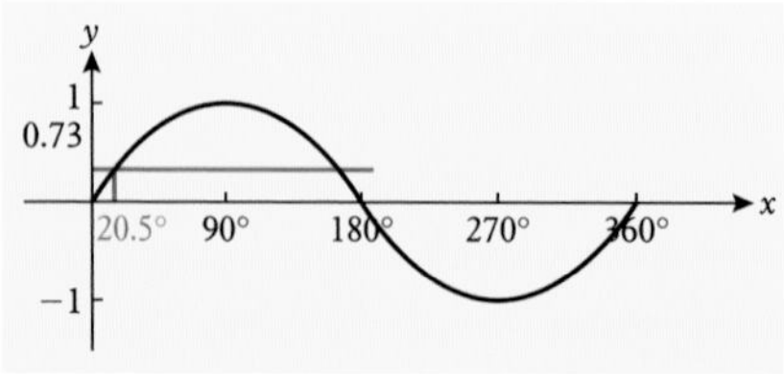

b 20.5°

c Red lines on diagram

d $180° - 20.5° = 159.5°$

3 Because the graph of $y = \sin x$ has only one trough in a 360° cycle.
$x = 270°$ is the solution for $0 \leqslant x \leqslant 360°$.

4 116.1°, 243.9°

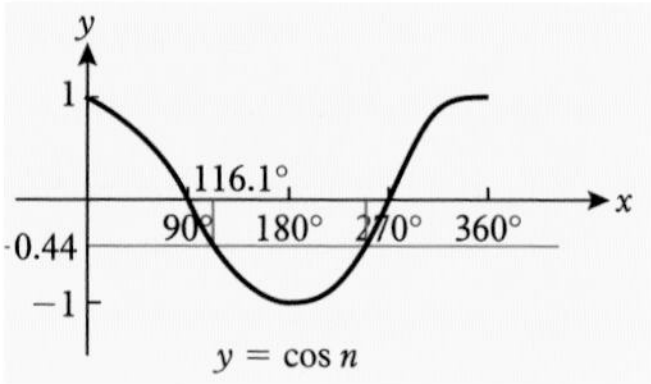

5 **a** 30°, 150°

b 150°, 210°

c 225°, 315°

d 45°, 315°

6 **a**

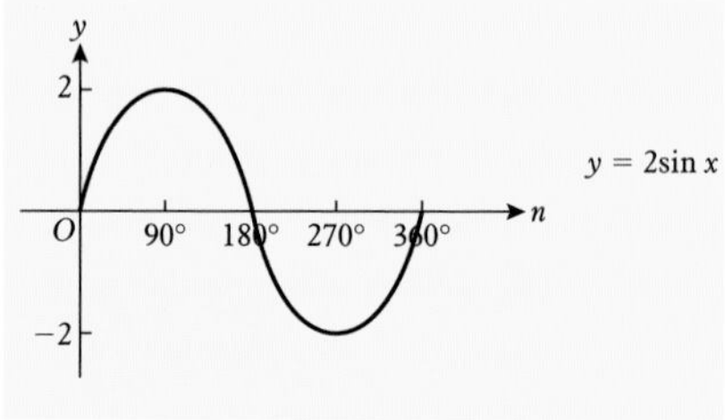

b

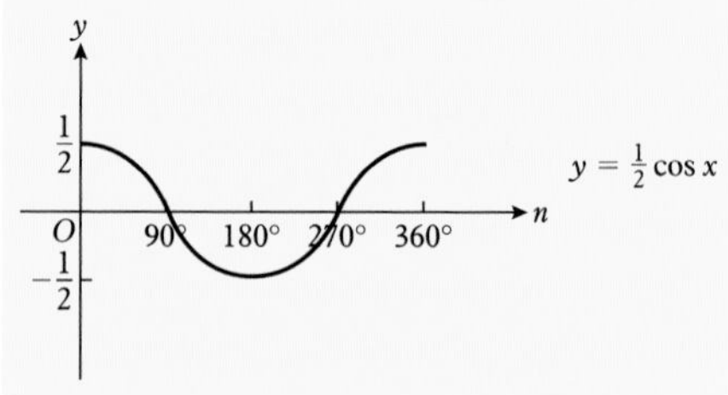

c

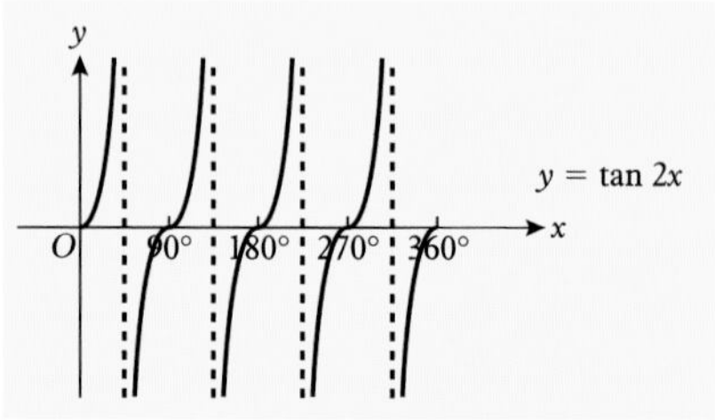

d

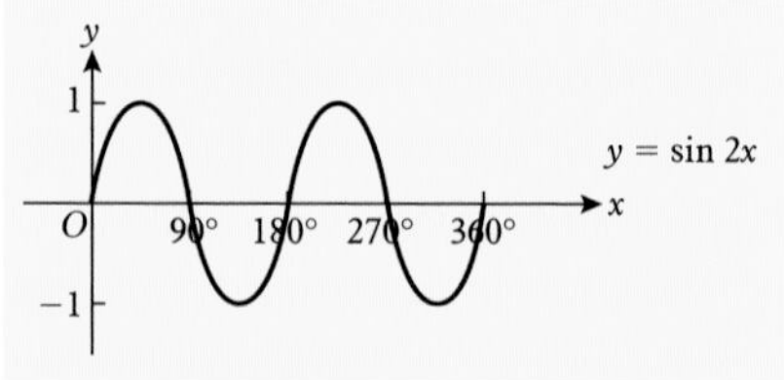

e

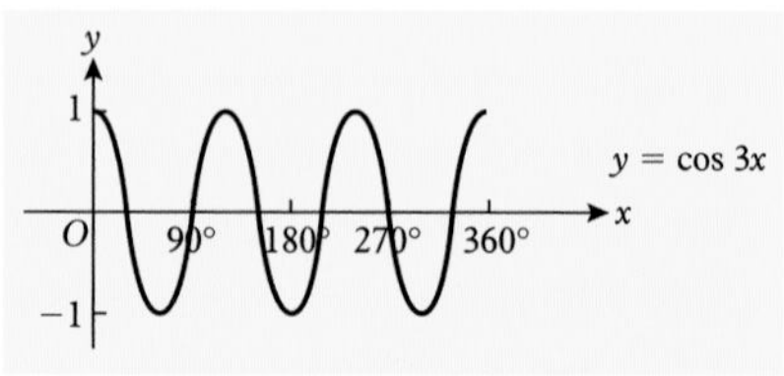

f

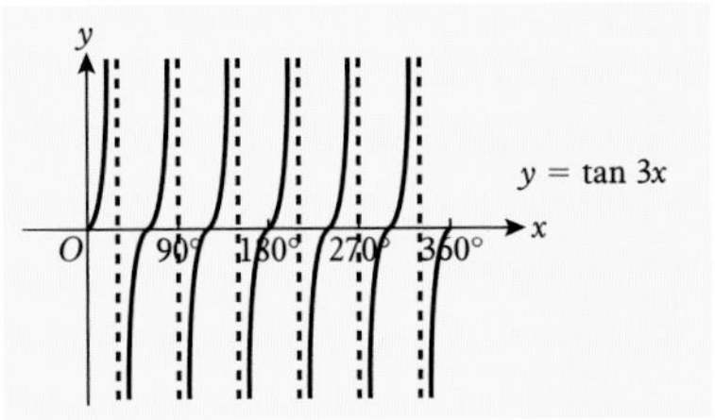

g

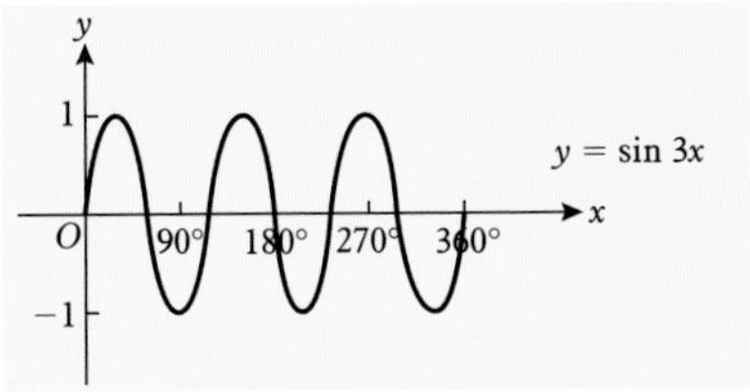

h

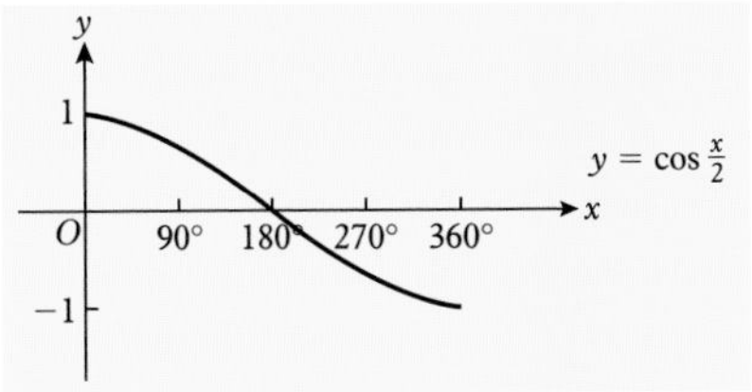

i

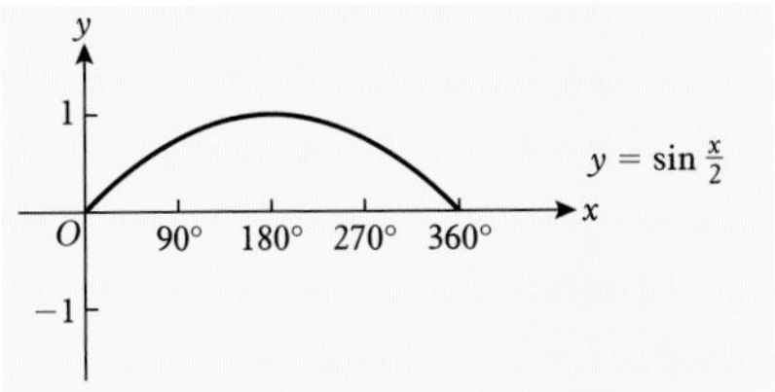

j

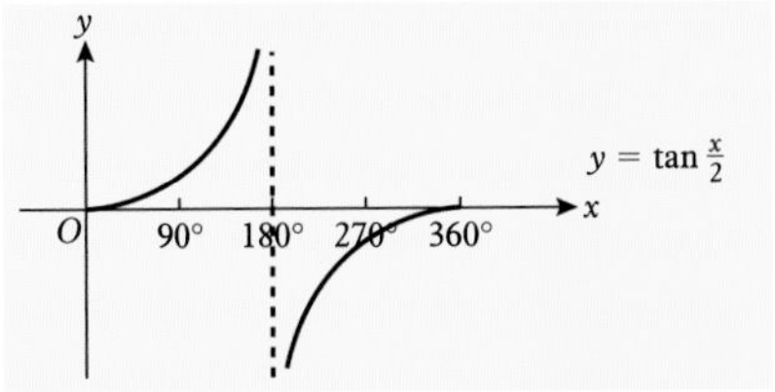

k

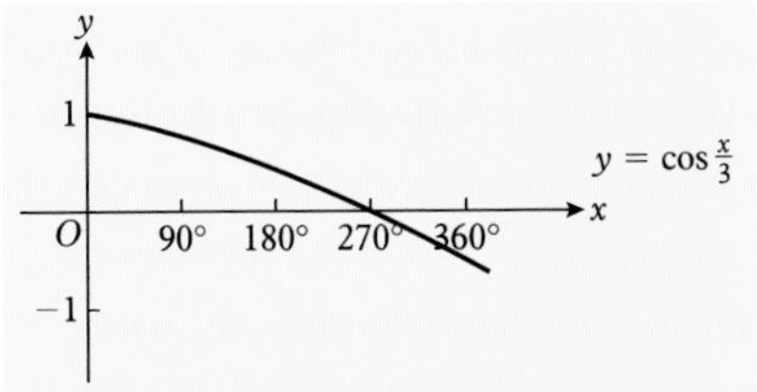

l

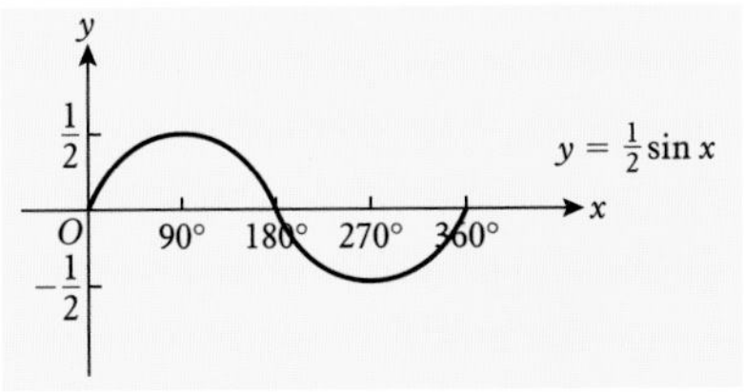

7 **a** 4

b −4

c

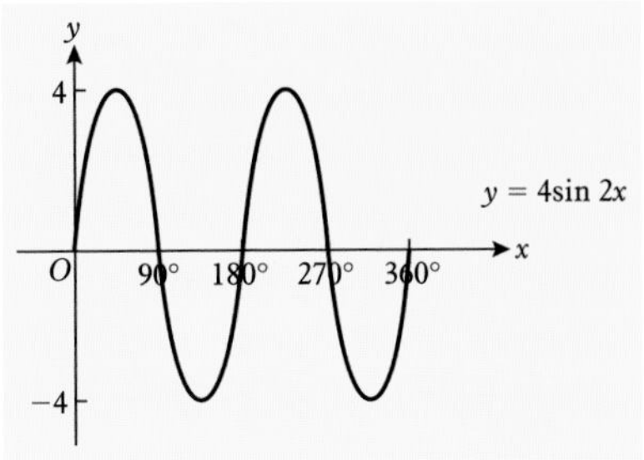

8 **a** 5

b −5

c

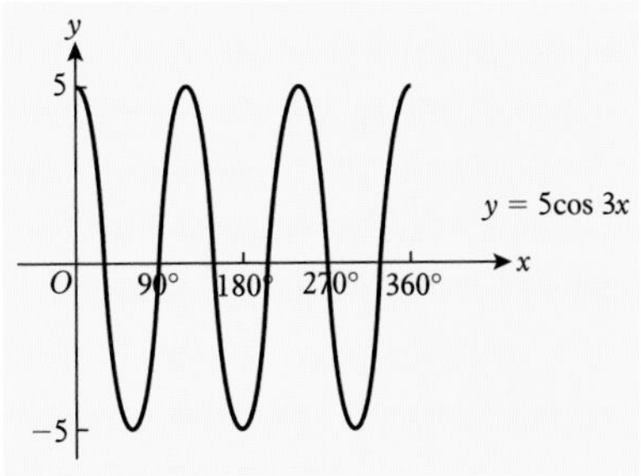

9 **a**

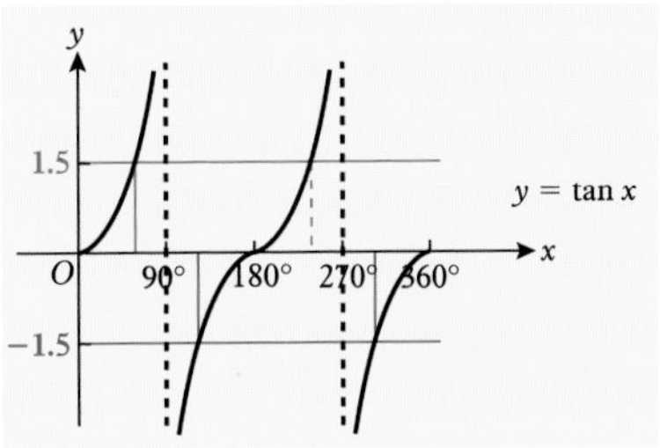

b Red lines on **a**

c $56.3° + 180° = 236.3°$ (see dashed red line on **a**)

d $180° - 56.3° = 123.7°$ (see green lines on **a**)
$360° - 56.3° = 303.7°$

10 **a** $\dfrac{a}{\sin 48°} = \dfrac{13}{\sin 37°}$

$a = 16.1$ cm

b $\dfrac{\sin b}{18} = \dfrac{\sin 36°}{13}$

$\sin b = 0.8138\ldots$

$b = 54.5°$ or $125.5°$

11 **a** $a^2 = 11^2 + 17^2 - 2 \times 11 \times 17 \times \cos 34°$

$a^2 = 99.939\,94\ldots$

$a = 10.0$ cm

b $\cos b = \dfrac{10^2 + 14^2 - 19^2}{2 \times 10 \times 14}$

$= -0.232\,14\ldots$

$b = 103.4°$

12 $\frac{1}{2} \times 23.6 \times 28.1 \times \sin 123°$
$= 278.1 \text{ cm}^2$

13 **a**

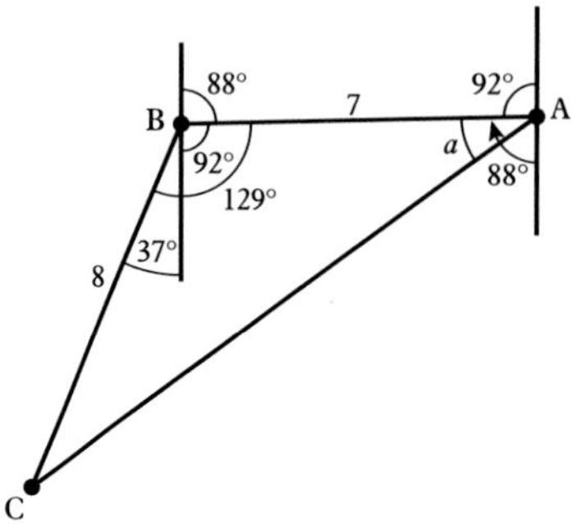

b $AC^2 = 7^2 + 8^2 - 2 \times 7 \times 8 \times \cos 129°$
$= 183.48\ldots$
$AC = 13.5$ miles (3 sf)

c $\frac{\sin a}{8} = \frac{\sin 129°}{AC}$
$\sin a = 0.458\,579\,8\ldots$
$a = 27.3°$ (152.7° is too big to exist in given triangle)
So bearing $= 360° - 92° - 27.3°$
$= 241°$

14 **a** $WY^2 = 12^2 + 14^2 - 2 \times 12 \times 14 \times \cos 130°$
$WY = 23.6$ cm (3 sf)

b $\frac{\sin \angle WYX}{12} = \frac{\sin 130°}{WY}$
$\angle WYX = 22.9°$ (157.1° is too big for triangle given)

c In $\triangle WYZ$: $YZ^2 = 5^2 + WY^2$ (Pythagoras)
$YZ = 24.1$ cm (3 sf)

d In $\triangle XWZ$: $XZ^2 = 5^2 + 12^2$
$XZ = 13$ cm

e In $\triangle XZY$:

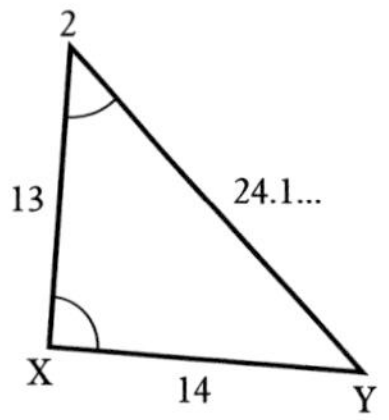

$\cos \angle XZY = \frac{13^2 + 24.1^2\ldots - 14^2}{2 \times 13 \times 24.1\ldots}$
$\angle XZY = 27.9°$ (3 sf)

f $\frac{\sin \angle ZXY}{24.1\ldots} = \frac{\sin \angle XZY}{14}$
$\angle ZXY = 53.6°$ or $126.4°$
but $\angle ZXY$ is obtuse so $\angle ZXY = 126.4°$.

Exam questions

1 **a** **(1)** (90, 1) **(2)** (270, −1)
b

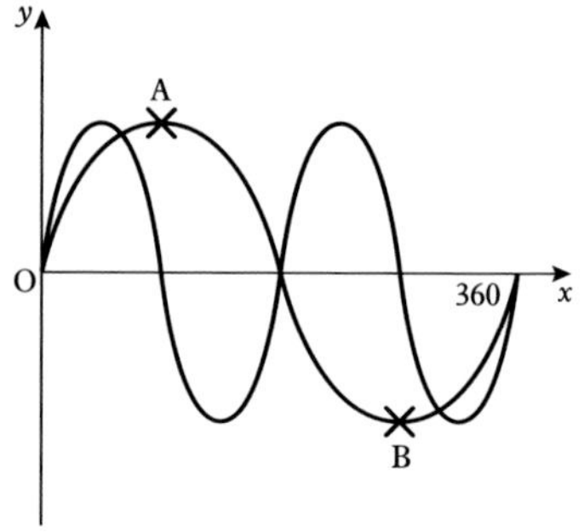

2 **a** 41.4 cm **b** 9.21 cm² **c** 22.3 cm²

3 **a**

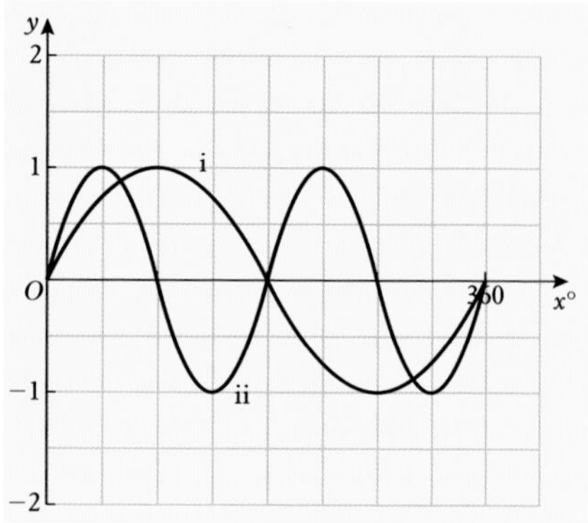

4 **a** $p = 3, q = 2$ **b** 120, 240
5 **a** 3 hs, 15 hs **b** 105, 165, 285, 345
6 **a** 11.7 cm **b** 42.5°
7 177 m
8 61.9°

CHAPTER 20

Questions

1 **a** 30 km/h
b 50 km/h
c The car is stopped. Glenda is at her destination and has stopped before returning home.
d $\frac{80}{1\frac{1}{4}} = 64$ km/h.

2 **a** $\frac{600}{1.5} = 400$
b ℓ/min², the rate of increase of the rate of flow. The flow rate is increasing at 400 litre/min every minute.
c Steady flow of 600 ℓ/min
d $-\frac{600}{2} = -300$ ℓ/min²
The negative sign means that the flow rate is decreasing.
e Volume of water passed through the pipe.

f $100\ \ell/\text{min} \times \frac{1}{2}\ \text{min} = 50$ litres

g $(\frac{1}{2} \times 3 \times 6 + 3 \times 6 + \frac{1}{2} \times 4 \times 6)$ squares
$= 39$ squares
$= 1950$ litres

3 **a**

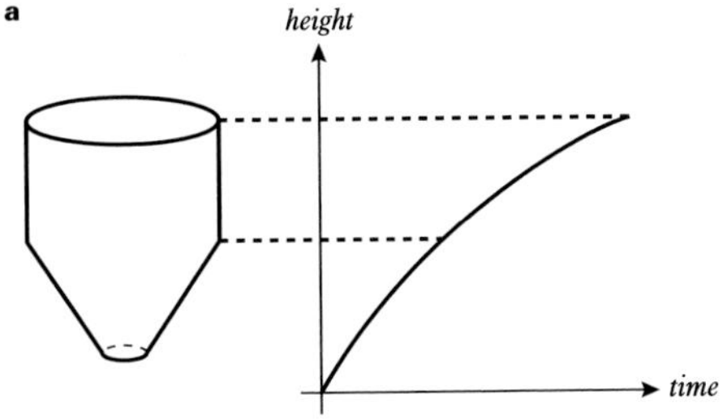

b

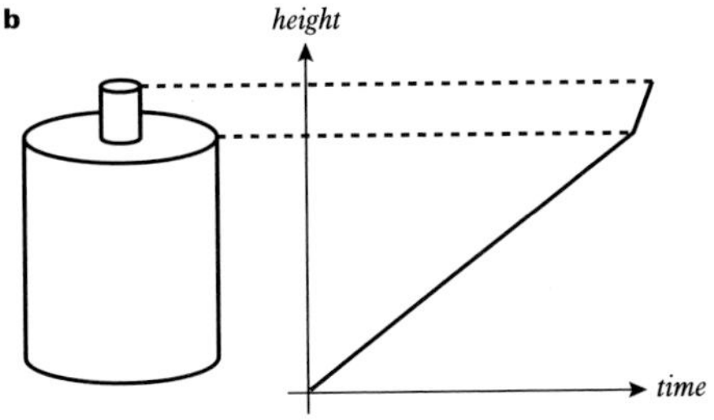

4 $n \times 2^t$

5 **a**

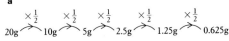

$20\text{g} \xrightarrow{\times \frac{1}{2}} 10\text{g} \xrightarrow{\times \frac{1}{2}} 5\text{g} \xrightarrow{\times \frac{1}{2}} 2.5\text{g} \xrightarrow{\times \frac{1}{2}} 1.25\text{g} \xrightarrow{\times \frac{1}{2}} 0.625\text{g}$

5 half lives are needed

b $m = 20 \times \left(\frac{1}{2}\right)^t$

6 **a**

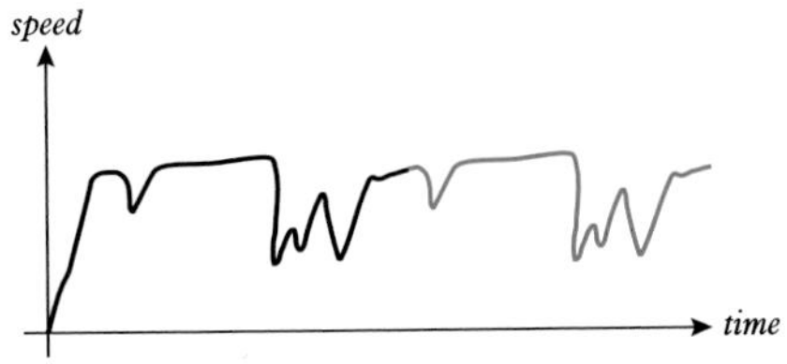

b Red part above.

7 **a** $V = pq^t$
$t = 0, v = 4000 \Rightarrow p = 4000$
curve passes through (2, 4410)
$4410 = 4000\,q^2 \qquad q^2 = 1.1025$
$q = 1.05$

b The initial investment

c 4000×1.05^6
$= £5360.38$ (nearest penny)

8 **a** 6 m

b 9 m (3 sin $30t°$ has maximum value of +3)

c 3 m (3 sin $30t°$ has maximum value of −3)

d $\sin (30t)° = 1$ when $30t = 90$
$t = 3$
So 3 am for first high tide.

e $\sin (30t)° = 1$ next when $30t = 450$
$t = 15$
So 3 pm for next high tide.

f $\sin (30t)° = -1$ when $30t = 270$
$t = 9$
So 9 am for first low tide.

g At 9 pm $d = 6 + 3 \sin (30 \times 21)$
$d = 3$ m (it's the next low tide)

Exam questions

1 **a**

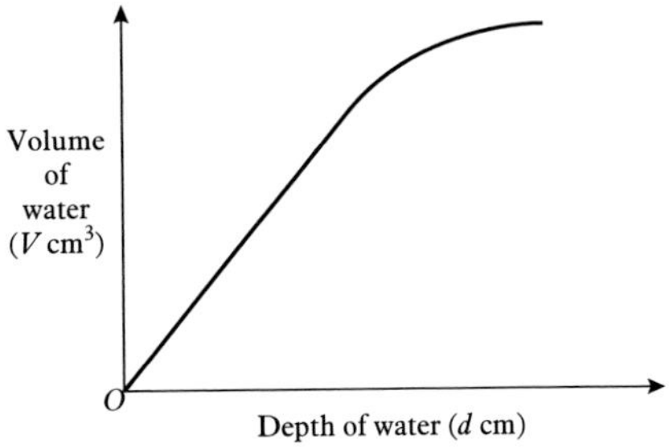

b 137 000 cm^3

2 **a** **(1)** pq^3 **(2)** 3^{x+y}

b $k = 7c$

c $p = 2.5, q = 4$

3 $p = 8, q = \frac{18}{8}, k = 27$

4 **a** 9.8 m/s **c** 11.2 m/s

b 35.4 km/h **d** 2.2 m/s^2